MAD SCIENCE

MAD SCIENCE

EINSTEIN'S FRIDGE, DEWAR'S FLASK, MACH'S SPEED, AND 362 OTHER INVENTIONS AND DISCOVERIES THAT MADE OUR WORLD

Edited by Randy Alfred
From the *WIRED* blog
This Day in Tech,
founded by Tony Long

Little, Brown and Company
New York Boston London

Little, Brown and Company
Hachette Book Group
237 Park Avenue, New York, NY 10017
littlebrown.com

First Edition: November 2012

Little, Brown and Company is a division of Hachette Book Group, Inc., and is celebrating its 175th anniversary in 2012. The Little, Brown name and logo are trademarks of Hachette Book Group, Inc.

The publisher is not responsible for websites (or their content) that are not owned by the publisher.

The Hachette Speakers Bureau provides a wide range of authors for speaking events. To find out more, go to hachettespeakersbureau.com or call (866) 376-6591.

The inspiration for this book, and much of its content, come from *Wired*'s blog "This Day in Tech."

Illustration credits: p. 5: NASA Marshall Space Flight Center; p. 23: U.S. Naval Historical Center Photograph; p. 31: NASA Marshall Space Flight Center; p. 36: Detroit Publishing Company Photograph Collection, Library of Congress; p. 50: NASA Marshall Space Flight Center; p. 63: U.S. Patent Office; p. 67: Bayer; p. 77: NASA Headquarters, "Great Images in NASA"; p. 84: Cornell University Library, Making of America Digital Collection; p. 87: Allcock's Porous Plaster; p. 92: George Grantham Bain Collection, Library of Congress; p. 100: Russell Lee/Library of Congress/Office of War Information Photograph Collection; p. 101: Attributed to Robert Edward Holloway; p. 114: National Center for Supercomputing Applications; p. 115: Rayovac; p. 118: National Oceanic and Atmospheric Administration; p. 121: U.S. Patent Office; p. 122: Gottscho-Schleisner Collection/Library of Congress; p. 129: Ezrdr/Wikimedia; p. 132: U.S. National Archives; p. 135: John Vachon/Office of War Information Photograph Collection/Library of Congress; p. 148: Illlustration, Henry Baker, 1756; p. 149: NASA Langley Research Center; p. 151: U.S. Bureau of Reclamation/Library of Congress; p. 152: Smithsonian Institution; p. 166: NASA; p. 168: Eadweard Muybridge/Library of Congress; p. 170: U.S. Patent Office; p. 188: Friedrich Zschäckel/German Federal Archives; p. 190: U.S. Patent and Trademark Office; p. 191: *How to Make and Operate Moving Pictures*, edited by Bernard E. Jones, Funk & Wagnalls, 1917; p. 194: NASA; p. 196: U.S. Patent Office; p. 197: U.S. Patent Office; p. 205: U.S. Geological Survey; p. 207: NASA; p. 233: Liudmila & Nelson; p. 236: U.S. National Archives; p. 239: U.S. Patent Office; p. 243: U.S. Navy; p. 248: NASA; p. 259: NASA; p. 262: C. H. Claudy/U.S. National Archives; p. 263: Painting, Carl Rakeman [1878–1965]/Federal Highway Administration; p. 272: U.S. National Archives; p. 276: National Cancer Institute; p. 296: Restored photolithograph/U.S. National Archives; p. 310: U.S. Army; p. 312: Library of Congress; p. 353: John Daniels; p. 360: NASA; p. 366: NASA, Galaxy Evolution Explorer image.

Library of Congress Cataloging-in-Publication Data

Mad science : Einstein's fridge, Dewar's flask, Mach's speed, and 362 other inventions and discoveries that made our world / edited by Randy Alfred ; from the WIRED blog This Day in Tech, founded by Tony Long. — First edition.
p. cm.
ISBN 978-0-316-20819-2
1. Science—Miscellanea. 2. Science—Blogs. I. Alfred, Randy, editor.
Q173.M18 2012
500—dc23   2012022234

10 9 8 7 6 5 4 3 2 1

RRD-C

Printed in the United States of America

For the founders of the feast:

Isaac Newton and Albert Einstein
Charles Babbage and Ada Lovelace
Charles Darwin and Alfred Russel Wallace
Henry David Thoreau and Rachel Carson

And my parents, Eleanore and Pete Alfred,
who showed me the rainbow

Give up your Ptolemy,
Rise up and follow me.

— COPERNICUS, AS GLOSSED
BY NANCY L. STARK

LIST OF CONTRIBUTORS

AA	Amy Ashcroft	JBJ	Jason B. Jones
AM	Alexis Madrigal	JM	Jess McNally
AW	Angela Watercutter	JP	Jason Paur
BK	Brandon Keim	JSL	John Scott Lewinski
BM	Betsy Mason	KB	Keith Barry
BXC	Brian X. Chen	KS	Kerstin Sjödén
CB	Chris Baker	KZ	Kim Zetter
CK	Chris Kohler	LG	Lisa Grossman
DC	Doug Cornelius	LK	Leander Kahney
DD	Daniel Dumas	LW	Lewis Wallace
DK	David Kravets	MC	Michael Calore
DS	Darryl Siry	NS	Noah Shachtman
DT	Dylan Tweney	PG	Priya Ganapati
DV	Danielle Venton	RA	Randy Alfred
EH	Evan Hansen	RS	Ryan Singel
EM	Erik Malinowski	SG	Scott Gilbertson
EVB	Eliot Van Buskirk	ST	Scott Thill
HH	Hugh Hart	TB	Tony Borroz
HL	Hadley Leggett	TL	Tony Long
JCA	John C Abell	ZR	Zach Rosenberg

MAD SCIENCE

January 1
1583: First New Year of Gregorian Calendar

The calendar established by Julius Caesar in 45 BCE was running ten days behind the real seasons of the year. Easter arrived too late in spring.

All because the Earth year is about eleven minutes short of 365¼ days. Without a calendar correction, Easter would eventually have fallen in summer, and Christmas in the spring. So Pope Gregory XIII appointed a commission. It proposed eliminating three leap years every four centuries (years ending in 00, unless divisible by 400). That would prevent further creep of the calendar against the seasons. But to reset the calendar immediately, ten days had to be eliminated. The pope decreed the new calendar should start in October 1582. The day after October 4 would not be October 5, but October 15.

This was just months away. Only Italy, Spain, and Portugal made deadline. Many people feared their lives were being shortened by ten days. The pious worried that saints might not listen to prayers that were ten days late. Everyone's birthday moved up ten days too, so 365 days would pass between one birthday and the next. Rents, interest, and wages had to be discounted for that October, since it now had only twenty-one days.

A mob in Frankfurt rioted against the pope and his mathematicians. France made the change in December. Parts of the Low Countries jumped from December 21, 1582, directly to January 1, 1583, skipping Christmas. Most Catholic countries adopted the Gregorian calendar by 1584. But Europe became a patchwork of calendars. You could cross a border and go backward or forward ten days. (Makes the international date line kid stuff.) The Julian calendar (see page 61) held on until 1752 in Britain and its colonies, and right through 1918 in Russia.

As a result, the old Soviet Union used to celebrate its October Revolution in November. —*RA*

Also January 1:

1801: Piazzi Discovers Ceres, First Known Asteroid
1845: Telegraph Helps Capture Murderer John Tawel
1915: Aspirin Tablets Replace Powdered Form
(see page 67)

Also Sixteenth Century:

May 4, 1536: C U @ the Piazza (see page 126)

1

January 2
1870: Building the Brooklyn Bridge

Construction begins on the Brooklyn Bridge. It's the first suspension bridge to use steel—rather than iron—cables.

Conceived in 1867 by famed bridge designer John Augustus Roebling, it was by far the world's longest suspension bridge, with a deck that connected Manhattan and Brooklyn suspended by cables hung from two neo-Gothic towers that pierced the skyline. Not precisely certain of the strength of his materials, Roebling designed the bridge to be six times stronger than it had to be. As a fail-safe, he added straight, diagonal cables to stiffen the superstructure. They make the bridge not a true suspension bridge (which has just vertical stringers, or suspender cables, hanging from huge, curved catenary cables) but a hybrid of suspension and cable-stayed design. That hybrid also gives the steel webwork its characteristic—and mesmerizing—crisscross appearance.

Roebling died of tetanus in 1869, the result of an injury sustained while surveying the bridge site. John's son Washington assumed the title of chief engineer, but tragedy struck again when he became ill with the bends (severe decompression sickness) after rapidly exiting one of the bridge's caissons. It fell to Washington's wife, Emily Warren Roebling, to supervise construction of the bridge. Though never formally trained as an engineer, Emily had studied alongside her husband and began her own research after he became bedridden. Historian David McCullough credits Emily with saving the project, and she rode alongside President Chester Arthur during the ceremonial opening of the bridge, in May 1883. The Brooklyn Bridge has over the years carried P.T. Barnum's elephants, light-rail, and six lanes of automotive traffic, as well as the thousands of pedestrians who left Lower Manhattan after the attacks of September 11, 2001.

But it's not for sale, no matter what that guy on the corner tells you. —*KB*

Also January 2:

1860: Mathematician Le Verrier "Discovers" the Planet Vulcan — Oops!

1959: Soviets' Luna 1 Probe Misses Moon, Goes into Solar Orbit

Also 1870:

February 9: Weather Service Forerunner Established (see page 40)

January 3
1957: Debut of the Electric Watch, a Space-Age Marvel

The Hamilton Electric 500 is announced at a press conference. It is the first battery-operated electric wristwatch and the first watch you never need to wind.

The 500 was made by the Hamilton Watch Company of Lancaster, Pennsylvania, which began developing the timepiece in 1946. It was still not completely ready eleven years later, but the company, feeling the pressure of competition and wanting badly to be the first out the door with this innovation, called the press conference and went into production anyway. It was an instant hit at a time when *progress* was everyone's watchword, so to speak, and all eyes looked expectantly to the future. And the 500 was, briefly, the watch of the future, with its ultramodern design and cutting-edge technology. But fundamental problems with the watch soon became apparent.

Battery life was relatively short, for one thing, so while winding was no longer necessary, frequent battery replacement—in some ways a more arduous chore—was. And *newer* doesn't always mean "better," which the 500 proved by being prone to failure, making it less reliable than the standard winding watch.

The watch's hands were driven by a complex wheel train. By the late 1960s, quartz-movement watches—which had many fewer parts—had arrived, and Hamilton ended production in 1969.

The 500 is now a highly prized collector's piece.—*TL*

Also January 3:
1888: Waxed Paper Drinking Straw Patented
1977: Apple Computer Incorporates

Also 1957:
September 4: The Start of the Short, Unhappy Life of the Edsel (see page 249)

3

January 4
1903: AC/DC Current Events
(or, Edison Fries an Elephant)

Thomas Edison stages a highly publicized electrocution of an elephant to demonstrate the danger of alternating current (although the only true danger it poses is to Edison's own direct-current patents).

Edison had established direct current as the standard for electricity distribution and was living large off the patent royalties, income he didn't want to lose just because George Westinghouse and Nikola Tesla showed up with alternating current. Edison's aggressive campaign to discredit the new current took the macabre form of a series of animal electrocutions using AC—a process he referred to snidely as getting Westinghoused. Stray dogs and cats were the most easily obtained subjects, but he also zapped a few cattle and horses.

Edison got his big chance when the Luna Park Zoo at Coney Island decided to terminate Topsy the elephant. The cranky female had squashed three handlers in three years—including one idiot who'd tried to feed her a lit cigarette. Someone suggested having the pachyderm "ride the lightning," a method American prisons had started using in 1890 to dispatch the condemned. Edison was happy to oblige. Topsy was restrained using a ship's hawser. Copper electrodes were attached to her feet. To stack the odds against Topsy even more, she was fed cyanide-laced carrots moments before a 6,600-volt AC charge slammed through her body. She was killed instantly, and Edison—in his own mind, anyway—had proved his point. A crowd of fifteen hundred witnessed Topsy's execution, which was filmed by Edison and released later that year as *Electrocuting an Elephant*.

But in the end, all Edison had to show for his efforts was a string of dead animals, including the unfortunate Topsy. DC fell out of favor as AC demonstrated its superiority and became the standard. — *TL*

Also January 4:

1643: Birth of Isaac Newton, Physicist (see page 107)
1809: Birth of Louis Braille, Inventor of Alphabet
    for Blind
1813: Birth of Isaac Pitman, Inventor of Shorthand
    Alphabet

Also 1903:

August 11: Instant Coffee, a Mixed Blessing
    (see page 225)

January 5
1972: Nixon Okays "Low-Cost" Space Shuttle

President Richard M. Nixon announces that NASA will develop a space shuttle system, touting its reliability, reusability, and low cost.

The Mercury and Gemini programs had put Americans into Earth orbit. Apollo had been to the moon seven times — landing four times — and would land twice more later in 1972.

But NASA wanted a reusable rocket ship to explore Earth's orbit and to supply and staff a space station. Nixon gave the go-ahead: "This system will center on a space vehicle that can shuttle repeatedly from Earth to orbit and back. It will revolutionize transportation into

First flight of space shuttle *Endeavour*, 1992.

near space, by routinizing it. It will take the astronomical costs out of astronautics." NASA director James Fletcher's remarks also referred to the shuttle's "modest budget" and reduced complexity. The plan was to make forty-eight flights a year at about $50 million per launch ($274 million in 2012 money).

From 1981 to 2011, the shuttles made a hundred and thirty-five space flights, averaging four or five missions a year. University of Colorado researcher Roger Pielke Jr. calculated in early 2005 that the shuttle program to date had cost $145 billion, or about $1.3 billion per flight. (Based on a 1995 midpoint, that's almost $2 billion per flight in 2012 dollars.)

The Apollo program cost a total of $19.4 billion from 1960 to 1973. That averages almost $2.2 billion each for the nine lunar missions. (Based on a 1967 midpoint, that would be about $15 billion each today.) So, space shuttle flights have certainly been less expensive than Apollo lunar missions. But even adjusting for inflation, and despite the program's copious achievements, shuttle launches cost seven times more than was estimated. — *RA*

Also January 5:

1943: Death of George Washington Carver, Inventor of Peanut Butter

1996: First Cellphone Bomb Kills Hamas Operative

Also 1972:

April 10: 140 Nations Sign Biological Weapons Treaty

November 29: Pong, a Game Any Drunk Can Play
(see page 335)

January 6
2000: Air Traffic Control System Suffers a Really Bad Air Day

Less than a week after Y2K passes without a global computer melt-down, a glitch in a 1960s computer at the air traffic control center in Washington, DC, slows and shuts down airlines in the Northeast.

Normally, when a flight is completed or a plane leaves a region's airspace, the flight plan is deleted. A software glitch on this day meant the computer was no longer deleting the flight plans. As new flight plans came in, the system overloaded, and it shut down. So controllers went old-school, using small strips of paper with the flight information typed out. These slips were then hand-carried from one controller to another. This sneakernet system works just fine and is still used today as a backup. But it's slower to walk across a room than to click a mouse, and the system ground to a crawl. Hundreds of flights from Boston, New York, Philadelphia, and Washington were directly affected, and that caused a ripple effect as waiting flights stacked up. Travelers around the country suffered nightmarish delays until the computer was fixed.

In the early days of aviation, pilots flew when conditions were good enough for them to see one another. By 1926, the United States started to implement rules for air traffic, and by 1930 the first radio-equipped control tower was installed in Cleveland. The first full-scale air traffic control center was established in Newark in 1936, the same year the federal government took over control of air traffic.

In the early days, controllers used a blackboard and maps with small models to estimate the locations of airplanes as reported by airports, the airlines, and radio operators along the routes. ATC modernized after World War II with the implementation of radar. U.S. controllers today combine radar, satellites, and good old-fashioned eyesight to keep track of more than eighty-five thousand flights a day. The Federal Aviation Administration plans to double air traffic capacity by switching from radar to GPS navigation. —*JP*

Also January 6:

1912: Birth of Tech Skeptic Jacques Ellul
1930: First Diesel Auto Trip, Indianapolis to New York

Also 2000:

June 10: London's "Wobbly Bridge" Opens, Briefly
June 24: President Goes Live on Net
August 17: Half of U.S. Homes Have Internet Access
October 11: Ozone Hole Exposes Chilean City
     (see page 286)

1851: Foucault Demonstrates Rotation of Earth

Scientists had been trying for two centuries to measure Earth's motion by dropping objects from towers. It didn't work: too quick, too crude, too many interfering factors.

Jean-Bernard-Léon Foucault had an insight. A pendulum hanging on a wire and swinging directly north and south would appear to the observer to slowly move its plane of oscillation as Earth turned underneath it. In January 1851, after weeks of work in the cellar of his home, Foucault hung an eleven-pound pendulum from a six-and-a-half-foot cable. He observed a small clockwise motion of the pendulum's apparent plane of oscillation. The pendulum was going straight back and forth, but the earth moved for Foucault. It was the first direct visual evidence of Earth's rotation that was not based on watching the stars circle in the sky. He arranged a demonstration for the scientists of Paris on February 3. He told them, "You are invited to see the earth turn." And so they did, as they watched Foucault's pendulum move on a thirty-six-foot wire at the Paris Observatory.

French president Louis-Napoléon was a science buff, and he arranged for Foucault to give a public demonstration of his remarkable pendulum on March 31. Under the lofty roof of the Panthéon in Paris, Foucault hung a sixty-two-pound brass sphere on a two-hundred-twenty-foot cable. A pointer attached to the bottom of the sphere traced patterns in sand on a low wood platform. The public was dazzled. President Louis-Napoléon soon became Emperor Napoléon III, and he gave Foucault the position of physicist attached to the Imperial Observatory. But the university-trained scientists of Paris sniffed at Foucault, seeing him as an unschooled upstart. They repeatedly turned him down for membership in the French Academy of Sciences, although they finally admitted him in 1865. — *RA*

Also January 7:

1785: Balloon Carries Two Passengers Across English Channel

1904: CQD Radio Distress Signal Adopted, but SOS Soon Prevails

1953: President Truman Announces U.S. H-bomb

Also 1851:

May 1: First World's Fair Opens at London's Crystal Palace

December 10: Birth of Librarian Melvil Dewey, Inventor of Dewey Decimal System

January 8
1942: *A Brief History of Time* Is Made Possible

British physicist Stephen Hawking is born.

Hawking was born in Oxford, where his parents had moved to escape the German Blitz on London. His website notes, in an interesting historical aside, that his birth came on the three hundredth anniversary of Galileo's death.

Though naturally predisposed to mathematics, young Hawking switched to studying physics, because University College at Oxford did not offer the other discipline. He turned out to be pretty good at physics too. From there, Hawking moved on to Cambridge to do his research in cosmology. It was while attending Cambridge that he developed the first signs of amyotrophic lateral sclerosis, an incurable neurological disease. The average ALS patient dies within a few years of diagnosis. Defying all odds, Hawking has managed to live with it for five decades. In fact, Hawking has done far more than simply live with ALS. Almost willfully ignoring his disability, which leaves him paralyzed and unable to speak or walk, Hawking has carved out a brilliant career as a theoretical physicist specializing in the study of the universe.

His *A Brief History of Time*, a bestseller on the subject, helped raise public consciousness regarding the nature of the universe, its possible beginnings, and its probable end.

The reason for the very slow progression of Hawking's disease remains a mystery. A special computer rigged to his wheelchair and operated by a blink switch attached to his glasses gives him some limited self-sufficiency, although he still requires around-the-clock nursing care.

In April 2007, Hawking realized his dream of taking a zero-gravity flight, making him the first quadriplegic ever to do so. He hopes to fly a suborbital flight on Virgin Galactic's SpaceShipTwo tourist service. The company's owner, Richard Branson, says he'll pick up the $200,000 tab for Hawking. —*TL*

Also January 8:

1790: President Washington Urges New Nation to Encourage Science (see page 53)
2004: Largest Ocean Liner, the RMS *Queen Mary 2*, Christened

Also 1942:

January 13: Ejection Seat Works, Pilot Elated (see page 13)

1643: Astronomer Sees Ashen Light of Venus

Italian astronomer Giovanni Riccioli discovers a faint glow on the night side of the planet Venus. Other astronomers over the ensuing centuries will also observe the Ashen Light, but it remains an astronomical mystery that defies conclusive explanation.

Riccioli was the lunar cartographer who named many of our moon's features, like the Sea of Tranquility and the Sea of Storms. The faint luminescence he saw on Venus has been witnessed many times since, but there have been others who looked for it and did *not* see it. Some chalk it up to observer error, distortion caused by Earth's atmosphere, and/or artifacts induced by telescope optics. But still: Four centuries of similar observations? Those who've seen the Ashen Light of Venus report it looks a lot like the reflected earthshine that sometimes casts a dull glow on the moon but not even that bright. It's most easily sighted when the dusk edge of the sunlight on Venus faces Earth.

The U.S. Pioneer mission and the Soviet Venera 11 and 12 landers looked for it without any luck. The Keck I telescope in Hawaii did spot a faint green glow consistent with the emissions from oxygen atoms, but it was too weak to account for all the amateur sightings over the years. Another possibility is multiple lightning flashes. During Venus flybys in 1998 and 1999, the Cassini spacecraft failed to detect the high-frequency radio noise that lightning would be expected to generate, like AM radio static during terrestrial thunderstorms. But other radiation detected on Venus is characteristic of lightning discharges. It's also possible the Ashen Light of Venus is caused by solar particles energizing the atmosphere, like the terrestrial aurora borealis and aurora australis—hence its evanescence.

Or it's some previously unknown combination of things we understand. Or something we don't understand at all. —*RA*

Also January 9:
1816: Davy's Safety Lamp First Used in a Mine
1839: Birth of Photography — Daguerre Displays His
Daguerreotypes (see page 233)

Also 1643:
January 4: Birth of Isaac Newton, Physicist
(see page 107)

January 10
1863: Take the Tube

London inaugurates the world's first subway service. Approximately forty thousand Londoners ride the trains the first day.

The original line ran from Paddington Station to Farringdon Street via Edgware Road, Baker Street, Portland Road (now Great Portland Street), Gower Street (now Euston Square), and King's Cross. It took the train eighteen minutes to make the three-and-three-quarter-mile journey. By 1880, the line was carrying forty million passengers a year. Despite the subway's success, it was not widely copied for more than three decades. The next metro subways to open were in Budapest and in Glasgow, in 1896. Boston's opened in 1897, Paris's in 1900, and New York's in 1904.

The London system relied on steam-driven trains, which made proper ventilation critical. There were several independent companies operating the trains, making logistics a nightmare. By the twentieth century, however, trains had been electrified and tunneling methods improved. The London system, which became known as the Underground, was largely consolidated by 1902 under the ownership of American tycoon Charles Yerkes. But it was 1933 before all the lines came under the control of a public corporation, the London Passenger Transport Board.

Governance of the system has successively passed to the London Transport Executive (1948), London Transport Board (1963), London Transport Executive (again; 1970), London Regional Transport (1984), and Transport for London (2000). During the Blitz in World War II, Underground stations were used as ad hoc air-raid shelters. Today the Underground comprises eleven lines (not counting the Dockland Lights Railway or the converted suburban rail lines of the London Overground) serving 270 stations in metropolitan London. The system carries roughly a billion passengers per year, making it one of the largest in the world. By comparison, New York City's subway system carries 1.6 billion passengers annually, and Tokyo's 3.1 billion. — *TL*

Also January 10:

1899: Electric Flashlight Patented
1901: Spindletop Well Produces First Texas Gusher
1949: 45 RPM Records Debut

Also 1863:

February 7: An Early Stab at Organizing the Elements
(see page 38)
December 19: Walton Patents Improvement to His
Invention of Linoleum

January 11
1922: Insulin Makes a Nice Shot

Insulin is used to treat diabetes in a human being for the first time.

During a clinical test at the University of Toronto, an injection of bovine insulin was administered to fourteen-year-old Leonard Thompson by surgeon Frederick Banting, who had been researching ways of extracting insulin from the pancreas. The boy suffered an allergic reaction to the first injection, but further work improved the extract. A second injection administered a few weeks later was successful. Banting was then able to produce large quantities of insulin, but the process remained impure until pharmaceutical maker Eli Lilly offered its assistance. Banting cut a deal with the drug company, and insulin came into common use.

For his work, Banting, along with collaborator John James Rickard Macleod, was awarded the 1923 Nobel Prize in Physiology or Medicine. Banting shared the credit—and half of his prize money—with his research assistant Charles Best, and Macleod shared his with coresearcher James Collip.—*TL*

Also January 11:

1902: *Popular Mechanics* Publishes First Issue
1911: Forerunner of Max Planck Institute Founded
   in Berlin

Also 1922:

April 2: Rorschach Dies, Leaving a Blot on His Name
   (see page 94)
June 16: First Controlled Horizontal Helicopter
   Flight in U.S.
August 4: U.S. Phones Shut Off for Bell's Funeral
   (see page 71)

January 12
1992 (or 1997): HAL of a Computer

HAL 9000, the master computer aboard the spaceship *Discovery* in the novel and film *2001: A Space Odyssey*, becomes operational. He will inspire millions of dreams—and some nightmares—of artificial intelligence.

Late in the film, HAL regresses to his infancy and begins an eerie recitation of bits of his earliest knowledge: "I am a HAL 9000 Computer, Production No. 3. I became operational at the H–A–L plant in Urbana, Illinois, on the twelfth of January, 1992." At least that's what HAL says in the 1968 film. Director Stanley Kubrick and author Arthur C. Clarke cowrote the screenplay, which was inspired by Clarke's 1950 short story "The Sentinel." Clarke then wrote a novelized version of the screenplay, in which he changed HAL's birth year to 1997.

Chapter 16 of the novel clearly states that HAL stands for "Heuristically programmed ALgorithmic computer." Many people, however, thought HAL was a one-letter-ahead cipher for IBM. In his book *The Lost Worlds of 2001*, Clarke dismissed that idea, claiming that it was embarrassing, given all the help IBM provided for the film: "We...would have changed the name had we spotted the coincidence." The IBM theory is further denied in the sequel, *2010*.

Urbana, Illinois, is home to the University of Illinois and—since 1986—the National Center for Supercomputing Applications, which developed the first web browser, Mosaic (see page 114).

The movie cost $10.5 million ($69 million in 2012 dollars) and premiered in New York City on April 3, 1968. The dazzling special effects did not impress all the critics: the *New York Times* described *2001* as "somewhere between hypnotic and immensely boring," while Pauline Kael deemed it "monumentally unimaginative." Kubrick promptly cut nineteen minutes from the film, and the smash hit debuted across the country three days later.—*RA*

Also January 12:

1665: Fermat's Last Breath

1967: It's Cold in Here — First Cryonic "Burial"

Also 1992:

May 5: Wolfenstein 3-D Shoots First-Person Shooter into Stardom (see page 127)

Also 1997:

February 10: IBM's Deep Blue Outmatches Chess Champ Kasparov (see page 41)

January 13
1942: Ejection Seat Works, Pilot Elated

At the height of World War II, German test pilot Helmut Schenck becomes the first person to use an ejection seat to successfully exit his aircraft in an emergency situation.

Schenck, testing a Heinkel He-280 jet fighter, was in tow behind a conventionally powered aircraft when his plane iced up, making it impossible for him to start the engines. He jettisoned his canopy and activated the seat, which catapulted him clear of the aircraft. Schenck was the first to use this method of exiting his aircraft in an actual emergency, although another Heinkel pilot had previously ejected successfully under test conditions.

Germany led the way in developing the ejection seat. The speed and g-forces generated by its newly developed jet planes made escape problematic for a pilot equipped with only a parachute. Exiting the aircraft by bailing out, common in propeller-driven planes, was exceedingly dangerous in a jet. The British also studied aircraft ejection during the interwar years but then set the project aside in favor of other pursuits. They didn't seriously revisit the subject until after World War II.

The Germans experimented with several types of ejection seat, or *Schleudersitzapparat*, "seat catapult device." Schenck's was activated by compressed gas; another type relied on a spring-operated mechanism, and a third used a propellant charge. Schenck's type of seat was later abandoned in favor of the propellant-charge version.

Ejection seats were eventually installed in several models of Luftwaffe jets but in only a few Me-262s, the most widely used German jet fighter of the war. From the time of Schenck's successful escape to the end of World War II, approximately sixty Luftwaffe airmen ejected from their planes in combat situations. — *TL*

Also January 13:

1908: First Pilot to Fly Kilometer Wins Big
    Aviation Prize
1928: First Home TV Sets Have Three-Inch Screens
2004: Astronomers Hold Back, Don't Panic World Over
    Asteroid That Misses

Also 1942:

February 23: Japanese Sub Shells Santa Barbara
    Refinery (see page 55)

January 14
1794: First Successful Cesarean in United States

Elizabeth Bennett delivers a daughter by cesarean section, becoming the first woman in the United States to give birth this way and survive. Her husband, Jesse, is the physician who performs the operation.

Elizabeth, struggling with a difficult labor and believing she would die, had asked her attending physician to perform a cesarean, in the hope of saving the baby. The doctor refused on moral grounds, so Jesse stepped in. Conditions were crude. The procedure was performed at home—a log cabin deep in the backwoods of Mason County, Virginia (now West Virginia). The operating table consisted of a couple of planks laid across two barrels. Jesse Bennett resorted to laudanum—lots of it—to knock out his wife. Despite these limitations, the surgery went smoothly. Bennett extracted a healthy girl, and he closed the incision.

The cesarean section was not new. What was new was the idea that both mother and child could survive the ordeal. The operation itself dated from antiquity but, with very few exceptions, was performed only when the mother was dead or dying. The first recorded cesarean in which both mother and child survived was done in Switzerland in 1500. That was also a husband-and-wife affair, although in this case the husband, Jacob Nufer, was a swine gelder, not a doctor.

Before the nineteenth century, the success rate for C-sections was very low. Times have sure changed: in the United States today, more than 30 percent of all births are by cesarean section.—*TL*

Also January 14:

1898: Death of Mathematician Charles Dodgson, AKA Lewis Carroll

1914: Ford Starts Using Chain-Driven Assembly Line to Build Model Ts

1943: Birth of U.S. Astronaut Shannon Lucid

Also 1794:

March 14: Eli Whitney Patents the Cotton Gin (see page 75)

March 27: U.S. Navy Established

January 15
2001: Enter Wikipedia

Wikipedia goes online. Doing research will never be the same again.

Wikipedia is the web's vast user-generated, crowd-sourced encyclopedia. The 2001 innovation is itself based on wiki software introduced in 1995 to let groups collaborate on a document. *Wiki* comes from the Hawaiian word meaning "fast." Wikipedia is the brainchild of Jimmy Wales and Larry Sanger. Wales gets credit for the open-to-all encyclopedia concept, while Sanger is said to have advocated using a wiki to accomplish this.

Most entries can be edited by anyone, so there's legitimate skepticism. What if the Wikipedian who wrote it is a liar? An idiot? Merely incompetent? Well, if you'd rather not screw up your term paper, you'd better cross-check the information. Wikipedia can help with that too: Just follow the links from the footnotes. And use a search engine to find independent verification.

But the wisdom of the crowd can result in pretty good work. Besides, after a couple of high-profile snafus, Wikipedia introduced some controls on the intellectual free-for-all. For instance, you can't anonymously edit pages about celebrities, some other quickly changing topics, and some specific emotionally charged topics. And upper-level editors use special software to keep an eye on recent changes.

Looking for a list of U.S. first ladies or Swedish kings? Likely no problem. Middle Eastern politics? Tread carefully. But there's no arguing that Wikipedia usually has the info we want, right there at the top of the search results. In the rush-rush atmosphere of the twenty-first century, there's plenty to be said for that. And plenty of people know it. The site ranks consistently in the top ten most visited worldwide.

As of 2012, Wikipedia boasts nearly 4 million articles in English alone. Counting everything that appears in 280 languages, it's 21 million entries from 85,000 active contributors. You could look it up. — *TL, RA*

Also January 15:

1919: Molasses Explosion and Flood Kills Twenty-One in Boston
1929: Birth of Martin Luther King Jr., Who Warned of Technological Amorality

Also 2001:

January 28: Hey, Don't Tampa with My Privacy
(see page 28)

January 16
1936: Day at the Races, and Your Nag in a Photo Finish

A photo-finish camera is installed at Florida's Hialeah race track. It marks the first use of the device for Thoroughbred horse racing, the sport with which it is most closely associated.

Photo-finish cameras are also used at track meets, auto races, and bicycle races—anywhere, in fact, where the winner is determined by who hits the finish line first. The photo-finish camera was originally a conventional still camera modified to handle rapid multiple imaging: its focal-plane shutter was replaced with a capping shutter, and it employed a vertical-slit view of the finish line. The camera was elevated to avoid blocking its view of any competitors.

With refinements, which resulted in a dramatic increase in the number of photographs taken per second, this remained the basic photo-finish technology until the advent of digital photography. A camera in the mid-twentieth century might shoot a hundred images per second. A modern digital camera can take a thousand images per second.

The first photo-finish camera was used at the 1912 Olympic Games in Stockholm. By the 1932 Olympic Games in Los Angeles, photo-finish technology had advanced to include the finish line and a chronograph reading on a single image. A version of this so-called Kirby camera went into use at Hialeah for the 1936 meeting.

In horse racing, the winning horse is the one whose nose hits the line first. In a race among humans, the winner is the one whose torso—any part of it—hits the line first, which is why you often see runners lunging at the tape in a close race.

Perhaps the most famous photo finish of all time occurred with a triple dead heat (three horses hitting the finish line simultaneously) in the 1944 Carter Handicap at Aqueduct Racetrack in New York. —*TL*

Also January 16:

1909: Ernest Shackleton's Expedition Locates Magnetic South Pole
1969: Two Soviet Spacecraft Dock and Transfer Cosmonauts

Also 1936:

May 12: August Dvorak Patents New Keyboard Arrangement, but QWERTY Still Reigns
July 7: Get a Grip — Phillips Screws Up the Toolbox (see page 190)

1985: Red Phone Boxes Are Put on Hold

Britain's famous red telephone boxes are officially retired by British Telecom, another victim of encroaching technology and corporate expediency.

Retired does not mean "completely gone," however. Although their numbers have dwindled over the years, Sir Giles Gilbert Scott's iconic telephone boxes are still kicking around Old Blighty. And they remain the phone box of choice on the island of Malta. Scott's design was chosen in a 1924 competition, and Scott himself originally recommended painting the boxes silver. The post office—which ran Britain's phone service at the time—chose red so the booths would be easier to see. Red was also, of course, the color (or should we say *colour*) of the post office letter boxes found on so many corners. Following privatization in 1984, the phone service switched to British Telecom, and the design of phone boxes became more utilitarian. Many of the red boxes were replaced, but a public outcry led to the preservation of a fair number, especially in areas of central London frequented by traditionalists and tourists.

The growing prevalence of cellphones in the past two decades has severely reduced demand for public phone boxes of any type. Some of the red beauties still in use have been converted to Internet kiosks, but smartphones are knocking the wind out of that market too. —*TL*

Also January 17:

1706: Birth of Benjamin Franklin, Inventor and
     Electricity Pioneer
1966: U.S. Accidentally Drops Three H-Bombs on
     Spanish Village; They Don't Explode

Also 1985:

March 11: First Public Net Links Consumers'
     Computers (see page 72)

January 18
1911: Clear the Deck...for an Aeroplane

A Curtiss biplane becomes the first airplane to perform a landing on a ship.

The plane, piloted by Eugene Ely, landed on a platform bolted to the armored cruiser USS *Pennsylvania* moored in San Francisco Bay. Ely had been flying for less than a year when he was approached by the U.S. Navy to help investigate military uses for aircraft. (Flying was such a new endeavor that mere months of experience qualified Ely for the risky attempt.) The previous November, he'd taken off in a Curtiss plane from a specially built eighty-three-foot wooden platform on the bow of the light cruiser USS *Birmingham* in Hampton Roads, Virginia. The plane didn't have a lot of speed when it taxied off the edge of the ship. The wheels actually dipped into the water, and spray splattered on Ely's goggles. Instead of making a triumphant circuit of the harbor and landing at the Norfolk Navy Yard, he quickly landed on a beach and counted his blessings. But Ely performed flawlessly in his 1911 shipboard *landing*. And within the hour, he took off from the deck of the *Pennsylvania* and returned safely to San Francisco.

He was not so lucky later in the year: he was killed during a flying exhibition in Macon, Georgia, just shy of his (some sources say) twenty-fifth birthday.

The military significance of Ely's shipboard landing and takeoff was staggering, leading directly to the development of the aircraft carrier, which remains, since World War II, the most dominant nonnuclear naval weapon. — *TL*

Also January 18:

1778: Captain Cook Lands in Hawaii
1888: Birth of Thomas Sopwith, Designer of World
    War I's Sopwith Camel Airplane
1933: Birth of Audio Engineer Ray Dolby, Inventor of
    Dolby Noise Reduction

Also 1911:

January 21: All Roads Lead to Monte Carlo...Rally
    (see page 21)
September 17: First Transcontinental Flight Takes
    Weeks (see page 262)

January 19
1883: Let There Be Light, All Over Town

Roselle, New Jersey, earns its place in tech history when the first electric lighting system employing overhead wires goes into service.

The system was built by Thomas Edison as part of an experiment to prove that an entire community could be lit by electricity from a shared, central generating station. A steam-driven generator sent the juice through the wires strung overhead to a store, the town's railway depot, forty or so houses, and a hundred and fifty streetlights. The First Presbyterian Church of Roselle made electrical and ecclesiastical history three months later when it installed a thirty-bulb electrolier and became the world's first church to be lit by electricity. The electric chandelier still hangs in the church. In the centennial year of 1983, a bronze-and-granite marker was dedicated at the original site of Edison's generator, at the corner of Locust Street and West First Avenue, in Roselle.

Edison, one of the most prolific inventors of all time, was known as the Wizard of Menlo Park. That predates, apparently, the age of visionaries, geniuses, and gurus. Among Edison's 1,093 U.S. patents are the incandescent lightbulb; the phonograph; the stock ticker; and the kinetoscope, an early film projector. (We could put an Edison patent on nearly every page of this book, but you'd lose interest pretty fast.) Whether Edison was the actual inventor of everything he patented remains debatable; in fact, doubtful (see pages 29, 192, 222, and 354). But there's no denying that the man knew what he was doing (see pages 4, 229, and 296). —*TL*

Also January 19:

1736: Birth of James Watt, Improver of the Steam Engine
1983: Apple's Lisa Debuts, First Commercial Computer with a Graphical User Interface

Also 1883:

May 24: Brooklyn Bridge Opens (see page 2)
November 18: Railroad Time Goes Coast to Coast (see page 324)

January 20
Inaugural Tech, from Steamboat to Twitter

When George Washington was inaugurated as the first U.S. president, in 1789, his words could be heard no farther than his unaided voice could project them, and they could be carried no faster than the decrees of an ancient Roman or Persian emperor had been.

The choices one had to hurry news on its way were horseback, sailing ship, and carrier pigeon. No railroads, no telegraph. Time and tech have changed how we see and hear the inauguration. Here are some firsts:

1801: Jefferson's inaugural address prompts a newspaper extra.

1817: Steamboats (see page 198) on the Potomac can carry news of Monroe's inauguration.

1837: Trains (see page 263) can carry word of Van Buren's inauguration, because the B&O Railroad began service from Washington in 1835.

1845: As Polk is sworn in, telegraph inventor Samuel Morse (see page 173) taps out the news to places as far away as Baltimore.

1857: Buchanan's inauguration is photographed (see page 233).

1885: News of Cleveland's inaugural can be spoken over long-distance telephone wires (see page 71).

1897: McKinley's inauguration is captured in moving pictures.

1905: Telephones are installed on the Capitol grounds for Theodore Roosevelt's inaugural.

1921: Harding rides to and from his inauguration in an automobile. Loudspeakers let people in the crowd actually hear the proceedings. An announcer on radio station KDKA in Pittsburgh (see page 274) reads an advance copy of the inaugural address.

1925: Coolidge's inauguration is broadcast live on twenty-four radio stations.

1929: Hoover's inaugural is recorded by talking newsreel (see page 73).

1949: President Truman is inaugurated on national TV.

1961: Kennedy's inaugural parade is on color TV (see page 86).

1981: Reagan's inauguration is the world's first telecast to include live, closed-caption subtitles for the hearing-impaired.

1997: Clinton's inaugural is webcast.

2009: Obama's inauguration is tweeted (see page 82).

Note: Inauguration Day was March 4 from 1793 through 1933. —RA

Also January 20:

1838: First Traveling Post Office as Britain Starts Sorting Mail Aboard Trains

1942: Nazi Wannsee Conference Unleashes "Final Solution," Genocide by Technology

1911: All Roads Lead to Monte Carlo... Rally

The first Monte Carlo Automobile Rally is held. Twenty-three cars starting from eleven different locations around Europe eventually converge on the tiny principality of Monaco.

The event was organized at the behest of Prince Albert I (great-grandfather of current Prince Albert II). Like many motoring contests of the time, it was seen primarily as a way for auto manufacturers to test new cars and new technologies. Remember, this was the era of the Model T, the first purpose-built race cars, and the earliest electric starters. Results of the hybrid event depended not only on driving time but also on judges' assessment of the automobiles' design and passenger comfort, as well as what condition the vehicles were in after covering a thousand kilometers of roads not really made for the horseless carriage. The arbitrary system provoked minor outrage, but the judges' decision stood.

Automobile dealer Henri Rougier won first place driving a Turcat-Méry 45-horsepower model. Second place went to a driver named Aspaigu in a Gobron, and third to Jules Beutler in a Martini. The rally was held the following year, but then not again until 1924, after which it became an annual event. World War II and its aftermath interrupted the yearly races, so there were no rallies from 1940 through 1948.

Winning the annual rally gives a car manufacturer a great deal of publicity and trust. Before Paddy Hopkirk won the rally in 1964, the Mini was seen as just a commuter car. After winning, the Mini Cooper had the rep of a world-beating performance car.

The Monte Carlo Rally isn't held on a closed track but on narrow cliffside roads to the north of the tiny principality. About a quarter of it is run at night. — *TB*

Also January 21:

1954: USS *Nautilus*, World's First Nuclear-Powered Submarine
1979: Neptune Moves Outside Pluto's Wacky Orbit
2008: Chief Marie Dies; So Does Eyak Language

Also 1911:

October 24: Birth of an Inventive Wyeth

January 22
1984: Dawn of the Mac

The Apple Macintosh personal computer is introduced to the world in a now-legendary TV commercial during Super Bowl XVIII.

The sixty-second spot, directed by Ridley Scott, featured a female athlete running through a dystopian landscape inspired by George Orwell's *1984* and then throwing a sledgehammer at a TV image of Big Brother (meant, in this case, to represent IBM). It ended with a promise: "Apple Computer will introduce Macintosh. And you'll see why 1984 won't be like *1984*." The first Mac computer went on sale two days later. It was a product of its time—underpowered and not very easy to use. But it did represent a sea change, a paradigm shift, whichever late-twentieth-century business cliché you care to use. It was the first consumer computer to feature a graphical user interface that could be called user-friendly, and it was the first— with the advent of the LaserWriter printer and Aldus PageMaker—to make desktop publishing a reality.

The Macintosh 128K (that was your RAM) screamed along at 8 MHz, featured two serial ports, and could accommodate one 3.5-inch floppy disk. It ran the Mac OS 1.0, came with a nine-inch black-and-white monitor, and sold for a cool $2,500 (about $5,500 in 2012 dollars). In a little under three months, Apple sold fifty thousand of these babies; not exactly an avalanche. But successively improved models did sell big and were enormously influential in goading other computer manufacturers to make their products more intuitive.

As for the ad itself, *Advertising Age* eventually selected it as the Commercial of the Decade, and it's credited with beginning the annual creative orgy of modern Super Bowl commercials. — *TL*

Also January 22:

1950: Innovative Automaker Preston Tucker Acquitted of Financial Fraud Charges
1980: Soviets Exile Nuclear Physicist Andrei Sakharov

Also 1984:

April 11: Shuttle Makes House Call, Repairs Satellite (see page 103)

1960: Journey to the Deepest Place on Earth

The diving submersible vessel *Trieste* descends to the floor of the Marianas Trench, the deepest known place on the planet.

The *Trieste*, a bathyscaph, was designed by Swiss scientist Auguste Piccard and built in Naples, Italy, as a deep-diving research vessel. After proving its mettle in the Mediterranean, the *Trieste* was purchased by the U.S. Navy and shipped to San Diego. Its dive into the Challenger Deep, the deepest part of the trench, was made, according to a U.S. Navy press release, "to demonstrate that the United States possesses the capability for manned exploration of the sea down to the deepest part of its floor." The vessel was equipped with air

Jacques Piccard (*right*) and Ernest Virgil with *Trieste* before its record-breaking dive.

tanks that could be flooded, like a submarine's, to help it dive. Additionally, it was weighted down with nine tons of iron shot. *Trieste* descended at a rate of three feet per second to a depth of 27,000 feet, after which its descent rate was halved.

It took U.S. Navy lieutenant Donald Walsh and Jacques Piccard (son of *Trieste*'s designer) four hours and forty-eight minutes to make the descent to a depth of 35,810 feet, roughly seven miles. The pressure there is nearly 17,000 pounds per square inch. Although an outer Plexiglas window cracked during the descent, *Trieste* handled the dive like a champ. After sitting on the bottom for twenty minutes, eating candy bars and shining a light that illuminated a shrimplike creature outside, Walsh and Piccard pumped out two tons of iron shot and began their ascent. The trip home was faster—three hours and seventeen minutes to the surface.

Trieste was sent to the Atlantic Ocean in 1963 to search for the nuclear attack submarine USS *Thresher*, which was lost with all hands during a deep-diving exercise. Upon the bathyscaph's return, it was retired to the U.S. Navy Museum in Washington. —*TL*

Also January 23:

1911: French Academy of Sciences Rejects Marie Curie (see page 357)

1978: Sweden Is First Country to Ban Aerosol Sprays to Save Ozone Layer

Also 1960:

May 9: Easy Birth Control Arrives, but There's a Catch (see page 131)

October 24: Soviet Rocket Explodes, Killing Top Engineers, Technicians

January 24
1935: First Canned Beer Sold

The first canned beer in the United States goes on sale in Richmond, Virginia.

The American Can Company began experimenting with canned beer in 1909. But the cans couldn't withstand the pressure from carbonation, and they exploded. As Prohibition ended, in 1933, the company developed a keg-lining technique, coating the inside of the can in the same way the inside of a keg is lined. American Can offered the beer maker Krueger a deal: if the beer flopped, the brewery wouldn't have to pay for the canning equipment. So Krueger's Cream Ale and Krueger's Finest Beer became the first beers sold to the public in cans. Canned beer was an immediate success. The public loved it, giving it a 91 percent approval rating.

Compared to glass, the cans were lightweight, cheap, and easy to stack and ship. Unlike bottles, cans didn't require you to pay deposits. By summer, Krueger was buying 180,000 cans a day from American Can. By the end of the year, thirty-seven breweries had followed Krueger's lead. The first cans were flat-topped and made of heavy-gauge steel. To open one, you had to punch a hole in the top with the sharp end of a church-key-style opener. Some breweries tried out cans with a conical top sealed with a bottle cap, but they didn't stack and ship as easily as flat-tops.

During World War II, canning was interrupted in order to save resources. Aluminum cans, cheaper and lighter still, were introduced in 1958.

Beyond their economy and convenience, cans are actually better for beer than glass bottles. The opaque can blocks light completely; light splits the B vitamin riboflavin, which then reacts with the beer's hops to form a molecule that smells like...skunk spray. —*DV*

Also January 24:

1848: Gold Discovered in California
1984: First Apple Macintosh Computer Goes on Sale
   (see page 22)

Also 1935:

February 2: First Polygraph Lie-Detector Evidence
   (see page 33)

January 25
1921: Robots First Czech In
1979: Robot Kills Human

1921: A play about robots premieres at the National Theater in Prague, Czechoslovakia.

R.U.R., by Karel Čapek, marks the first use of the word *robot* to describe an artificial person. Čapek invented the term, basing it on the Czech word meaning "forced labor." The robots in Čapek's play are not mechanical men made of metal. Instead, they are molded out of a chemical batter, and they look exactly like humans.

In the play, over the course of just fifteen years, the price of a robot has dropped from $10,000 to $150. (In today's money, that's $128,000 down to $1,900.) Each robot "can do the work of two and a half human laborers," so humans could be free to have "no other task, no other work, no other cares" than perfecting themselves. However, the robots come to realize that even though they have "no passion, no history, no soul," they are stronger and smarter than humans. They kill every human but one.

The play explores themes that would later become staples of robot science fiction, including freedom, love, and destruction. Although many of Čapek's other works were more famous during his lifetime, *R.U.R.* is the one he is best known for today.

1979: In a remarkable coincidence, the first recorded human death caused by a robot occurred on the fifty-eighth anniversary of the play's premiere.

A twenty-five-year-old Ford Motor assembly-line worker was killed on the job in a Flat Rock, Michigan, casting plant. Robert Williams died instantly when the robot's arm slammed him as he was gathering parts in a storage facility, where the robot also retrieved parts. Williams's family was later awarded $10 million in damages. The jury agreed that the robot struck Williams in the head because it lacked safety measures, such as an alarm that would go off when the robot approached a person. —*TL, DK*

Also January 25:

1627: Birth of Physicist-Chemist Robert Boyle

1945: Grand Rapids, Michigan, Becomes First U.S. City to Fluoridate Drinking Water

Also 1921:

December 9: Leaded Gasoline Shown to Reduce Knocking

Also 1979:

March 28: Reactor Meltdown at Three Mile Island (see page 89)

January 26
1983: Spreadsheet as Easy as 1-2-3

Lotus begins selling its spreadsheet application for Microsoft DOS, called 1-2-3.

Lotus 1-2-3 was not the first spreadsheet application—it was preceded by VisiCalc. But 1-2-3 became the most popular, boosting sales of IBM PCs and PC clones, all of which ran DOS, and facilitating the rapid rise of Microsoft's operating system. The built-in charting and graphing capabilities of 1-2-3 helped it outsell VisiCalc. Lotus 1-2-3 quickly came to dominate the business-software market in the mid- and late 1980s.

Spreadsheet software, which seems commonplace today, was a major breakthrough for personal computing. It made it easy to keep track of columns of numbers, such as sales receipts, paychecks, expenses, or even athletic records. But the real power of the spreadsheet was the ability it gave businesspeople to run quick and easy what-if calculations. What if we lowered the price of our widgets by ten dollars? What if mortgage rates drop to 5 percent and we refinance? What if we laid off five thousand workers and shuttered our Kalamazoo plant, then outsourced manufacturing to a Chinese company for half the cost? Technology pundit John C. Dvorak says the what-if society turned corporate executives into slavish devotees of spreadsheet scenarios, no longer making decisions based on what customers actually want. But there's no doubt that the spreadsheet has transformed American business and the economy.

Lotus 1-2-3's reign lasted nearly five years, dwindling only when the company failed to transition from DOS to the increasingly Windows-centric world of the late 1980s and early 1990s. Microsoft Excel was much easier to learn than the forbiddingly austere text screen of Lotus's product, and Excel started outselling 1-2-3.

Lotus went on to create another incredibly successful business application, Lotus Notes, still used by many companies today. Lotus has been a division of IBM since 1995. —*DT*

Also January 26:

1700: Northwest Quake Unleashes Transpacific Tsunami
2006: End of an Era — the Last Western Union Telegram

Also 1983:

June 13: Pioneer 10 Reaches an End...and a Beginning (see page 166)
June 18: Sally Ride Becomes First American Woman in Space

1888: National Geographic Society Gets Going

A small cadre of businessmen, explorers, scientists, and scholars incorporates the National Geographic Society.

What began as a small, elite society for "the increase and diffusion of geographic knowledge" is now one of the world's largest nonprofit scientific and educational institutions. Its mission today is broader: "to inspire people to care about the planet." Founding president Gardiner Green Hubbard was the father-in-law and early financial backer of inventor Alexander Graham Bell (see page 71), another founding member and the society's second president. The society's *National Geographic* magazine first appeared just nine months after that founding meeting. It started as a drab scholarly journal sent to 165 charter members.

National Geographic's hallmark photojournalism began as an editor's desperate attempt to fill eleven blank pages of the January 1905 issue before it went to press. Gilbert Grosvenor gambled with a photo spread on Lhasa, Tibet. Members loved it. *National Geographic* has been a constant pioneer in photojournalism, photographic technology, and color printing. It was the first U.S. publication to establish a color-photo lab (in 1920), first to publish color underwater photographs (in 1927), first to print an all-color issue (in 1962), and first to print a hologram (in 1984).

Membership revenue has provided funding for more than nine thousand grants for research and exploration, including Robert Peary's expedition to the North Pole, Hiram Bingham's excavation of Machu Picchu, Jacques-Yves Cousteau's underwater exploration (see page 164), Louis and Mary Leakey's research on human evolution in Africa, and Dian Fossey's and Jane Goodall's studies of gorillas and chimpanzees, respectively. The society's yellow-bordered flagship publication is published in thirty-two languages and sent to eight million subscribers worldwide. NGS also produces films, books, DVDs, music, and games; runs a website; and has a television channel that reaches 270 million households in 166 countries.

That's increase and diffusion for you. —*AA*

Also January 27:

1910: Death of Thomas Crapper, Improver and Popularizer of the Flush Toilet

1950: Antibiotic Terramycin Announced in *Science*

1967: Three Astronauts Die in Apollo Capsule Fire

Also 1888:

August 12: Berta Benz Takes Boys on First Auto Road Trip (see page 226)

September 4: Eastman Patents Roll-Film Camera, Registers Kodak Name

January 28
2001: Hey, Don't Tampa with My Privacy

When the Baltimore Ravens and the New York Giants face off in Tampa, Florida, in Super Bowl XXXV, officials debut a new video technology that has nothing to do with instant replay. Facial-recognition surveillance cameras point at tens of thousands of fans entering the game.

The idea: catch known con artists or terrorists. The system set off alarms in the press before the big game, which one magazine called the Snooper Bowl. But there were no touchdowns for facial recognition that day. Not a single bad guy was caught.

Undeterred, Tampa police deployed the system on a busy street; a year later, they were forced to admit that they hadn't snagged a single fugitive and had mostly given up on the multimillion-dollar system. Tampa kept on with $8 million in federal grants to improve the system. Tampa patrolmen now use digital cameras to take pictures of citizens at traffic stops and compare them against a database of 7.5 million mug shots. About five hundred people have been arrested thanks to the system. But facial recognition on a mass scale remains an engineer's challenge and a civil libertarian's nightmare. Photo angles and lighting complicate the task of matching faces to photos. New 3-D solutions that take facial depth into account may overcome those problems. And the Department of Homeland Security hands out millions of dollars to cities and airports for more security cameras.

In the meantime, you can turn the technology on your friends and family using facial-recognition tools in photo-sharing sites like Picasa and Flickr. —*RS*

Also January 28:
1807: Flickering Gaslight Illuminates London's
    Pall Mall
1938: Race-Car Driver Bernd Rosemeyer Dies in
    Record-Setting Time Trial
1986: Space Shuttle *Challenger* Explodes
    (see page 103)

Also 2001:
October 23: Now Hear This ... the iPod Arrives
    (see page 298)
December 3: Segway Starts Rolling (see page 97)

January 29
1895: Electrifying!

Charles Proteus Steinmetz receives a patent for a "system of distribution by alternating currents." His engineering work makes it practical to build a widespread power grid for use in lighting and machinery alike.

Steinmetz fled Germany during an antisocialist crackdown and arrived in the United States in 1889. A brilliant mathematician, he figured out the law governing the power loss, or hysteresis, caused by the reversing magnetic fields of AC circuits. Steinmetz garnered instant fame among his peers, and the constant in his equation remains in use today. He also developed mathematical models for predicting the performance of complex circuits, so electrical engineers no longer had to build every system before they could determine how it would perform.

Thomas Edison was still working only with direct-current electricity (see page 4), but George Westinghouse (see page 66) had bought Nikola Tesla's patents for alternating current. Westinghouse rival General Electric placed its bet on AC and new employee Steinmetz. Building on his own work and Tesla's (see page 192), Steinmetz completed a system that let AC be used not just for lighting but for running multiphase motors "without the necessity of installing special multiphase generators for the motors, or running special circuits." Steinmetz was ready to electrify the nation—and the world.

After retirement, he still consulted for GE on difficult problems. Once, he painstakingly traced the problem in a nonfunctioning apparatus to the element that wasn't working and then marked it with chalk. When he submitted a bill for $10,000, GE asked him to itemize the charges.

He sent this invoice:

Making chalk mark: $1
Knowing where to place it: $9,999

—*RA*

Also January 29:
1901: Birth of TV Pioneer Allen B. DuMont
1964: *Dr. Strangelove* Premieres
1998: Tobacco Exec Admits Nicotine Is Addictive

Also 1895:
May 7: First Calculating Machine That Can Multiply
   (see page 129)
November 5: George Selden Receives First U.S.
   Automobile Patent

January 30
1975: Rubik Applies for Patent on Magic Cube

Ernö Rubik files for a patent on his twisty toy cube.

Rubik, who'd been schooled in sculpture and architecture, taught interior design at a Budapest art college. His initial interest in building the cube was structural: how the little cubies, or cubelets, could move without the big cube falling apart. Rubber bands didn't work, so he carved them to interlock with one another. He also applied different-colored paper to each of the big cube's six sides. As Rubik twisted his bright little bauble, the shifting colors pleased his design sense. But when he tried to put the colors back in order, he found it wasn't all that easy. Random twisting, he figured, would take a lifetime. (Spoiler alert: Partial solution ahead.) Rubik hit on the rubric of starting by aligning the corner cubes, but it still took a few weeks to solve the puzzle.

He applied for a Hungarian patent and arranged for a small Budapest co-op to produce the toy. Four other inventors in three different countries held similar patents, but the laurels would be Rubik's. Ideal Toy bought exclusive rights to the Magic Cube in 1979 and changed the name to Rubik's Cube to provide some trademark protection.

Omni magazine wrote about the cube in late 1980, and a slew of publicity followed. Rubik's Cube became the mega-fad of the early '80s. More than 300 million cubes have been sold. That's nothing, of course, compared to the 43,252,003,274,489,856,000 different possible permutations of the classic cube — enough to cover the planet with 273 layers of cubes, each with a unique arrangement of colors.

Speaking of large numbers, Ernö Rubik became the first self-made millionaire from the Communist bloc, or, in this case, block. — *RA*

Also January 30:
1790: The Lifeboat, an Idea Whose Time Has Come
1894: Charles King Patents Pneumatic Hammer

Also 1975:
April 4: Bill Gates, Paul Allen Form a Little Partnership
   (see page 96)

January 31
1958: First U.S. Satellite Discovers Van Allen Belt

The United States enters the space age with the successful launch of the Explorer 1 satellite. Data from the satellite confirms the existence of a radiation belt girdling Earth.

Explorer 1 (officially Satellite 1958 Alpha) blasted into orbit from Cape Canaveral atop a Jupiter-C rocket. Caltech and the Jet Propulsion Laboratory carried out the project for the U.S. Army (before NASA was founded). The Jupiter-C was a modified version of Wernher von Braun's Redstone ballistic missile, itself a direct descendant of another von Braun production, the German V-2 rocket.

Explorer was already on the drawing board but development was accelerated dramatically following the Soviet Union's successful launch of Sputnik 1 the previous October (see page 279). In all, it took eighty-four days to modify the rocket and to design and build the satellite.

Installation of Explorer 1 on its launch vehicle.

Explorer 1 was tiny, weighing thirty pounds fully loaded. More than half the weight was instrumentation, including a cosmic-ray-detection package, a variety of temperature sensors, and a microphone for picking up micrometeorite impacts. It was a model of simplicity, and it worked. Astrophysicist James Van Allen designed the onboard equipment that helped detect and return data about the Earth-circling radiation belts that now bear his name. Trapped within these two croissant-shaped belts, which run from roughly 125 to 620 miles above Earth, are radioactive particles capable of penetrating about one millimeter of lead. Radiation can damage the solar cells, integrated circuits, and sensors necessary to satellite operation, and astronauts who pass through the field may run a slightly higher than normal risk of developing cancer.

In response, NASA has taken to turning off sensors as they pass through the Van Allen belt, and it has improved the protective housing for sensitive instruments. The belt has not seriously impeded human travel in space. — *TL*

Also January 31:

1769: Birth of Pioneer Parachutist André-Jacques Garnerin

1881: Birth of Irving Langmuir, Inventor of Atomic-Hydrogen Welding Torch

Also 1958:

February 3: *Silent Spring* Seeks Its Voice (see page 34)

February 1
1951: TV Shows Atomic Blast, Live

For the first time, television viewers witness the detonation of an atomic bomb live, as KTLA in Los Angeles broadcasts the blinding light produced by a nuclear device dropped on Frenchman Flat, Nevada.

One of a hundred aboveground nuclear tests conducted between 1951 and 1962 in the Nevada desert, this A-bomb test found its way into history when a camera crew that had secretly taken position on top of a Las Vegas hotel focused on the blast. The images were relayed to the station's transmitter on Mount Wilson Observatory about two hundred miles away, and early-bird viewers saw their television screens fill with white light at 5:30 in the morning.

Witnessing the blast firsthand was KTLA reporter Stan Chambers. In a YouTube (see page 46) interview, Chambers described how station manager Klaus Landsberg pulled off the unauthorized broadcast: "We couldn't get near the field, because it was all top secret. Klaus sent a crew to Las Vegas and put them on top of one of the hotels....They kept the camera open for the flash of light that would come on when the blast went off." Los Angeles viewers tuned in for the one-off event. "We had a rating that was very large for 5:30 in the morning," Chambers recalled. "That one flash. You just see this blinding white light. It didn't seem real. We didn't have videotape. You couldn't say, 'Let's look at it again.'"

In 1952, KTLA set up the first live national feed for a Nevada atomic bomb explosion. That one was carried by the major networks. The tests became so commonplace that watching mushroom clouds turned into a Las Vegas tourist attraction. —*HH*

Also February 1:

1893: Edison Opens America's First Film-Production Studio
2003: Space Shuttle *Columbia* Lost

Also 1951:

April 5: Birth of Dean Kamen, Pied Piper of Technology (see page 97)
May 8: DuPont Debuts Dacron

A polygraph machine (sometimes known as a lie detector) is used for the first time to bring a conviction in court.

Criminal justice systems in many societies have long believed that you can spot a liar based on several physiological reactions to questioning. An increase in blood pressure and heart rate, dry mouth, perspiration—all are believed to suggest the likelihood of guilt. These factors are present in someone feeling anxiety, and, well, why would you feel anxiety unless you were lying? The polygraph measures and records these reactions, but of course the method is not exactly foolproof. Some people get anxious easily and fold at the knees without any real provocation. Others are as cool under duress as the proverbial cucumber.

Nevertheless, on February 2, 1935, Leonarde Keeler, a detective and coinventor of the Keeler polygraph, tested his invention on two suspected criminals in Portage, Wisconsin. The results of these tests were admitted as evidence in court, and both suspects were convicted of assault. Case closed.

But probably not.—*TL*

Also February 2:

1046: Monks Note Cold Weather, Start of Medieval "Little Ice Age"
1923: Leaded Gasoline Goes on Sale

Also 1935:

February 28: Sheer Bliss (see page 60)

February 3
1958: *Silent Spring* Seeks Its Voice

Rachel Carson writes to *New Yorker* editor E.B. White suggesting an article about the danger of pesticides. It's the genesis of her pioneering book *Silent Spring*.

Carson was already a successful scientist and author. She'd earned a master's in zoology and worked for the Fish and Wildlife Service. She'd written for *The New Yorker* and *Atlantic Monthly* and had authored bestselling books, including *The Sea Around Us*.

The proposed magazine article grew into a book. *Silent Spring* called forth the image of a spring without birds...and birdsong. Carson pointed out that inadequately tested pesticides were killing hundreds or possibly thousands of beneficial species. Not only did the chemicals often not work against their intended targets, Carson wrote, but they became concentrated: small animals and poisoned vegetation were eaten by other animals, who were eaten by larger animals, and so on up the food chain. And neither the interactions of the multiple chemicals nor their possible effects on humans, pets, and farm animals had been properly studied. *The New Yorker* started serializing *Silent Spring* in June 1962, and it was published in book form later that year. With its warning that it is arrogant to believe humans can totally control nature, *Silent Spring* is probably the most influential environmental book of the twentieth century. Still in print today, it stands with the previous century's *Walden* (see page 223) as a founding volume of eco-awareness. Carson galvanized the modern environmental movement. Despite the chemical industry's massive counterattack on the book, *Silent Spring* helped bring about the Clean Air Act, the Clean Water Act, the Occupational Safety and Health Act, and the founding of the Environmental Protection Agency—all within a decade.

Carson succumbed to breast cancer eighteen months after *Silent Spring* was published. She was fifty-six.—*RA*

Also February 3:

1468: Death of Johannes Gutenberg (see page 133)
1984: First Human Birth from Transplanted Embryo

Also 1958:

July 29: Ike Inks Space Law; NASA Born (see page 212)

February 4
2004: You've Got a Friend in the Facebook

Some college dudes unveil a website only Harvard University people can use. By 2011, it's worth $50 billion, and for hundreds of millions of people, the site now called Facebook is so integral to daily life that, for all intents and purposes, it is the Internet.

Trying to remember a time before Facebook? That'd be difficult for many people. Friendster was the social media leader in early 2004. Google was still months away from an IPO. YouTube (see page 46) was a year away, and Twitter (see page 82) would have been borderline indescribable.

The speed with which Facebook zoomed from dorm project to tech behemoth makes it seem like an overnight success. But Facebook's rise has not been without drama over the boundaries of privacy, the definition of *friend,* and the etiquette of blocking your mother.

And then there were the lawsuits over whose idea it really was, because success has many parents. Regardless, as *Wired*'s Fred Vogelstein argued, Mark Zuckerberg built the company, "and in Silicon Valley, at least, that's all that matters." Indeed, the concept of a social network wasn't new. Zuckerberg himself had a profile on Friendster, and as Facebook grew, the dominant player was MySpace. Facebook gained early currency as an elite destination for top-college students. It became a global sensation by reversing course and opening up to anyone. Astonishingly, when Facebook allowed mom and dad in, the cool kids didn't bolt. Instead, they bolted from MySpace. Will Facebook still be the dominant force in social networking in 2020? That'll depend as much on how much we want to continue sharing as on how the firm operates. After all, it's "always depended on the kindness of strangers."—*JCA*

Also February 4:

1915: Experiments Prove Poor Diet Causes Pellagra
1998: Belgian Artist-Provocateur Plants Cream Pie in
    Bill Gates's Face

Also 2004:

June 21: SpaceShipOne — First Privately Financed
    Manned Craft Reaches Edge of Space
    (see page 254)
August 13: Adam Curry Launches First Podcast, *Daily
    Source Code*
December 25: Next Stop, Titan, Saturn's Largest Moon
    (see page 361)

February 5
1840: Rat-A-Tat-Tat, You're Dead

Hiram Maxim, inventor of the machine gun, is born.

Maxim machine gun, 1898–1901.

Although multishot weapons existed in one form or another for centuries, Maxim's gun, which made its debut in 1881, is considered the first true machine gun. The key to the Maxim gun was that it eliminated the need for hand power, relying instead on the recoil of the previously fired bullet to reload the chamber. That exponentially increased the rate of fire achieved by earlier weapons, such as the Gatling gun. Maxim's other innovation was the introduction of a water-cooling system to reduce barrel overheating during extended firing.

By World War I, the Maxim gun had been adopted in one form or another by all the major combatants, and it was put to use with deadly effect in the trenches. The Maxim gun also saw action in aerial combat; variations were mounted on both Allied and German aircraft (see page 110). It was ironic, perhaps, that Maxim died only a few days after the Battle of the Somme, during which German machine gunners had mowed down thousands of attacking British infantrymen.

Technology had advanced far beyond Maxim's original gun by World War II, but his basic innovations were incorporated into newer, even deadlier designs. — *TL*

Also February 5:

1897: Indiana Legislature Nearly Declares $\pi = 3$
1940: Birth of H.R. Giger, Cyborg Surrealist Artist
1999: Millions View Victoria's Secret Online Fashion
    Webcast

Also 1840:

May 6: U.K. Issues World's First Adhesive Postage
    Stamps (see page 128)

February 6
1959: Titan Launches; Cold War Heats Up

The United States successfully test-fires its first Titan 1 intercontinental ballistic missile. The threat of global nuclear holocaust moves from the plausible to the likely.

The Titan 1 was not the first ICBM; both the United States and the Soviet Union deployed ICBMs earlier in the 1950s. But the Titan represented a new generation: a liquid-fueled rocket with greater range and a more powerful payload that upped the ante in the Cold War.

The Titan that the U.S. Air Force successfully launched from Cape Canaveral featured a two-stage liquid rocket capable of delivering a four-megaton warhead to a target eight thousand miles away. Puny by today's standards, four megatons nevertheless dwarfed the destructive power of the A-bombs dropped on Japan (see page 220). The Titan's range meant that, firing from its home turf, the United States was now capable of hitting targets in Eastern Europe, the western Soviet Union, and the far eastern Soviet Union.

The first squadron of Titan 1s was declared operational in April 1962. The missiles were stored in protective underground silos but had to be brought to the surface for firing. The Titan 2, which followed in the mid-1960s, could be launched directly from its silo. Though developed as a vehicle for delivering nuclear warheads to targets thousands of miles away, the Titan also proved effective as a launch platform for NASA. The Titan 2 was used extensively during the Gemini program (see page 243), before being replaced for Apollo (see page 203) by the far more powerful Saturn 5.

The Cold War is now history, and various treaties have led to the reduction of nuclear arsenals in both the United States and Russia. But the ICBM (which can be launched from silos, mobile launchers, or submarines) is still around, and still lethal. — *TL*

Also February 6:

1959: Texas Instruments Files to Patent Integrated Circuit (see page 257)

1971: Astronaut Alan Shepard Plays Golf on the Moon

Also 1959:

May 28: COBOL, a New Language for Business (see page 150)

July 24: Moscow Exhibit of U.S. Kitchen Ignites Nixon-Khrushchev Debate on Technology

February 7
1863: An Early Stab at Organizing the Elements

British chemist John Newlands organizes the known elements, listing them in a table determined by atomic weight according to what he provisionally calls his law of octaves. It is not an instant hit.

Newlands noticed, as he cataloged the elements sequentially, based on Stanislao Cannizzaro's atomic-weight system, that elements with similar properties tended to appear in regular intervals of eight, reminding him of the perfect eighth, or octave, in music. He called his explanatory paper "The Law of Octaves, and the Causes of Numerical Relations Among the Atomic Weights."

He arranged the elements by weight (like Cannizzaro) *and* by shared characteristics, grouping elements with similar properties on shared lines of his table. This required some fudging on Newlands's part and ultimately resulted in some inaccuracies. Nevertheless, Newlands defended his org chart, saying that no other method for cataloging the elements was workable.

Newlands's table was initially dismissed by the English Chemical Society as irrelevant. It wasn't until the Russian chemist Dmitry Mendeleyev published his own periodic table of the elements, in 1869, that Newlands's achievement began to be appreciated. Still, it would be another eighteen years before the Royal Society got around to awarding Newlands the Davy Medal in recognition of his work.

And it wasn't until 1913 that Henry Moseley established that the properties of the elements varied periodically according to atomic number, not atomic weight. — *TL*

Also February 7:

1984: Astronauts McCandless and Stewart Perform First Untethered Spacewalk

2000: "Mafiaboy" Hacker Overwhelms Major Websites with Denial-of-Service Attacks

Also 1863:

January 10: London Underground Opens (see page 10)

October 14: Alfred Nobel Patents Detonator for Nitroglycerin

February 8
1865: Mendel Reads Paper Founding Genetics

Gregor Mendel reads his first paper on genetics to the local scientific organization.

From 1856 to 1863, Mendel grew 28,000 pea plants in his monastery's garden. He kept careful records of his crossbreeding experiments, recording each individual plant's height, pod shape, flower location and color, and seed shape and color. He presented his literally seminal research at the Nature Research Society of Brünn (now Brno, Czech Republic) on February 8 and March 8, 1865. The papers introduced the concepts of dominant and recessive factors. He also postulated his two laws of heredity:

The Law of Segregation: An organism inherits two factors from its parents but contributes only one to its offspring.

The Law of Independent Assortment: Different traits are sorted separately from one another.

Taken together, these new concepts explained why crossbreeding pea plants that have purple flowers (dominant factor) with plants that have white flowers (recessive factor) yields three-quarters purple-flowered plants and one-quarter white-flowered in the next generation.

Mendel published his lectures as "Experiments on Plant Hybridization" in 1866. The methodical monk sent reprints to forty leading biologists. Only one responded. Charles Darwin (see page 363) never read his copy, this even though both Darwin and natural-selection codiscoverer Alfred Russel Wallace (see page 184) had acknowledged they could not explain how traits of successful organisms in one generation were passed on to the progeny.

Mendel's paper was cited a mere three times over the next thirty-five years. He died in 1884. It was 1900 before biologists realized their current research on heredity was merely reproducing, so to speak, Mendel's much earlier work. Near-simultaneous publications by three different botanists credited Mendel. An English translation of Mendel's 1865 paper finally appeared in 1901.

Mendel is acknowledged today as the founder of genetics and the scientist who first uncovered the mechanism that had eluded Darwin and Wallace. —*RA*

Also February 8:

1672: Isaac Newton Reads His First Paper on Optics Before the Royal Society
1924: First Execution by Gas Chamber

Also 1865:

September 28: England Gets Its First Woman Physician, the Hard Way (see page 273)
December 26: James Mason Patents Coffee Percolator

February 9
1870: Feds Get on Top of the Weather

President Ulysses S. Grant signs a bill creating what we now call the National Weather Service.

It had been obvious for centuries that weather in North America generally moves from west to east or from southwest to northeast. But other than looking upwind, you couldn't use that knowledge to predict the weather. You needed to move weather reports downwind faster than the weather itself was moving. The telegraph (see page 173) finally made that possible. In 1849, the Smithsonian Institution began supplying weather instruments to telegraph companies. Volunteer observers submitted observations to the Smithsonian, which tracked the movement of storms across the country.

Several states established their own weather services, but Congress thought the nation needed a centralized weather office with military precision. The War Department assigned the new function to the Division of Telegrams and Reports for the Benefit of Commerce. The network went online on November 1, 1870. At 7:35 a.m., observers at twenty-four stations in the eastern United States began taking synchronized readings and telegraphing them to the division's headquarters in Washington, DC. To head the unit, the U.S. Army hired Cleveland Abbe, a private forecaster who (his name notwithstanding) operated out of Cincinnati. He made his first official forecasts in February 1871.

A forecast looked like this:

Probabilities: It is probable that the low pressure in Missouri will make itself felt decidedly tomorrow with northerly winds and clouds on the lakes, and brisk southerly winds on the Gulf.

The weather division was renamed the U.S. Weather Bureau and transferred to civilian control as part of the Agriculture Department in 1891. President Franklin D. Roosevelt moved it to the Commerce Department in 1940.

The bureau was renamed the National Weather Service in 1970, when it joined the Commerce Department's newly created National Oceanic and Atmospheric Administration. —*RA*

Also February 9:

1883: Birth of Garnet Carter, Inventor of Miniature Golf
1969: Boeing 747 Jumbo Jet's First Test Flight

Also 1870:

February 26: New York City Blows Subway Opportunity (see page 58)

February 10
1996: Checkmate!

IBM's Deep Blue becomes the first computer to win a chess game against a reigning human world champion.

Had grand master Garry Kasparov gone on to lose the whole match, it would have stoked the fears of those who believed in the imminent arrival of a dystopian world where man is ruled by his inventions. But Kasparov, who became the world's youngest grand master in 1985, at the age of twenty-two, recovered his equilibrium after his initial stumble. He won the next game, drew twice, then took games five and six to win the match, 4–2. Kasparov lost a rematch to Deep Blue the following year—his first match loss ever to any kind of opponent. He managed a 3–3 draw in 2001 against Deep Junior, an entirely different software program.

Aside from their stunt value, these man-versus-computer matches have changed the way that chess is played, and not necessarily for the better. "We don't work at chess anymore," complained grand master Evgeny Bareev. "We just look at the stupid computer, we follow the latest games and find small improvements. We have lost depth."

Others, however, are more philosophical: "Cars can outrun us, but that hasn't stopped us from having foot races," said U.S. grand master Maurice Ashley. "Even if a computer is the best player on the planet, I'll still want to go around the corner, set up the chess pieces, and try to kick your butt." — *TL*

Also February 10:

1957: Styrofoam Ice Chest Invented
1961: New Niagara Falls Hydroelectric Power Plant
    Starts

Also 1996:

July 23: Stand By … High-Definition TV Is on the Air
    (see page 206)

February 11
2005: This Guy's No Dummy

Samuel Alderson, inventor of the automotive crash-test dummy, dies. His creation saved countless lives…and amused millions along the way.

Alderson studied physics under J. Robert Oppenheimer and Ernest O. Lawrence and worked on missile-guidance systems during World War II. He built dummies for the military to test jet-ejection seats (see page 13) and parachutes. He also engineered one for NASA to test the Apollo lunar-module splashdown.

Those dummies matched the size, shape, and weight of pilots and astronauts, had joints to mimic human biomechanics, and contained scientific instruments to measure acceleration and impact forces. Alderson adapted one to test automobile safety in 1960, but he was a few years ahead of his time. At that point, automotive engineers tested their cars using cadavers, with unsatisfactory results: The stiffs were, well, stiff. Also, no two cadavers were alike, and after a couple of tests, they degraded rapidly (to say the least). That made it difficult to generate consistent and reproducible results.

Alderson's first production-model auto dummy hit the road in 1968. The dummy featured a steel ribcage, articulated joints, and a flexible spine. Engineers at General Motors combined elements of Alderson's dummies with those from another company to create a dynasty of Hybrid dummies: men, women, children, and infants. They're used in testing seat belts, air bags, and other safety features that the National Highway Traffic Safety Administration figures have saved more than 300,000 lives since 1960.

The *New York Times* obituary said Alderson's "cultural legacy includes Vince and Larry, the ubiquitous dummy stars of highway safety advertisements in the 1980s and '90s; the television cartoon *Incredible Crash Dummies;* and the pop group Crash Test Dummies."

In case you're wondering, Alderson, like the inventor of the three-point seat belt (see page 193), died of natural causes. He was ninety.—*RA*

Also February 11:

1751: First U.S. Hospital Opens in Philadelphia

1847: Birth of Thomas Edison (pages 4, 19, 222, 229, 296)

1939: Lise Meitner Publishes Discovery of Atomic Nuclei Splitting in Uranium Reactions

Also 2005:

February 15: YouTube and Your Fifteen Minutes of Fame (see page 46)

February 16: Kyoto Protocol on Greenhouse Gases Goes into Effect

February 12
1878: A Face-Saving Invention from Harvard

Fred Thayer, captain of Harvard University's baseball team, receives a patent for his inspired invention: the catcher's mask.

Getting hit in the face with a baseball is no fun at all. Only a complete idiot would willingly squat behind home plate without wearing a mask (as well as a chest protector, shin guards, and, yes, if appropriate, a cup). But when the game was in its infancy and pitchers didn't throw a hundred miles an hour, catchers used zero protection.

The curveball was getting popular, and some catchers had trouble handling it. When Harvard catcher James Tyng started acting head-shy, Thayer looked for ways to fix things. Taking his inspiration from the fencing mask, Thayer designed a mask that added a forehead pad and chin rest to help absorb the impact of the ball. He also replaced the fine wire mesh with a birdcage face protector that improved visibility. Thayer and Tyng experimented with the catcher's mask away from the diamond before Tyng finally used it in a regular-season game, on April 12, 1877. It was, pardon the pun, a hit. Gushed the *Harvard Crimson*:

> [T]he new mask was proved a complete success, since it entirely protects the face and head and adds greatly to the confidence of the catcher, who need not feel that he is every moment in danger of a lifelong injury.

Tyng originally wore the mask only on a two-strike count, when catching the third strike cleanly was necessary to record the out. But he soon saw the advantage of wearing it all the time. Nobody argued.

Word spread quickly throughout baseball, and soon even major-league catchers adopted the mask. The Spalding sporting-goods firm began selling Thayer's Patent Harvard Catcher's Mask for $3 each (figure $70 in 2012 dollars).

Thayer's original is on display in the National Baseball Hall of Fame. — *TL*

Also February 12:

1809: Birth of Charles Darwin (see pages 184, 363)
1908: Automobiles Start Great Race from New York
   to Paris

Also 1878:

April 21: Firehouse Pole Invented (see page 113)

February 13
1990: Seeing the Earth as Others See Us

The first picture of our solar system taken from deep space is sent to Earth by Voyager 1.

The image captures the sun and six planets, including Earth, in a single frame. The sun appears much as any other star would look from Earth, and the planets are barely visible dots.

It was Voyager's final look back before it passed beyond the planets and headed for the edge of our solar system.

Voyager 1, launched September 5, 1977, is an interplanetary probe designed to collect and transmit various data for as long as the craft can function. Even though it was launched two weeks after its sister probe, Voyager 2, Voyager 1 was placed on a faster trajectory, and it quickly overtook its mate.

The launch of the probes was timed to coincide with a rare alignment of the solar system's four largest planets—Jupiter, Saturn, Uranus, and Neptune. The spacecraft used the planetary configuration for gravitational assists from one planet to the next. The technique cut the probe's flight time to Neptune from thirty years to twelve. Both Voyagers sent back spectacular images of these enormous gaseous spheres and their key moons.

After leaving the planets behind, Voyager 1 shifted to transmitting data from the outer solar system, including results of plasma-wave experiments conducted to locate the heliopause—the boundary of interstellar space.

Nuclear batteries on board are expected to keep Voyager 1 functioning and transmitting data until 2020, at which point the craft will be forty-three years old and thirteen billion miles from Earth—the most distant man-made object in the universe. —*TL*

Also February 13:

1633: Galileo Arrives in Rome for Heresy Trial (see page 256)

1895: Lumière Brothers Patent Cinématographe (movie camera and projector)

2004: Astronomers Announce Discovery of Largest White Dwarf Star

Also 1990:

April 24: NASA Launches Hubble Telescope

May 22: Microsoft Launches Windows 3.0

August 18: Psychologist B.F. Skinner Dies. Is His Coffin a Skinner Box?

1929: Al Capone's .45-Caliber Valentine

Gangland slaying takes a quantum leap when mobsters working for Al Capone use the cutting-edge technology of the day — the Thompson submachine gun — to wipe out a rival gang.

It's a reminder of the darker side of tech. The St. Valentine's Day massacre wasn't the first time a mobster used a tommy gun in a rub-out, but the slaughter of seven men was unprecedented, even by jaded Chicago standards.

Capone ordered the massacre to wipe out George "Bugs" Moran and his North Side Gang, who were muscling in on Capone's Prohibition-era bootlegging operations. The plan was to lure Moran and as many of his men as possible to a garage at 2122 North Clark Street and take care of business there. A shipment of smuggled alcohol is usually cited as the bait, but what actually brought Moran's men there remains unclear.

Moran himself was a no-show. A Capone lookout mistakenly identified one guy as Moran, and Capone's men closed in. Two of them, disguised as cops, disarmed Moran's mugs, who probably suspected some kind of shakedown. The "cops" had the gangsters line up against the back wall of the garage as if they were going to be frisked. They were: with .45-caliber slugs from a couple of Thompsons brought in by two plain-clothed killers. Although the tommy guns provided plenty of firepower, these were professionals. The executioners used shotguns to seal the deal.

The police made no arrests even though there was no doubt who lay behind the St. Valentine's Day massacre. And although Capone failed to physically eliminate Moran, the damage was done. Moran lost power and eventually control of the North Side.

The Thompson went on to be used effectively during World War II but will forever be identified as the Mob's favorite weapon. — *TL*

Also February 14:

1838: Birth of Margaret Knight, Inventor of Paper-Bag-Making Machine

1989: First GPS Satellite Launched

Also 1929:

March 29: President Hoover Gets a Telephone Inside Oval Office Instead of Anteroom

September 25: Doolittle Proves Instrument Flying Works from Takeoff to Landing

February 15
2005: YouTube and Your Fifteen Minutes of Fame

Technology further erodes the notion of private life as the YouTube.com domain starts up.

The video-sharing site, founded by Steve Chen, Chad Hurley, and Jawed Karim, has become one of the most visited websites worldwide; as of 2011, it trailed only Facebook, Google (YouTube's parent company), and Google's Gmail.

With its slogan Broadcast Yourself sounding a clarion call to exhibitionists the world over, YouTube was an instant hit, encouraging individuals to submit not only personal videos but movie clips, TV clips, and music videos as well. Its success attracted the attention of Google; in 2006, less than two years after YouTube was founded, Google shelled out $1.65 billion to acquire it.

YouTube's reach is enormous. Although the company doesn't make exact numbers available, YouTube (which is available in thirty languages) boasts hundreds of millions of registered users and has a constantly expanding video archive— twenty-four hours of video are added every minute. The site's popularity has not gone unnoticed by American political candidates, Democrats and Republicans alike, including those running for president. One candidate, Republican Ron Paul, began his quest for the White House with a campaign limited almost exclusively to the Internet. YouTube can backfire as a political tool, however. At least one candidate, U.S. senator George Allen (R-VA), was undone by it—he lost his 2006 reelection bid after a clip showing him allegedly making racist remarks was repeatedly played on YouTube.

One of the biggest stumbling blocks for the YouTube business model is potential copyright infringement. Registered users routinely submit clips from concerts and movies and TV shows, drawing the ire of (and lawsuits from) a variety of original-content providers.

YouTube has also come under fire for both its censorship and its failure to screen carefully for material that many would consider disgusting or hateful. —*TL*

Also February 15:
1809: Birth of Cyrus McCormick, Inventor of
    Mechanical Grain Reaper
1995: FBI Busts Computer Hacker Kevin Mitnick

Also 2005:
February 11: Inventor of Auto Crash-Test Dummy Dies
    (see page 42)
March 3: Fossett Flies Nonstop Solo Around Globe

February 16
1978: Bulletin Board Goes Electronic

Ward Christensen and Randy Suess launch the first public dial-up bulletin-board system.

It was several decades before the hardware or the network caught up, but the seeds of today's online communities were sown when the two launched CBBS, the computerized bulletin-board system.

Reportedly conceived when Christensen was trapped in his Chicago home during the Great Blizzard of 1978, CBBS took its basic premise from the community bulletin boards that adorned the entrance of libraries, schools, and supermarkets.

The notion of digital meeting places in mind, Christensen and Suess set out to create the software and managed to go from idea to working bulletin board in just a month. The two developers announced their creation in the November 1978 issue of *Byte* magazine. The article created a stir among hobbyists and hackers, and it wasn't long before others began building CBBS clones. By the mid-1980s, BBSs supported an active community with three magazines devoted to covering the latest in the proto-online world.

The original homebrewed Internet, CBBS was primitive but quickly proved revolutionary. Sure, connecting to someone on a BBS meant dialing into a phone line through your clunky, first-generation personal computer, typing a message in your monochromatic terminal, and waiting days or weeks before you (hopefully) got a reply. Dialing outside your area entailed steep long-distance phone charges. But holy cow, look! You've got digital friends.

Because of its complexity, limitations, and slowness, the system was largely populated by computer enthusiasts willing to shell out big bucks for the fastest modems. So, like the early web that came after it, early BBSs consisted mostly of technical postings, software downloads, and primitive online games.

BBSs still thrive in Taiwan, where they're an extremely popular form of communication for young people. —*SG*

Also February 16:

1968: First 911 Call Made, in Haleyville, Alabama
2005: Kyoto Protocol on Greenhouse Gases
    Goes into Effect

Also 1978:

January 23: Sweden Is First Country to Ban Aerosol
    Sprays to Save Ozone Layer
June 26: First Dedicated Oceanographic
    Satellite in Orbit
August 11: First Atlantic Balloon Crossing Takes Off

February 17
1818: Proto-Bicycle Gets Things Rolling

A minor German noble patents a two-wheeled, foot-powered vehicle. It looks almost like a modern bicycle, minus some key components.

A couple of bad oat crops had caused horses to starve, which got Baron Karl Christian Ludwig von Drais de Sauerbrun thinking about how you could get around quickly without a horse. His first attempt was a four-wheeled vehicle with a treadmill crankshaft between the rear wheels. That got nowhere, so he invented a two-wheeler on a frame that looked much like a modern bicycle frame, with a seat and front-wheel steering. It didn't have a chain drive; it didn't even have pedals. You drove the thing with your feet, much like a scooter. You also stopped it with your feet: it had no brakes.

Von Drais's *Laufmaschine*, or running machine, bested 9 miles an hour on its first trip, June 12, 1817, near Mannheim. He patented the invention the next year, but better weather and falling oat prices dimmed its future as a practical replacement for the horse.

Some towns fined the machines' users for riding on public roadways. The two-wheelers needed paved or at least smooth surfaces, of which there weren't many. It was easy to fall off the contraption, and people's leather shoes weren't as durable as a horse's iron shoes. And the *Laufmaschine* faced competition from another new invention: the railroads. So the utilitarian-inspired mechanical horse became a fancy toy for aristocrats and the rising bourgeoisie. The French called it a *draisine;* the English a hobby horse. The devices were often graced with equine figureheads or even carved dragons and elephants.

A *draisine* today is a hand- or foot-propelled railcar used for track maintenance. For the modern bicycle, we thank Ernest Michaux (pedals and brakes), Harry John Lawson (chain transmission), and Robert W. Thompson (pneumatic tire, see page 346). — *RA*

Also February 17:

1864: Confederate *H.L. Hunley* Is First Submarine to Sink Enemy Ship

1972: VW Beetle Surpasses Ford Model T as Most Popular Car Ever

Also 1818:

January 1: Mary Shelley's *Frankenstein* Published

December 24: Birth of Physicist James Joule, Discoverer of Heat-Work Relationship

February 18
1838: Physicist Machs His Entrance

Ernst Mach is born in what's now the Czech Republic. His most memorable work in aerodynamics will be the understanding of supersonic speeds, leading to the measure that bears his name.

Mach explained his understandings of supersonic flow in a paper published in 1887. The paper included the first photograph showing the shock waves that form when an object moves at supersonic speeds.

Mach didn't become a measure until 1929, thirteen years after the physicist's death. A Mach number is the ratio of the speed of an object to the speed of sound in the surrounding medium. Mach 1.0 means that the speed of the object is equal to the speed of sound in that medium. Pilots who fly near the speed of sound use the Mach number because it gives the aerodynamic condition of the aircraft independent of changes in air density due to altitude, temperature, or humidity. Primarily because air density decreases as altitude increases, an aircraft "feels" less air the higher it flies. So the aerodynamic forces acting on an aircraft traveling 700 miles an hour at sea level are very different from the forces acting on an aircraft traveling the same speed at 50,000 feet. The airplane at sea level is traveling slower than the speed of sound, at Mach 0.92. The aircraft flying at 50,000 feet will be flying at Mach 1.06, faster than the speed of sound. High-speed pilots and engineers describe velocity by Mach number because airliner performance largely depends on Mach, not on miles-per-hour speed. An airliner at Mach 0.85 at 10,000 feet will behave much the same as at Mach 0.85 at 30,000 feet. Likewise, a fighter jet's behavior at Mach 1.4 will be essentially the same whether the craft is at 10,000 feet or 30,000 feet.

The first time an aircraft flew faster than Mach 1.0 was October 14, 1947, when Captain Chuck Yeager flew a Bell X-1 to Mach 1.07. —*JP*

Also February 18:

1898: Birth of Enzo Ferrari
1913: Radiochemist Frederick Soddy Coins *Isotope*
1930: Pluto Discovered (see page 238)

Also 1838:

March 12: Birth of W.H. Perkin, Synthesizer of Mauve Dye

February 19
1986: Mir, the Little Space Station That Could

Mir, the first modern-era space station, is launched by the Soviet Union.

Space shuttle *Atlantis* connected to Russia's Mir space station.

Mir (Russian for both "peace" and "world") came after the U.S. Skylab and Soviet Salyut stations and before the current International Space Station. Mir was assembled in stages; seven of them, counting the original core module. The station became the centerpiece of the Soviet (and later Russian Federation) space program.

The Soviets devoted more time than the Americans did to studying the effects on humans of long-term stays in microgravity. Cosmonauts were typically aloft for much longer periods than astronauts, and Valery Polyakov spent a record 437-plus days aboard Mir (see page 83).

With the Cold War over, the Russians began welcoming U.S. shuttle crews to Mir as part of the collaborative groundwork for the coming ISS. Seven shuttle missions docked with Mir, their crews remaining aboard for varying lengths of time. As part of this collaboration, NASA agreed to pay for the installation of the Spektr and Priroda modules, the last two pieces of the Mir puzzle. A special docking module to accommodate the U.S. shuttles was also added. So, in a sense, Mir can claim to be the first international space station.

Attempts were made to keep Mir aloft using private money, but all the plans (including one that would have converted it into an orbiting movie studio) came to nothing. Burdened by its financial commitments to the ISS, Russia decided to end Mir's life. The station's orbit was gradually decelerated, and on March 23, 2001, Mir came flaming through the atmosphere on its way to a watery grave in the South Pacific. Mir spent 5,511 days in orbit, more than fifteen years. —*TL*

Also February 19:

1473: Birth of Nicolaus Copernicus
2002: Odyssey Spacecraft Turns Its Cameras on Mars

Also 1986:

January 28: Space Shuttle *Challenger* Explodes
March 11: NFL Adopts Instant Replay
April 26: Chernobyl Nuclear Plant Suffers Meltdown

February 20
1962: John Glenn Orbits Earth

John Glenn isn't the first American in space—Alan Shepard beat him by eleven months—but he is the first to orbit the planet, which he does three times aboard the *Friendship 7*. It's a watershed moment for the U.S. space program.

Glenn was a Marine pilot who flew more than a hundred combat missions in World War II and the Korean War (sometimes with baseball great Ted Williams as his wingman). He became a test pilot in the 1950s and joined NASA as a Project Mercury astronaut in 1959. By then he had logged over nine thousand hours of flying time and was among several astronauts selected to provide input for the design of the Mercury space capsule. (He would also have a say in the interior configuration of the early Apollo capsules.)

On February 20, Glenn got to inspect his handiwork up close and personal, from atop an Atlas rocket booster at Cape Canaveral. *Friendship 7* lifted off shortly before 10:00 a.m. and made three orbits of Earth at an altitude of 162 statute miles. From liftoff to splashdown, Glenn was aloft for a total of four hours, fifty-five minutes, and twenty-three seconds. He was the second human, after the Soviet Union's Yury Gagarin, to orbit Earth.

Following his retirement as an astronaut in 1965, Glenn went into private business. He was elected to the U.S. Senate in 1974 as an Ohio Republican, and he served until 1999. He wasn't through with space, however. Glenn returned to the heavens once more, in 1998, as a crewmember aboard the space shuttle *Discovery.—TL*

Also February 20:

1792: U.S. Post Office Established

1934: Ernest O. Lawrence Patents Cyclotron Subatomic Particle Accelerator

Also 1962:

May 23: Doctors Reattach Youngster's Arm (see page 145)

February 21
1947: *Take a Polaroid* Enters the English Language

Inventor Edwin Land demonstrates the first instant camera. The camera would become better known by his company's name: Polaroid.

Using developer contained in a sac layered with the film and photographic paper, the first Polaroid camera could produce a black-and-white photo in sixty seconds. After the exposure was made, the packet was pulled out of the camera, which squeezed the chemical from its sac and began the developing process. (Later models included the photographic fixer, but the original Polaroids required hand-fixing with a little gizmo.) Land introduced Polacolor film in 1963, allowing color Polaroids.

Edwin Herbert Land was a prolific inventor, second only to Thomas Edison in the number of patents received. His Polaroid Corporation produced polarizers for a variety of scientific and commercial applications. Polaroid became immensely profitable during World War II, making high-quality optics for military use. Land and his Polaroid engineers also helped design the optics for the Cold War's U-2 spy plane.

Land was legendary for his eccentric and exhausting work habits. Like the archetypical mad scientist, he locked himself in the lab for days on end, stopping only to eat and often not bothering to change clothes. When Land was on one of these jags, he scheduled his assistants in shifts so they could keep up with him.

The Polaroid Land camera remained in production until 1983. The Polaroid Corporation itself fell on hard times after Land's death and filed for bankruptcy protection in 2001. The company announced in 2008 that it was closing the facilities that had produced its namesake instant cameras and would concentrate instead on making digital cameras and portable printers. —*TL*

Also February 21:
1866: Birth of August von Wassermann, Discoverer of
  Syphilis Test
1931: Alka-Seltzer Debuts

Also 1947:
July 6: The AK-47, an All-Purpose Killer (see page 189)

Tech Presidents' Day—George, Tom, and Abe

Three U.S. presidents were surprisingly inventive.

WASHINGTON'S ADVICE

January 8, 1790: During the very first State of the Union address, President George Washington urges the young nation to import "useful inventions from abroad" but encourage homegrown genius to flourish by offering patent protection for inventors. Washington was trained as a surveyor, and he attached great importance to the study of science and literature. He was looking to the country's economic future and its military security.

In response to Washington's request, Congress passed a patent act. Washington signed it into law on April 10, and on July 31, 1790, the United States granted its first patent to Samuel Hopkins for his process of making potassium carbonate (see page 214).

JEFFERSON'S FOSSIL

March 10, 1797: Thomas Jefferson presents a scientific paper that's considered the first American contribution to vertebrate paleontology.

Jefferson, the new vice president of the United States, was president of the American Philosophical Society, a distinguished association founded by Ben Franklin and others in 1745. For his presidential address to the group, he read a paper, "A Memoir on the Discovery of Certain Bones of a Quadruped of the Clawed Kind in the Western Parts of Virginia." The bones belonged to an extinct, ox-size clawed sloth of the genus *Megalonyx*. A French naturalist in 1822 assigned the sloth the name *Megalonyx jeffersonii*. The Sage of Monticello also invented a wheel-shaped mechanical cipher-decipher machine; automatic double doors; and improvements to plows, sundials, clocks, beds, bookstands, and several devices that copied writing.

Despite all this, and despite his position on the board that oversaw the first U.S. patent law, Jefferson held no patents himself. He was an open-source kind of guy; he believed it was necessary for inventors to be rewarded, but he distrusted a system that could be abused to keep needed innovations from reaching public use.

President John F. Kennedy paid tribute to Jefferson at a 1962 White House dinner honoring all forty-nine living American Nobel Prize recipients. He told the august assemblage, "I think this is the most extraordinary collection of talent and of human knowledge that has ever been gathered together at the White House— with the possible exception of when Thomas Jefferson dined alone."

LINCOLN'S PATENT

March 10, 1849: Abraham Lincoln files for a patent, starting a process that would make him the only U.S. president to patent an invention. Lincoln had worked on riverboats, and he once engineered a stranded flatboat off a dam by shifting cargo and drilling a temporary hole to let out bilge water. As a boat passenger on another occasion, he observed a captain use planks and empty barrels to lift his stranded vessel off a river sandbar.

Inspired, Lincoln set to work to design a system of inflatable, rubberized cloth bags that could, theoretically, be built into or added onto any boat. It was a complex arrangement of ropes, pulleys, spars, and sacks. His patent application called it "a new and improved manner of combining adjustable buoyant air chambers with a steamboat or other vessel for the purpose of enabling their draught of water to be readily lessened to enable them to pass over bars, or through shallow water, without discharging their cargoes." The U.S. Patent and Trademark Office says patent 6469 is the only patent held by an American president. Lincoln's law partner observed, "The invention was never applied to any vessel, so far as I ever learned."

But a dozen years later, inventor Lincoln would be steering the nation through treacherous shoals. — *TL, RA*

Also February 22:

1828: Wöhler Reports Making Urea, First Synthetic Organic Compound
1857: Physicist Heinrich Hertz Enters the Cycle of Life
1997: Dolly the Sheep, First Cloned Mammal, Is Announced

Also 1790:

May 8: France Authorizes Metric System (see page 130)

Also 1797:

October 22: A.-J. Garnerin Jumps from Hot Air Balloon with First Silk Parachute

Also 1849:

April 10: Safety Pin Patented (see page 102)

February 23
1942: Invasion! Japanese Sub Attacks California!

A Japanese long-range submarine surfaces off the California coast and uses its five-and-a-half-inch deck gun to shell an oil refinery near Santa Barbara.

The attack lasted about twenty minutes but caused little damage to the Ellwood refinery. It did stoke fears, which had existed since the raid on Pearl Harbor ten weeks earlier, that the Japanese were preparing for a full-scale invasion of the West Coast.

Japan's imperial high command envisioned nothing of the sort, lacking both the military capacity and a strategic reason for invasion. The *I-17* was on combat patrol along the Pacific Coast, and five days after shelling the refinery, it torpedoed an American tanker off Cape Mendocino. Commander Nishino Kozo, skipper of the *I-17*, was familiar with the Ellwood refinery, having docked there as the captain of an oil tanker before the war. A *Parade* magazine article in 1982 suggested that Kozo staged the attack on his own initiative, in retaliation for a slight he had suffered during a prewar visit to Ellwood.

Kozo's gunnery display scared already skittish Americans. On the night following *I-17*'s shelling of the refinery, trigger-happy antiaircraft gunners in Los Angeles spotted some UFOs and lit up the sky with tracer ammunition for a couple of hours.

The *I-17* was a B1-class submarine: 350 feet long with 2,200 tons of surface displacement, and by far the largest combat sub to see service during World War II. By comparison, Germany's largest long-range combat U-boat, the IXD, was 70 feet shorter and displaced barely 1,600 tons when surfaced.

American coastal defenses were poorly organized in early 1942, and Kozo was able to take advantage of that fact. German U-boat commanders on the East Coast discovered the same thing, with devastating effect on Allied shipping. — *TL*

Also February 23:
1941: Seaborg's UC Berkeley Team Isolates Plutonium
1987: "Quintessential" Supernova Bursts on Scene

Also 1942:
July 18: World's First Operational Jet Fighter Takes Wing (see page 201)

February 24
1664: Steam Power Is Newcomen In

Thomas Newcomen, creator of the first practical steam engine, is born in Devon, England.

Flooding was a major problem in the local mines, and water often had to be pumped out by human or animal power. It was for this task that Newcomen created the first practical steam engine, the prototype that made the Industrial Revolution possible (see page 198).

Newcomen's engine used a vertical brass cylinder with a piston connected to a rocking beam. A copper boiler below the cylinder heated the water to boiling. When the piston was at the top of its range of motion, water was sprayed into the cylinder. That cooled the insides, condensing the steam within. This formed a vacuum, pulling the piston down. The boiler, still on, then reheated the steam, driving the piston up again. Repeating this process caused the rocker beam to move up and down like a seesaw. Newcomen's design attached the working end of the beam (opposite the piston) to chains that descended to pumps located deep in the mine.

The first one was installed in 1712. Simplicity and effectiveness made the Newcomen engine popular. More than 110 were in operation by 1773.

Newcomen wasn't the only one to stumble on the basic principle. Thomas Savery had invented a device in 1698 that used a vacuum and atmospheric pressure to suck water, but it didn't work well in practice. After Savery's death, a joint stock company was set up to issue licenses for both designs. The Newcomen design eventually faded away, mainly because the machine was expensive to operate. The James Watt design that replaced it in the 1770s added a condenser cylinder to reduce heat loss and increase fuel efficiency.

Newcomen himself finally ran out of steam and died at home in 1729. — *PG*

Also February 24:
1582: Papal Bull Creates Gregorian Calendar
    (see page 1)
1949: First Rocket to Reach Space, 100 KM Up

Also 1664:
May 9: Robert Hooke Observes a Giant Spot on Jupiter

Samuel Colt receives a U.S. patent for the first revolver.

The inspiration for Colt's refinement of the one-shot pistol apparently came while he was a young hand aboard the sailing ship *Corvo*. The rotation of the helmsman's wheel, or perhaps the turning of a capstan, got him thinking about the possibility of a revolving chamber rapidly feeding bullets into a pistol. He carved a wooden prototype, and after his sailing days were over, he produced a metal revolving cylinder based on the model. It proved to be a remarkably simple and effective method for delivering five or six bullets in rapid succession without reloading.

After receiving his patent, Colt began manufacturing revolvers—two sidearms and a rifled model—in Paterson, New Jersey. Despite the guns' demonstrated efficiency, sales lagged, and Colt nearly lost his shirt. It wasn't until 1845, when U.S. troops fighting Indians in Texas came back with stories of the weapon's effectiveness, that Colt began mass production in a Connecticut plant. Prior to the outbreak of the American Civil War, Colt shipped his firearms nationwide. With secession and the birth of the Confederacy, he stopped all deliveries to the South and supplied only Union troops during the war.

Colt's simple shipboard invention revolutionized the firearms industry, changed the nature of warfare, and made him insanely wealthy. His worth at the time of his death, in 1862, was estimated at $15 million, about $340 million in 2012 dollars. — *TL*

Also February 25:

1723: Mathematician-Astronomer-Architect Christopher Wren Dies

1837: Davenport Patents Electric Motor (see page 333)

1919: Oregon Taxes Gasoline 1¢ per Gallon, First in U.S.

Also 1836:

July 13: U.S. Issues Patent No. 1 after 9,957 Unnumbered Patents

December 15: U.S. Patent Office Burns, Destroying 10,000 Records

February 26
1870: New York City Blows Subway Opportunity

Inventor Alfred Ely Beach opens New York City's first subway line, a pneumatic demonstration project in a three-hundred-foot tunnel under Broadway.

Beach had permission to build a package-delivery tunnel under Broadway, but he secretly began work on a passenger-transit demonstration instead. A rush of air from a massive blower propelled the car. "When the blower is in motion, an enormous volume of air is driven through the tunnel, which drives the car before it like a boat before the wind," Beach wrote. After only fifty-eight days of construction, Beach's subway opened as a demonstration on February 26, 1870. Passengers entered the railway through a luxurious station in the basement of Devlin's clothing store. The price of admission was a small donation to a home for orphans of Union soldiers and sailors from the Civil War.

The railway wasn't actually operational on its opening day, because of an engine failure, but within a week, passengers began taking the short journey under Broadway from Warren Street to Murray Street and back. When Beach couldn't get state funding for a complete subway network, he blamed infamous Tammany Hall ruler Boss Tweed. Wealthy Broadway landowners also helped scuttle the plan, fearing the tunnels would damage their buildings' foundations.

But even if he'd gotten state approval, the financial panic of 1873 and the depression that followed would have prevented Beach from building his system. Because they cost less than underground lines, elevated lines gained popularity. The IRT didn't begin underground public transit service until thirty-four years after Beach's demonstration line opened.

No elements of Beach's subway remain (despite its appearance in *Ghostbusters 2*). The station burned in 1898, and the tunnel was destroyed in 1912 during construction of a BMT tunnel. Today's City Hall station occupies the former tunnel's footprint. — *KB*

Also February 26:
1935: Radar, the Invention That Saved Britain
1991: Birth of the Web (see page 114)

Also 1870:
January 2: Brooklyn Bridge Construction Begins
    (see page 2)

February 27
1812: Rage, Rage Against the Industrial Age

The poet Lord Byron gives an impassioned speech urging the House of Lords not to make the crimes of the Luddites a capital offense. It doesn't work. But the term *luddite* persists today for one who ignores, declines, opposes, refuses, or resists technology.

The Luddites were workers, especially croppers and weavers, who saw growing industrialization as a threat to their livelihoods. They took their name from the folkloric figure of Ned Ludd, who was said to have smashed a couple of stocking frames (knitting machines) in the late 1770s. The Frame Breaking Act made those convicted of machine breaking—the willful destruction of mechanized looms, cloth-finishing machinery, and other devices that eliminated jobs—subject to the death penalty.

The movement began in early 1811; Nottingham factory owners and craft employers received letters calling on them not to install the new machines. When the letters were ignored, the movement spread across England. As the salaries of apprenticed workers were cut and jobs were lost, violence began. Early attacks came at night, with bands of workers breaking into locked factories to smash the machines. In one three-week period, they destroyed more than two hundred stocking frames.

Even with the imposition of capital punishment and an increase in police protection, the attacks continued and escalated in ferocity. Wheat prices soared in early 1812, and out-of-work craftsmen unable to feed their families became desperate. In clashes around the country, scores of Luddites were killed, as were a number of mill and factory owners. Other Luddites were arrested, convicted, and executed, including Abraham Charlston, a twelve-year-old boy. By mid-1812, the Luddites had effectively been broken, although sporadic attacks continued in England for several more years.

Byron's daughter became the first computer programmer (see page 158). Ned Ludd's name lives on. — *TL*

Also February 27:

1932: Neutron Discovered; A-Bomb on the Way

1940: Carbon 14 Discovered; Dating on the Way

Also 1812:

April 26: Birth of Alfred Krupp, Germany's Cannon King

June 24: First Successful Steam Locomotive (see page 177)

February 28
1935: Sheer Bliss

Nylon is produced for the first time.

Credit for the creation of this versatile synthetic goes to Wallace Carothers, a chemist who headed up DuPont's experimental station laboratory in Delaware. Nylon is a synthetic fiber made from coal, water, and air. Its first demonstrated use was as a toothbrush bristle, in 1938. With the coming of World War II, however, nylon turned up almost everywhere: in parachutes, flak vests, combat uniforms, and tires, among other things. It also became a staple in fabrics, carpets, and ropes. In its solid form, nylon is used as an engineering material. But its most celebrated use, perhaps, is in women's stockings, where it has helped fuel the erotic fantasies of young men for several generations.

Carothers didn't live to see his discovery put to any practical use. He killed himself using cyanide in 1937 at age forty-one. — *TL*

Also February 28:
1561: "Father of Surgery" Explains the Head Wound

Also 1935:
May 13: Enter the Parking Meter (see page 135)

February 29
45 BCE: Julius Caesar Takes the Leap

Roman dictator for life Julius Caesar, alarmed that the calendar is getting out of whack with the seasons, adds ten days to the calendar year and an extra day to the month of February every four years.

Caesar was reforming a calendar based on 355 days. An occasional leap month was supposed to be inserted to align the calendar with the seasons, but the Roman religious officials in charge of minding the calendar had been asleep at the switch, chronologically speaking. Caesar consulted with Egypt's top astronomers, who told him the year was 365¼ days long, so Caesar added ten days to the standard year and a leap day every four years.

Although Caesar decreed the new calendar in 46 BCE, that particular year had to have three extra months, for a total of fifteen months, to make up for the accumulated discrepancy. The first add-one-day leap year was 45 BCE. The Romans didn't number the days of the months but used an idiosyncratic system of calends, nones, and ides—and we all know what happened to ol' J.C. on the ides of March, 44 BCE.

The new Julian leap day was added on the day preceding the sixth day before the calends (first day) of March, or February 24 (more or less) by modern nomenclature. We're noting the first leap day on the date of our modern leap day using the same convenient logic by which the United States celebrates Washington's birthday on the third Monday in February, even though George himself observed it on February 22 (or he did after Britain and its colonies changed their calendar, in 1752) and even though he was actually born on February 11 in the Julian—or Old Style—calendar.

Though the Julian calendar was more accurate than its predecessor, it wasn't really as accurate as it needed to be. That's because an Earth year is not 365¼ days but 365 days, 5 hours, 48 minutes, and 46 seconds. This became known in the second century, but no corrections were made again until 1582 (see page 1).—*RA*

Also February 29:

1860: Birth of Herman Hollerith, Inventor of Punch-Card Tabulator

Also BCE:

April 24, 1184 BCE: Original Trojan Horse (see page 116)
May 28, 585 BCE: Predicted Solar Eclipse in Asia Minor Ends a Battle

March 1
1896: Becquerel Discovers Spontaneous Radioactivity

Radioactivity is discovered accidentally by French physicist Antoine Henri Becquerel.

Becquerel was investigating German colleague Wilhelm Roentgen's work (see page 314) on phosphorescence in uranium salts when he made his discovery.

While conducting an experiment using photographic plates, Becquerel found plates that were already fully exposed *before* being subjected to bright sunlight. After further investigation, he concluded that fluorescent uranium salts that had been placed next to the photographic plates (which were wrapped in thick black paper) emitted their own nuclear radiation. The uranium did not depend on the sun or other external light source to excite it.

Radioactivity is triggered by the spontaneous disintegration of atomic nuclei, resulting in radiant energy in the form of alpha, beta, or gamma rays.

Becquerel, the scion of a distinguished scientific family, shared the 1903 Nobel Prize in Physics with Pierre and Marie Curie (see page 357) for their combined work in the field of radioactivity. He was also elected permanent secretary of l'Académie des Sciences in 1908, the year he died. He's remembered today in a metric unit: the becquerel is the radiation caused by one nuclear disintegration per second, or approximately twenty-seven picocuries. —*TL*

Also March 1:
1966: Soviet Probe Venera 3 Lands on Venus, First
    Human Contact with Another Planet

Also 1896:
December 12: Marconi Demonstrates Radio
    (see page 348)

March 2
1887: Birth of the Master Locksmith

Harry Soref is born. The inventor will miniaturize the bank vault and put it into the everyday padlock.

Soref had an idea to improve padlocks at little expense. Most padlocks of the time had cheap metal casings that you could easily bust open with a hammer. Security? Ha! Building a padlock from thicker steel would have been expensive. Instead, Soref applied the laminated design of bank vaults and battleships: he used multiple layers. At the scale of a padlock, layers made of thin pieces of scrap steel would do the trick frugally.

Patent drawing for the original Master Lock.

Soref tried to interest big hardware companies in the idea, but engineers thought the construction process was too cumbersome. So with backing from a couple of friends, Soref established the Master Lock Company in 1921 and began building the little devils himself. His small Milwaukee shop had five employees, a drill press, a grinder, and a punch press. The locks — patented in 1924 — were tough, and the company prospered. Corporate lore says Soref taught Harry Houdini how to hide keys under his tongue and between his fingers.

Milwaukee was famous for its beer, but Prohibition was in force. When the growing firm needed larger quarters, it moved into the shut-down Pabst brewery. Master Lock sent a famous shipment of 147,600 padlocks to federal agents in New York City in 1928. The irony that speakeasies and distilleries were shut down and secured with locks made in a former brewery was noted.

The American Association of Master Locksmiths in 1931 awarded Soref the only gold medal it has ever bestowed. Soref died in 1957 and never saw Master Lock's famous 1974 Super Bowl commercial: it featured a high-powered rifle shooting a hole through a sturdy Master Lock *without* forcing it open. — *RA*

Also March 2:

1949: B-50 Flies Around World Without Landing
1969: Concorde Takes to the Skies
1995: Yahoo! Incorporated

Also 1887:

May 2: Celluloid-Film Patent Ignites Long Legal Battle
(see page 124)

March 3
1919: U.S. Starts International Airmail Service

The United States starts international airmail delivery by flying sixty letters from Vancouver, British Columbia, to Seattle.

The earliest airmail, at least in a sense, was flown on balloons (see page 231). The very first balloon flight in the United States is said to have carried a letter from tech-positive President George Washington (see page 53) to the owner of whatever property the balloon might land on.

The first U.S. Post Office airmail pilots, in 1918, were paid $4,000 a year ($61,000 in 2012 money). The following year, on March 3, pilot Eddie Hubbard and airplane builder Bill Boeing carried sixty letters over the border from Vancouver to Seattle in a Boeing Model C.

Flying was dangerous. More than half of the first forty postal pilots died in air crashes, most of them weather-related. When pilots Leon Smith and Ham Lee refused direct orders to take off during some nasty weather in 1919, both were fired. All the other pilots went on strike. The strike ended when new rules specified that a postal field manager had to fly a brief inspection flight to check the weather. If a manager didn't know how to fly a plane, he had to sit with the pilot during the test flight. With their own skins on the line, the managers quickly learned to be reasonable about balancing the weather and the schedule.

Postal planes started flying transcontinental relays in 1920. Pilots flew by day, when they could be guided by visible landmarks. At dusk they transferred bags of mail to overnight trains. The following morning, the mail was put on other planes for another day's flight. The system cut transcontinental mail-delivery time from four and a half days to one and a half.

Night flights became practical in 1924, thanks to a transcontinental airway of rotating beacon lights and lighted airstrips. Cross-country airmail cost 24¢ — more than $3 in today's cash. — *RA*

Also March 3:

1879: Birth of Nutritionist Elmer McCollum, First to Name Vitamins with Letters
2005: Fossett Flies Nonstop Solo Around Globe

Also 1919:

January 15: Morass of Molasses Mucks Up Boston
February 25: Oregon Taxes Gasoline 1¢ per Gallon, First in U.S.
June 15: First Nonstop Flight Crosses Atlantic

March 4
1877: The Microphone Sounds Much Better

Emile Berliner invents a new kind of microphone.

Alexander Graham Bell had already invented his telephone (see page 71), but without Berliner's carbon-disk or carbon-button microphone, telephones sounded terrible. Bell's microphone involved suspending a diaphragm above a pool of electrified liquid. Berliner added a layer of carbon particles between two contacts, one of which acted as a diaphragm for catching sound waves. Diaphragm motion varied the pressure on the carbon particles, varying the electricity that passed between the contacts.

His improvements allowed the signal to be amplified to compensate for electrical resistance in the wire. Without that amplification, the telephone would have remained a mere local curiosity, rather than transforming the world long-distance. Berliner's microphone also converted sound waves into electricity more accurately. It became commonplace in telephones, and even radio, until the condenser microphone appeared in the 1920s.

Bell bought Berliner's patent for $50,000 ($1.2 million in 2012 money) and in 1878 began manufacturing telephones using the technology. But the U.S. Supreme Court ruled in 1892 that Thomas Edison, and not Berliner, invented the carbon microphone. Neither could really claim total credit. The idea of transmitting speech by varying the current between two contacts had appeared in published works as early as 1854.

Berliner felt Edison had stolen his idea, but Berliner did receive ample credit for another crucial invention: disk records. They replaced Edison's phonograph cylinders, which took up much more space and were difficult to duplicate.

Berliner's company used a logo of a dog cocking its ear toward a record player. Modified versions of "His Master's Voice" have since been used by record companies around the world, including RCA and retail entertainment chain HMV.

We don't know Emile Berliner's first words transmitted by microphone, but they were probably not *Ich bin ein Berliner.* —EVB

Also March 4:

1887: Gottlieb Daimler Tests Benzine Motor Carriage
1890: Scotland's Forth Bridge Opens
1962: Nuclear Power Plant Starts in Antarctica

Also 1877:

August 14: Nicolaus Otto Patents Internal
    Combustion Engine
August 15: Edison Coins *Hello* as Telephone Greeting
    (see page 229)

March 5
1872: Westinghouse Gives Railroads a Brake

George Westinghouse Jr. patents the automatic railroad air brake.

Before the air brake, railroad engineers stopped trains by cutting power, braking their locomotives, and using whistles to signal their brakemen. The brakemen turned the brakes in one car, jumped to the next to set the brakes there, then jumped to the next, and so on. The system was dangerous (many brakemen died or were maimed), imprecise (the train might stop before or after the station), and unreliable (the train sometimes didn't stop until it ran into another train or something else on the tracks). Accidents were frequent and deadly (see page 121).

Westinghouse's first version of the device, the straight or direct air brake, used air hoses to connect the cars. When the engineer applied the brakes, air pressure turned the brakes on in each car of the train. Of course, if the hoses leaked or were disconnected, the train lost braking power.

With air brake 2.0, Westinghouse turned things around. Air pressure kept the brakes *off*. The engineer *reduced* pressure to apply the brakes. This built-in safeguard meant the brakes' losing pressure stopped the train automatically. That held true whether pressure dropped due to leakage or to cars coming unhitched; loose cars would brake to a stop. The system debuted in 1872 on the Pennsylvania Railroad. Automatic air brakes were soon adopted around the world. They were safer and more precise, and now that trains could be reliably stopped, railroads could operate at higher speeds. Air brakes are also used today on trucks, buses, and even amusement-park rides.

Westinghouse also invented electrical signals that saved lives by keeping two trains from occupying the same block of track. He bought Nikola Tesla's patents for alternating current and demonstrated its superiority over Thomas Edison's direct current (see page 4). And he founded the Westinghouse company. — *RA*

Also March 5:

1904: Physicist Nikola Tesla Expounds on Ball Lightning (see page 192)
1975: Silicon Valley's Homebrew Computer Club Holds First Meeting

Also 1872:

June 4: Robert Chesebrough Patents Process for Making Vaseline Petroleum Jelly

Felix Hoffmann, a young pharmacist working for the German pharmaceutical company Bayer, patents a new pain reliever. The trademark name is Aspirin.

Hoffmann, who was said to have been seeking an effective pain reliever for his father's rheumatism, successfully synthesized acetylsalicylic acid in August 1897. It would later be marketed as Aspirin—*a* for "acetyl" and *-spirin* for *Spiraea*, genus of the source plant for salicylic acid, the pain-relieving agent. That August also saw Hoffmann synthesize heroin, which he accomplished accidentally while attempting to acetylate morphine to produce codeine. That discovery didn't pan out like aspirin.

An early Bayer Aspirin bottle.

The benefits of salicylic acid as a pain reliever and fever reducer had been recognized since antiquity. Extracted from willow bark, it was commonly found in salves and teas. Unfortunately, it also caused stomach irritation.

Bayer's patent application was rejected in Germany because Hoffmann had not actually invented acetylsalicylic acid; a French and a German chemist had each synthesized it separately decades earlier. But the U.S. Patent Office issued a patent because Hoffmann was the first to synthesize it in a stable, usable form. Bayer began an aggressive worldwide marketing campaign, the German patent office eventually came around, and Bayer AG still holds the rights to the trade name Aspirin in more than eighty countries. In the United States, the word is often used generically to refer to almost any brand of acetylsalicylic acid.

Although aspirin is not without its side effects—Reye's syndrome is associated with a reaction to acetylsalicylic acid—it remains one of the world's most widely used pain relievers.—*TL*

Also March 6:

1930: Birds Eye Test-Markets First Mass-Produced Frozen Foods

1937: Birth of Valentina Tereshkova, First Woman in Space

1992: Michelangelo Computer Virus Pretty Much Fizzles

Also 1899:

March 14: Count Zeppelin Patents the Dirigible

August 23: First Ship-to-Shore Radio Signal to a U.S. Station

September 13: New Yorker Is First U.S. Pedestrian Killed by Car

March 7
1897: First Morning of the Cornflake

Dr. John Kellogg, believing that a strict diet benefits the patients at his sanitarium in Battle Creek, Michigan, serves up the world's first cornflakes.

Kellogg was a Seventh-Day Adventist and a passionate adherent of the healthy-living tenets of the church, which embraced a holistic approach to health in the days when there were no antibiotics and few effective drugs of any kind. The cornflakes he served at his Battle Creek Sanitarium (he coined the word *sanitarium* himself) were sugarless. His brother Will recognized and exploited cornflakes' commercial potential. He added sugar and sold the stuff as breakfast food, deeply offending the sensibilities of John, who sued to stop Will from marketing the cereal.

John lost. And Will marketed as few had done before. His signature became the company logo, and he devoted millions to advertising. In 1911, in New York City's Times Square, he put up the world's largest electric sign: eighty feet by a hundred and sixty feet, with the letter *K* standing more than sixty feet high. John Kellogg, meanwhile, used the royalties to subsidize his sanitarium, which he billed as a "place where people learn to stay well." It attracted a mainly wealthy clientele, who tended to check in for several weeks at a time.

John Kellogg also introduced new techniques for abdominal surgery. The sanitarium operated at its original site until 1942. The main building became a medical facility for veterans of World War II and the Korean War. — *TL*

Also March 7:

1912: Roald Amundsen Formally Claims South Pole for Norway

1929: Fleming Names His Mold Juice Penicillin

Also 1897:

February 5: Indiana Legislature Nearly Declares $\pi = 3$

June 12: The Swiss Army Gets Its Own Knife (see page 165)

March 8
1955: The Mother of All Operating Systems

Computer pioneer Doug Ross demonstrates the Director tape for MIT's Whirlwind machine. It's a new idea: a permanent set of instructions for how the computer should operate.

MIT's Whirlwind computer was the first digital computer that could display real-time text and graphics on a video terminal (in those days, just a large oscilloscope screen). Whirlwind processed data with 4,500 vacuum tubes—not chips, not even individual transistors. The Whirlwind occupied 3,300 square feet and was the fastest digital computer of its time. It also pioneered magnetic core memory for RAM.

The Director set of programming instructions was punched on paper tape and is regarded as the predecessor of operating systems in computers. The Director was designed to issue commands to the four-year-old Whirlwind machine. The idea eliminated the need for manual intervention in reading the tapes for different problems during a computing session.

The Director tape communicated with the computer through a separate input reader. That means tapes with various problems to be computed could be spliced together and run one after the other. Thanks to the Director tape on a separate input, each problem was recognized and appropriately processed. By eliminating the need to repeat the instructions on the tape with each and every problem, you could make a complete run by pushing a single button.

Programmers John Frankovich and Frank Helwig wrote the first Director tape program, and lead programmer Doug Ross demonstrated it in 1955. The Director tape was probably the first example of a Job Control Language–driven operating system. JCL is a scripting language used on mainframe operating systems to instruct them how to run a batch job or start a subsystem.

The Whirlwind is credited with influencing most of the computers of the 1960s.—*PG*

Also March 8:

1817: René Laennec Records First Use of Stethoscope to Listen to Heart and Lungs
1918: First Case of Killer Flu — Spanish Influenza Will Kill at Least 20 Million

Also 1955:

October 25: First Domestic Microwave Oven (see page 300)
November 5: Time-Travel Day in *Back to the Future*

March 9
1454: This Man Is a Continent... or Two

Amerigo Vespucci is born in Florence, Italy. He'll give his name to two continents.

Vespucci worked in Seville, Spain, at the time of Christopher Columbus. Not content to sit on the sidelines when fame and fortune awaited, he outfitted his own expeditions to seek a short trade route to India. On his second voyage to what we now call South America, Vespucci made a major breakthrough. He went south along the eastern coast and then farther south. He was off Patagonia, within four hundred miles of Tierra del Fuego, and the coast was like nothing previously known to Europeans. Vespucci was convinced it wasn't Asia, but a new continent.

He made more voyages to what was soon called the New World. A popular account appeared in a pamphlet, "The Four Voyages of Amerigo." It achieved widespread circulation, thanks to the growth of a relatively new technology, the printing press.

German cartographer Martin Waldseemüller reissued the pamphlet in 1507 with an introduction that suggested "calling this part... America, after Amerigo." Waldseemüller included a map on which the name America makes its earliest appearance. The map was popular. The name caught on, and it spread. Gerardus Mercator's 1538 world map included both *North* and *South* America.

Though Vespucci reached America after Columbus (and others), it's not unjust that two continents are named in his honor. He seems to have originated the idea that the new lands were not merely offshore islands of Asia. He reorganized the data, he shifted the paradigm, and he deserves the eponym.

We do not often refer to the New World as Columbia. Nor do we call it Ericsonia or Cabotland. Nor is our nation's name (and we should be grateful for this) Waldseemüller or the United States of Vespucci. —*RA*

Also March 9:

1862: USS *Monitor* Fights CSS *Virginia* in First Naval
   Battle of Ironclads
1945: American B-29s Bomb Tokyo, Killing 100,000

Also Fifteenth Century:

March 19, 1474: Venice Enacts a Patently Original Idea
   (see page 80)

March 10
1876: "Mr. Watson, Come Here"

Alexander Graham Bell makes the first telephone call in his Boston laboratory, summoning his assistant, Thomas A. Watson, from the next room.

Bell's journal, now at the Library of Congress, contains the following entry for March 10, 1876:

> I then shouted into M [the mouthpiece] the following sentence: "Mr. Watson, come here—I want to see you." To my delight he came and declared that he had heard and understood what I said.
>
> I asked him to repeat the words. He answered, "You said 'Mr. Watson—come here—I want to see you.'" We then changed places and I listened at S [the speaker] while Mr. Watson read a few passages from a book into the mouthpiece M. It was certainly the case that articulate sounds proceeded from S. The effect was loud but indistinct and muffled.

Watson's journal, however, says the famous quote was "Mr. Watson come here I want you." That disagreement is trifling compared to the long controversy over whether Bell truly invented the telephone. Another inventor, Elisha Gray, was working on a similar device, and recent books claim that Bell not only stole Gray's ideas but may even have bribed a patent examiner to let him sneak a look at Gray's filing.

Years of litigation ensued, but Bell's patents ultimately withstood challenges from Gray and others—perhaps by right, perhaps by virtue of Bell's bigger backers and better barristers. In that respect, the controversy recalls the patent battle over the telegraph and foreshadows later squabbles over the automobile, the airplane, the spreadsheet, online shopping carts, web-auction software, and the look and feel of operating systems.

One thing we know for sure: The telephone did not interrupt Mr. Watson's dinner that night with a special offer for home repairs or timeshare vacations in Florida.—*RA*

Also March 10:

1797: Jefferson the Paleontologist (see page 53)
1849: Lincoln the Inventor (see page 54)
2000: Dot-Com Bubble Ends, Pop Goes the NASDAQ!

Also 1876:

June 25: Indians May Have Had Better Guns at
Custer's Last Stand
August 8: Edison Patents Mimeograph (see page 222)

March 11
1985: First Public Net Links Consumers' Computers

The nation's first local, public packet-switching network opens for business.

Hooking into the world's network of interconnected computers isn't a notable event these days, especially now that millions of us have always-on connections in mobile devices that are rarely beyond arm's reach. But things were different when the Southern New England Telephone Company turned on ConnNet. It was the first local, public packet-switching network in the United States.

Customers in Connecticut could connect and reach NewsNet, the National Library of Medicine, CompuServe (see page 269), and Dow Jones News Retrieval. Companies could rent dedicated lines and get service from 4,800 to an astonishing 56,000 bits per second. Employees of subscribing companies could dial in from home to log in to their office mainframe. But computers using dial-up connections pulled down only 300 to 1,200 bits per second. (If you have a 5Mbps connection now, you are downloading more than 4,000 times as fast as the fastest ConnNet dial-up.)

ConnNet was not technically the first public Internet service provider, however. It was instead part of a global network using the X.25 protocol, which was rendered obsolete in the 1990s by the more popular Internet protocol, or IP.

Southern New England Telephone Company was a pioneer in bringing the latest telecommunications technology to consumers. It opened the first commercial phone exchange in the world in New Haven, Connecticut, on January 28, 1878, with twenty-one subscribers. The company also printed the world's first phone book.

The FCC's 2010 national broadband plan sets a goal for 2020 of 100 million Americans having 100Mbps broadband connections. That's more than 300,000 times faster than the basic 300bps service of 1985. — *RS*

Also March 11:
105 CE: "Invention" of Paper Shown to Chinese Emperor
1986: NFL Adopts Instant Replay

Also 1985:
March 15: Symbolics.com Is First Dot-Com (see page 76)

1923: Talkies Talk...on Their Own

Radio pioneer Lee de Forest demonstrates his Phonofilm movie process, bringing the world of synchronized sound to the movies.

Inventors who'd tried to link the phonograph and the moving picture found it nearly impossible to synchronize the sound with moving lips on the screen. The first sound films had recorded musical accompaniment but still used full-screen dialogue titles. They weren't talkies.

De Forest's technical advance was synchronizing sound and motion by placing the sound recording directly on the film in an optical soundtrack. Analog blips of light represented sound frequency and volume. It was the prototype of the optical sound-on-film process used from the 1930s onward, with continued improvements like high fidelity and stereo, until digital sound began to replace it in the 1990s. De Forest equipped thirty theaters around the world with Phonofilm. He didn't have a big budget for film production and couldn't seriously interest Hollywood in his invention. De Forest solved the sound-synch issue, but his fidelity was subpar even by 1920s standards.

The movie that introduced most of the public to talkies, 1927's *The Jazz Singer*, used the Warner Bros. Vitaphone. That system was essentially a phonograph hooked up to a projector and piped into loudspeakers. Meant to showcase Al Jolson's singing, the film instead amazed audiences with bits of spoken dialogue.

Fox Movietone, RCA, AT&T, and German companies were also developing sound-on-film, and their systems soon left de Forest's low-fi Phonofilm in the acoustical dust. De Forest had likewise been unable to capitalize on his earlier work in vacuum-tube technology and voice radio, spending years fighting lawsuits and even staving off charges of criminal fraud.

The Academy of Motion Picture Arts and Sciences honored de Forest in 1959 with a special Oscar for the "pioneer invention which brought sound to the motion picture."—*RA*

Also March 12:

1790: Birth of J.F. Daniell, Inventor of Practical Electric Battery
1838: Birth of W.H. Perkin, Synthesizer of Mauve Dye

Also 1923:

November 20: Morgan Patents the Traffic Light That GE Will Mass-Produce (see page 326)

March 13
1842: Henry Shrapnel Dies, but His Name Lives On

Henry Shrapnel, inventor of the long-range artillery shell that bears his name, dies.

Shrapnel, a British lieutenant, was serving in the Royal Artillery in the mid-1780s when he perfected his shell. A shrapnel shell, unlike a conventional high-explosive artillery round, was designed as an antipersonnel weapon. The projectile was packed with fragments — often sharp metal, lead balls, or nails — and detonated in midair, spraying enemy troops in the vicinity with what the British quickly christened *shrapnel.*

Shrapnel combined two existing weapons technologies, the canister shot and the delayed-action fuse. Canister shot, in use since the 1400s, burst upon leaving the gun's muzzle and was originally used in small arms at close range against infantry. Shrapnel's refinement carried the shell intact to the enemy's lines, where it detonated above the heads of the troops, to devastating effect.

The British army, not quick to embrace innovation, did not adopt Shrapnel's invention until 1803. The shell saw early action against the Dutch in Suriname but really came into its own after the Duke of Wellington demonstrated its effectiveness against Napoléon's army at several engagements, including the Battle of Waterloo. Henry Shrapnel, by then a captain, was rewarded with a promotion to major and then to lieutenant colonel. In 1814, the British government awarded him a lifetime annual stipend of £1,200 (about $100,000 in today's money). Shrapnel also worked on improvements in howitzers and mortars. He ended his military career a major general.

The shrapnel shell was quickly adopted by the armies of all Europe's great powers. Modern arsenals still employ shells that use canister-shot projectiles based on the shrapnel principle, but the nature of ordnance has obviously changed.

As for the word itself, *shrapnel* has long been used generically to refer to any shell fragment. — *TL*

Also March 13:

1781: William Herschel Discovers Planet Uranus
1964: Bystander Effect — No One Helps as Kitty
Genovese Is Slain

Also 1842:

March 30: Crawford Long Removes a Tumor Under
Ether (see page 275)
September 20: Birth of James Dewar, Inventor of
Thermos Bottle (see page 265)

March 14
1794: Whitney's Cotton Gin Patent Not Worth Much

Eli Whitney patents his new invention: a machine that quickly separates cotton seeds from cotton fibers. The cotton gin was the little engine that could—and did—transform the economy of the South and change the course of American history.

Short-staple cotton has sticky green seeds that are the devil to remove from the fluffy cotton bolls. Separating the seeds was time-consuming and labor-intensive, even with slave labor. Whitney figured he could make himself rich by inventing a machine to remove those seeds. His gin (short for *engine*) used a crank to rotate a cylinder with spiked teeth. The cotton was pulled through small slots to separate the seeds from the fibrous lint, which was pulled off the spikes by a rotating brush.

Whitney and his partner Phineas Miller chose a flawed business model. Rather than selling cotton gins to growers, they decided to run the gins themselves and charge growers for cleaning the cotton. Specifically, they took two bales out of every five they cleaned. That 40 percent charge motivated growers to tweak Whitney's design and build their own cotton gins. Miller brought suit against the pirate versions, but a loophole in the patent law prevented the partners from collecting any money until after the law was revised, in 1800.

Whitney and Miller finally decided to license the invention, and they collected some fees. But the patent expired in 1807. Whitney concluded, "An invention can be so valuable as to be worthless to the inventor."

Mechanization of the cotton harvest pressured the growers to cultivate more land, and the expansion of slavery led to the Civil War.

Whitney died in 1825, wealthy not from the cotton gin but from a technology that eventually played a big part in who won the Civil War: mass-produced muskets with interchangeable parts. —*RA*

Also March 14:

1879: Birth of Albert Einstein (see page 327)
1899: Count Zeppelin Patents the Dirigible
1988: 3/14 = Pi Day First Celebrated

Also 1794:

January 14: First Successful Cesarean in U.S.
(see page 14)

March 15
1985: Dot-Com Revolution Starts with a Whimper

Symbolics, a Massachusetts computer company, registers Symbolics.com, the Internet's first domain name.

In those early days, even before AOL, the Internet was a noncommercial medium that only eggheads and propellerheads used. It was more of a military and academic tool than today's vast public network. Nobody seemed in a terrible hurry to get a domain. Only five were registered during all of 1985. As you'd expect, the first hundred are packed with computer companies. Apple registered the sixty-fourth domain, in 1987. Microsoft waited until 1991.

IBM and Sun registered on the same March day in 1986, which was the year Intel and AMD joined the cool crowd. That was fourteen months ahead of even Cisco Systems, whose future tagline would be Empowering the Internet Generation.

Symbolics, though, was conceived at the MIT Artificial Intelligence Laboratory, the renowned academic incubator. One employee, a former member of the lab, created the LISP machine—the world's first computer workstation. Symbolics was known for developing what was considered the best computing platform for AI software. Others know Symbolics for its software, which, among other things, helped create scenes in *Star Trek III: The Search for Spock*.

By 1985 Symbolics was riding high. Things then turned south. The firm went into a freefall: founders were fired, buyers panicked, real estate investments turned bad, and the inexorable march of the PC trampled the company into near oblivion.

Symbolics still exists, but in a very diminished form—and at an entirely new (and less snazzy) address: symbolics-dks.com.

Symbolics.com changed hands for the first time in 2009, when it was bought by a domain-aggregation company whose owner was five years old when the address was first registered. —*JCA*

Also March 15:

1854: Birth of Pioneer Immunologist
   Emil von Behring

Also 1985:

October 18: Nintendo Entertainment
   System Launches (see page 293)

March 16
1926: Goddard Launches Rocketry

Robert Goddard's folly becomes fact with the first successful launch of a liquid-fueled rocket. It took place at Goddard's aunt Effie's farm in Auburn, Massachusetts.

Goddard has long been considered the father of modern rocketry, but it wasn't always that way. His ideas of sustainable rocket-powered flight were ridiculed by some colleagues early on and laughed at in the press. But he was appreciated abroad, especially in Germany, where the rocket's potential as a weapon was accepted and eventually realized. Wernher von Braun, who helped develop the V-2 rocket used against England during World War II (and which had its first test

Robert Goddard tows an early rocket behind a Ford Model A truck, c. 1931.

flight sixteen years to the day after the launch at Effie's farm), described Goddard as an early influence.

Goddard's great technical achievement was devising a method that radically increased the fuel efficiency of the rocket engine, to the point where a heavy mass could be propelled upward with minimum fuel expenditure. This, more than any other factor, is what made Goddard's vision of interplanetary travel feasible.

The press had been very hard on Goddard, and the *New York Times* ran an editorial in 1920 openly mocking his belief that humans could reach the moon. But the *Times* eventually came around. On July 17, 1969, the day after Apollo 11 left for the moon, the paper ran this belated retraction:

"Further investigation and experimentation have confirmed the findings of Isaac Newton in the 17th century, and it is now definitely established that a rocket can function in a vacuum as well as in an atmosphere. The *Times* regrets the error." — *TL*

Also March 16:

1802: U.S. Army Corps of Engineers Founded

Also 1926:

April 20: Warner Bros. Announces Vitaphone Talking Movies (see page 73)
April 28: Schrödinger Writes to Einstein About Wave Mechanics (see page 120)

March 17
1953: Airplane Black Box Is Born

After several high-profile crashes of pioneering de Havilland Comet airliners go unexplained, Australian researcher David Warren invents a device to record cockpit noise and instruments during flight.

Warren, a researcher at the Aeronautical Research Laboratories in Melbourne, Australia, believed recording instrument readings and pilots' voices could help determine the cause of a crash — and help prevent future crashes. He called his device a flight memory unit. Early versions recorded up to four hours of data on a steel foil.

The Australian aviation community initially rejected the device, for privacy reasons.

Eventually, British officials accepted the idea, and Warren began producing the flight-data recorder, or FDR, in a crash- and fireproof container and selling it to airlines around the world. After a 1960 crash whose cause could not be determined, Australia became the first country to require all commercial airplanes to use the devices.

Early recorders logged navigational heading, altitude, airspeed, vertical accelerations, and time. Today's FDR records many more parameters, including throttle and flight-control positions. The extra data lets investigators re-create most of the pilot-controlled activity in the moments leading up to a crash. Digital simulations of accidents now analyze both the problems leading to a crash and pilots' responses to them.

Modern black boxes are actually bright orange for visibility. They must be able to resist fire and piercing, withstand the pressure of being submerged twenty thousand feet under the ocean, and survive a 3,400g crash-impact test. To assist its recovery, the device emits a locator-beacon signal for up to thirty days. Modern FDRs use solid-state memory that can be downloaded almost instantly.

Future flight recorders might transmit flight data to ground stations in real time, which would eliminate the need to physically find the black box. — *JP*

Also March 17:

1845: Rubber Band Invented
1948: Birth of William Gibson, Father of Cyberspace

Also 1953:

April 13: CIA Okays Mind-Control Tests; *Casino Royale* Published (see page 105)

March 18
1931: The Schick Hits the Fans

The first practical electric shavers go on sale. They're a cut above their clumsy predecessors.

Inventor Jacob Schick wanted a more comfortable way to shave. He devised rough plans, so to speak, for putting a shaving head at the end of a flexible cable and powering it by external motor. He sent the idea to manufacturers, who quickly rejected it. Schick went into business for himself and, with exquisite timing, put his first electric shavers on the market in 1929, right on schedule for the stock market collapse and ensuing Great Depression. The early models didn't sell well. The design was a clumsy contraption with a heavy motor connected by a metal cable to the reciprocating shaving head.

So Schick got rid of the flexible cable and put a small electric motor inside the unit with the shaving head. The entire apparatus was encased in sleek, black Bakelite and fit comfortably in your hand. An electric appliance cord supplied power to the motor, which had to be kick-started by a turn-wheel switch on the unit. The new models went on sale in 1931 in New York City for $25 each (that's almost $380 in 2012 dollars). They didn't shave closer than a wet steel blade, but they were convenient. And when people considered the cost of blades, shaving cream, and other appurtenances needed for a wet shave, the razor didn't seem that expensive after all. Especially when prices dropped as Remington, Sunbeam, Philips, Zenith, and even Gillette entered the market.

Schick died rich, but young. Even though he maintained that a man who shaved correctly every day would live to a hundred and twenty, he didn't make it to sixty. —*RA*

Also March 18:

1662: Pascal Starts World's First Bus Service in Paris
1987: Superconductivity Debate Earns Meeting the
   Name "Woodstock for Physics"

Also 1931:

May 27: Wind Tunnel Lets Airplanes "Fly" on Ground
   (see page 149)
October 27: Dutch Elm Disease Discovered in
   Northeast U.S.

March 19
1474: Venice Enacts a Patently Original Idea

Venice passes the first-known written law to grant and protect patents.

The craft guilds, especially those of Venice's lucrative glassblowing trades, already had restrictions, but the senate was hoping to attract foreign innovators as well. So it gave the new law force throughout Venice's far-flung territories:

> Any person...who makes any new and ingenious contrivance, not made heretofore in our dominion, shall, as soon as it is perfected so that it can be used and exercised, give notice of the same to our [State Judicial Office], it being forbidden up to 10 years for any other person in any territory and place of ours to make a contrivance in the form and resemblance thereof, without the consent and license of the author.

Violators could incur a fine of 100 ducats (perhaps $7,000 in 2012 money) and immediate destruction of the offending contrivance. The government reserved the right "to take and use for its needs any of the said contrivances and instruments," provided "no one other than the inventors shall operate them." The senate itself could also issue special patents good for up to twenty-five years.

The U.S. Constitution empowered Congress to "promote the progress of science and useful arts, by securing for limited times to authors and inventors the exclusive right to their respective writings and discoveries," and the Patents Act became U.S. law in 1790 (see page 214). Global patent agreements began with the Paris Convention of 1883.

Spurred by digital technology and the Internet, intellectual-property law has grown contentious in recent decades. Not only are there conflicting claims as to who owns what, but a lively global debate rages over what constitutes an invention or original work, and even whether such protections are needed in the first place.—*RA*

Also March 19:

1979: House Proceedings Air Live on C-SPAN
1981: Two Technicians Killed in Ground Test of Space
Shuttle *Columbia*

Also Fifteenth Century:

June 1, 1495: First Scotch Whisky Recorded (see page 154)

March 20
1800: Volta's Battery Shows Potential

Italian physicist Alessandro Volta reports that he's developed a reliable source of electrical current. He's invented the wet-cell battery.

Volta had already devised the electrophorus to create static electric charges and discovered methane before he became a professor of physics at the University of Pavia in 1779. Volta theorized that electrical current was caused by the contact of dissimilar metals amid moisture. He went on to build a stack of alternating copper and zinc disks. Each pair of disks was separated from the next by cardboard that had been soaked in salty water. This voltaic pile produced continuous electrical current. Volta wrote about it to Joseph Banks, head of the Royal Society in London, who shared news of the invention with other scientists.

The results were literally electrifying and nearly immediate. Within weeks, William Nicholson and Anthony Carlisle had discovered that electrical current could decompose water into hydrogen and oxygen. Humphry Davy soon used this newly discovered process of electrolysis to isolate potassium, sodium, and calcium.

Emperor Napoléon of France, who also ruled much of Italy at the time, was impressed by Volta's inventions. He made him first a knight, then a senator, and finally a count.

Count Volta's name lives on, of course, as a unit of measure. In 1881, the volt was officially established as an electrical potential of one joule per coulomb of charge, or the electromotive force that will cause a current of one ampere to flow through a resistance of one ohm. It is used around the world.

(Note: West Africa's Volta River is not named for the scientist. It comes from the Portuguese word for "turn," perhaps because the river was a turnaround point for explorers, or simply because of its own twists and turns.) — *RA*

Also March 20:
1995: Poison Gas Wreaks Tokyo Subway Terror

Also 1800:
March 19: Alexander von Humboldt Captures Electric Eels in South America

March 21
2006: Twitter Takes Flight

Jack Dorsey sends the world's first (nonautomated) tweet:

```
inviting co-workers
```

Twitter is still derided by some as overhyped, a vast wasteland of "I am eating a ham sandwich" irrelevancies and 140-character non sequiturs, almost as big a time sink as Facebook (see page 35). But it has also been credited with accelerating revolutions, spawning new forms of literature, acting as a real-time watercooler for sharing snark during collective experiences like the Oscars, and generally operating as humanity's de facto announcement and early-warning system for... everything.

Not bad for a service that started as the Twttr, a project that Dorsey had thought up five years earlier but that never quite "solidified" until March 2006. Dorsey pitched a now-defunct Silicon Valley company called Odeo about his idea for a status-sharing service based on SMS. That approach got him two key partners — Odeo executives Biz Stone and Evan Williams. The three started a company named Obvious, which would become Twitter.

Dorsey, who would make the act of reflexively recording one's status de rigueur, kept marvelous historical records of his little project. Coding began March 13, and eight days later a machine-generated tweet was issued from Dorsey's account (no. 12):

```
just setting up my twttr
```

While some folks consider this the world's first tweet, Dorsey and Twitter are adamant that "inviting co-workers" (meaning his Odeo colleagues) was the first.

As of 2012, there were more than half a *billion* accounts. The company is thought by some to be worth perhaps $10 billion. —*JCA*

Also March 21:
1999: Around the World in Twenty Days, in a Balloon

Also 2006:
August 24: Pluto Demoted from Planetary Status
    (see page 238)
December 12: Baiji Yangtze Freshwater Dolphin
    Declared Extinct

March 22
1995: Longest Human Space Adventure Ends

Cosmonaut Valery Polyakov returns to Earth after the longest-ever stay in space by a human. He spent just over 437 days in the Mir space station.

Thanks to a strenuous workout regimen, he returned to Earth looking "big and strong" and "like he could wrestle a bear," in the words of NASA astronaut Norman Thagard. Polyakov, a medical doctor, said that he volunteered for the extra-long mission to prove that the human body could survive microgravity long enough to make a trip to Mars. When he got back on terra firma, he took pains to show that he was no worse for the zero-g wear.

"[W]hen his capsule landed in Kazakhstan he walked from it to a nearby chair, a tremendous achievement," Philip Baker wrote in his book *The Story of Manned Space Stations*. "He also stole a cigarette from a friend nearby, but could hardly be blamed for that. He sipped a small brandy and inwardly celebrated his mission. His record still stands, and it is unlikely to be broken until man ventures to Mars." Reportedly, his first statement back on Earth, spoken to a fellow cosmonaut, was "We can fly to Mars."

Polyakov's mission did not get off to an auspicious start. When the cosmonauts who dropped him off did a flyby to take pictures of Mir, they grazed the space station with their craft. Luckily, no major damage was done. The rest of Polyakov's mission wasn't that eventful. Scientists monitoring his mood and intellectual performance observed some impairment during his first three weeks in space and in the first two weeks following his return. In both instances, he soon bounced back to normal.

At the time, Polyakov also held the record for most *cumulative* time in space, but he has been surpassed by Sergei Krikalyov, with more than 803 days. —*AM*

Also March 22:

1935: Primitive, Low-Definition TV Shown in Berlin
1983: Pentagon Awards Contract for High-Mobility
Multipurpose Wheeled Vehicle, the HMV

Also 1995:

August 24: Say Hello to Windows 95
October 11: "We're Trashing the Ozone Layer"
(see page 286)
October 23: Judge Okays First Wiretap of Computer
Network

March 23
1857: Mr. Otis Gives You a Lift

Attention, shoppers: The first commercial elevator goes safely up and down in a New York City department store. It makes it possible to fill cities with skyscrapers.

Original Otis elevator.

The secret of Elisha Graves Otis's success wasn't so much that he could make a platform go up and down, which isn't really that big an engineering achievement. There had already been steam and hydraulic elevators in use here and there for a couple of years before Otis stepped up.

No: Otis's achievement was making an elevator that not only went up and down but also couldn't go into a free fall. If the elevator started dropping too fast, a governor automatically engaged the brakes and stopped it cold. To convince the public of his invention's safety, Otis gave a dramatic demonstration at the New York Crystal Palace, a grand exhibition hall built for the 1853 World's Fair. The company recounts this milestone in its history:

Perched on a hoisting platform high above the crowd at New York's Crystal Palace, a pragmatic mechanic shocked the crowd when he dramatically cut the only rope suspending the platform on which he was standing. The platform dropped a few inches, but then came to a stop. His revolutionary new safety brake had worked, stopping the platform from crashing to the ground. "All safe, gentlemen!" the man proclaimed.

Also March 23:

1983: Reagan Taunts Soviets with Star Wars Plan
1989: Scientists Skeptical about Cold-Fusion Announcement

Also 1857:

February 22: Physicist Heinrich Hertz Enters Cycle of Life

Otis's demonstration had the desired effect. He sold seven elevators that year, and fifteen the next. By 1873 there were two thousand Otis elevators in use. They expanded to Europe and Russia. Otis Elevator got the commissions for the Eiffel Tower, the Empire State Building, and the World Trade Center.

But the very first commercial installation was on March 23, 1857, at a five-story department store at Broadway and Broome Street in what is now New York City's SoHo district. — *JCA*

March 24
1882: Koch Pinpoints TB Bacillus

German physician Robert Koch announces his discovery of the tuberculosis bacillus, the cause of a scourge responsible for one in seven deaths during the mid-nineteenth century.

Koch turned to the study of infectious diseases while still in medical school at the University of Göttingen. There, he was influenced by anatomist Jakob Henle, an advocate of germ theory, which posited that communicable disease was transmitted through microorganisms.

Despite the work of other prominent microbiologists, including Joseph Lister and Louis Pasteur, scientists' prevailing view for much of the nineteenth century was still that diseases arose spontaneously within an individual. Koch, piggybacking on the work of his predecessors and making huge contributions of his own, played a key role in finally debunking that theory.

Koch worked on tuberculosis at the Imperial Health Bureau in Berlin. By 1882, he had isolated the bacillus and published his definitive paper on the subject. Besides discovering the TB germ, he also isolated the infectious bacilli for anthrax and cholera. The cholera work took him away from tuberculosis for a few years, but he returned to it as a professor of hygiene at the University of Berlin. Koch developed tuberculin, an extract of proteins from TB bacilli. He believed it would result in a cure for tuberculosis, but his claims proved to be exaggerated, which damaged his reputation for a time. The damage was not lasting, however, owing to Koch's many achievements that changed attitudes and approaches to the treatment of infectious diseases. Tuberculin proved useful too—not as a cure for the disease, but as a test for exposure to TB.

Koch was awarded a Nobel Prize in 1905 for his work in tuberculosis. He also laid down the conditions, known as Koch's postulates, that must be met before a specific bacterium can be said to cause a specific disease. — *TL*

Also March 24:
1976: President Ford Orders Swine Flu Shots for All
1989: *Exxon Valdez* Spill Causes Environmental Catastrophe
2001: Apple Unleashes Mac OS X

Also 1882:
December 22: Electric Christmas Tree Lights Invented (see page 358)

March 25
1954: RCA TVs Get the Color for Money

RCA begins production of its first color-TV set for consumers, the CT-100. It's destined to become a costly classic.

The RCA set had a fifteen-inch screen and sold for $1,000, which is about $8,500 in 2012 dollars. That's more than enough to take your pick today of fifty- to sixty-inch plasma screens with up to sixteen times the screen area of the 1954 model.

Admiral and Westinghouse sets had beaten RCA to the market by months and weeks, respectively, and they were expensive too. The Westinghouse went for $1,295. But the RCA standard—which was compatible with black-and-white broadcasts—came to define the market. Few families wanted to clutter the living room with one box for color and another for black-and-white. Compatible color required packing two sets of circuits into one TV console. That complexity not only added to the cost but also resulted in an image that was often blurry and ridden with ghosts. *Consumer Reports* warned the model was fit only for what these days we'd call early adopters: "Only an inveterate (and well-heeled) experimenter should let the advertisements seduce him into being 'among the very first' to own a color TV set." A 1954 *New York Times* headline should sound familiar to modern ears: "Set Buying Lags—Public Seen Awaiting Larger Screens, Lower Prices."

So RCA rolled out its twenty-one-inch 21CT55 in November 1954 at "just" $895 (over $7,000 today). Despite the price, the company was apparently losing money on every set it sold. It would take years of price drops and technical improvements before color TV was no longer a plaything of the rich.

In a sixty-four-gadget playoff bracket in 2007, *Wired* magazine readers named the RCA CT-100 the Greatest Gadget of All Time. —*RA*

Also March 25:

1916: Ishi, Last Survivor of Yahi American Indian Tribe, Dies in San Francisco
1995: First Internet Wiki Makes Fast Work of Collaboration (see page 15)

Also 1954:

June 27: Soviets Open World's First Nuclear Power Plant (see page 180)
July 15: Boeing 707 Makes First Flight

March 26
1845: A Sticky Application for an Old Problem

Drs. Horace Harrell Day and William H. Shecut patent an adhesive medicated plaster; in other words, a wound dressing that sticks on its own. It is the forerunner of the Band-Aid.

Samuel Gross had reported on using adhesive medicated plasters in a Philadelphia medical journal in 1830. Day and Shecut's innovation was to dissolve rubber in a solvent and then paint it on fabric. After obtaining a patent for the improved process, they sold the rights to Dr. Thomas Allcock, who sold it under the name Allcock's Porous Plaster. Dr. John Maynard advanced the idea in 1848. He took gun cotton (also known as nitrocellulose), dissolved it in sulfuric ether (regular ether, in modern parlance), brushed it on the skin, then covered it with cotton strips. Not exactly convenient or portable.

Advertisement for the forerunner of the Band-Aid.

Robert Wood Johnson and George J. Seabury came up with an improvement in 1874 that would stick for more than a century. They developed a medicated adhesive plaster with a rubber base. Johnson left Seabury and set up a partnership in 1885 with his brothers. That company became Johnson & Johnson. The Johnson brothers' factory in New Brunswick, New Jersey, shipped antiseptic surgical dressings to doctors and hospitals. The dressings were made of cotton and gauze and were individually wrapped. The company soon perfected a technique to sterilize the bandages.

It wasn't until 1920 that J&J created its most famous product. Earle Dickson, a newlywed employee, wanted a simple dressing for his klutzy wife, who suffered lots of cuts and burns in the kitchen. Dickson pre-assembled patches of gauze on adhesive tape for her. He showed the idea to his employer, and the company was soon marketing the Band-Aid for consumer use. — *RA*

Also March 26:

1850: Birth of Edward Bellamy, Socialist Science-Fiction Author

1999: Melissa Worm Wreaks Havoc on Internet

Also 1845:

August 28: *Scientific American* Debuts

December 10: Pneumatic Tire Patented Ahead of Its Time (see page 346)

March 27
1933: Just One Word: *Plastics*

Two British research chemists miss an important detail...and make polyethylene.

Reginald Gibson and Eric Fawcett worked at Imperial Chemical Industries' research laboratory at Winnington, Cheshire. When they attempted to react ethylene and benzaldehyde under high pressure, they produced a waxy lump of what the British call polythene. The experiment could not be successfully repeated; unbeknownst to the researchers, oxygen had leaked into their apparatus and catalyzed the reaction. Using better equipment two years later, ICI scientists M.W. Perrin and J.C. Swallow detected a leak. It took several months before they figured out that it was trace oxygen in the ethylene that played the key role in the reaction.

American chemist Carl Marvel actually made polyethylene by a different method before the ICI team, in the early 1930s. He had just ignored it, because "nobody thought polyethylene was good for anything." ICI, however, had plenty of ideas for it. The chemical conglomerate obtained its first patents in 1936 and quietly put the new plastic into production in 1938.

Polyethylene was a military secret during World War II. It was used to insulate cables on newly developed radar devices. Large-scale commercial polyethylene production began after the war, creating a plethora of plastic kitchenware, toys, containers, and packaging. Polyethylene achieved wide use because of its versatility and low cost. It now competes with other plastics like polyfluoroethylene and polypropylene. Low-density polyethylene (LDPE) has lots of branched polymer chains, which make it more flexible for use in plastic bags, films, and packaging materials. High-density polyethylene (HDPE) has long, straight polymer chains, which make it more durable for use in containers, plumbing, and other parts and fittings.

Too much of all this plastic stuff, alas, finds its way into the oceans as floating pollution. — *RA*

Also March 27:
1794: U.S. Navy Established
1977: Canary Island Runway Collision Kills 583 in Worst Airline Accident

Also 1933:
June 6: First Drive-In Movie Theater Opens (see page 159)
July 22: Wiley Post Completes First Solo Flight Around the World

March 28
1979: Meltdown at Three Mile Island Chills Nuclear Energy in U.S.

Equipment malfunction and human error lead to a partial reactor meltdown at the Three Mile Island nuclear power plant near Middletown, Pennsylvania. It is the most serious accident ever involving a U.S. commercial nuclear facility.

Things started going wrong at 4:00 a.m., when a malfunction caused the main feedwater pumps to stop running, leading to an overheating of the plant's TMI-2 reactor. Although the reactor shut down automatically, an emergency relief valve failed to re-close as the pressure decreased. That failure caused coolant to leak out and the reactor's core to overheat.

Operators were unaware of the leak and erroneously assumed that the core was properly cooled. They attempted to relieve the pressure on the core by reducing the flow of coolant—exactly the wrong thing to do. Overheating caused the zirconium tubes where the nuclear fuel pellets were stored to rupture, and the pellets began to melt.

Only the structural integrity of the containment building's walls prevented Three Mile Island from becoming a Chernobyl-style catastrophe. There was no rupture, meaning all the damage was contained within the facility.

No deaths or injuries have ever been attributed to the Three Mile Island meltdown. But it did chill the growth of nuclear power in the United States and lead to tighter safety and design regulations, with more rigorous oversight by the Nuclear Regulatory Commission.—*TL*

Also March 28:
1910: First Successful Seaplane Flight
1995: Craig Newmark Announces Original
    Craigslist Online

Also 1979:
July 11: Look Out Below! Here Comes Skylab!
    (see page 194)

March 29
1941: Radio Stations Shuffle Frequencies on Moving Day

Americans wake up on Saturday morning to discover that Jack Benny, Bob Hope, and other radio stars of the day no longer occupy their familiar spots on the dial. In a massive shuffle beginning at 3:00 a.m. eastern time, radio stations have engineered a game of musical chairs, and 80 percent of North America's AM frequencies are reassigned to new channels.

This so-called Moving Day resulted from the North American Radio Broadcasting Agreement, negotiated by the United States, Canada, and Mexico. The pact extended the AM broadcast band from 1500 kHz to 1600 kHz (mostly called kilocycles rather than kilohertz in those days). Designed to implement radio standardization throughout the Western Hemisphere, the agreement followed a futile 1939 attempt to squash Mexican "border blaster" stations, which for three decades would continue to overpower U.S. stations with extremely strong signals.

The agreement established clear-channel frequencies — which afford more protection from electromagnetic sky-wave interference at night — across the radio dial, and the reordering shifted most existing AM stations' frequencies to create bandwidth for those new clear-channel station allocations. The new broadcast order also reserved 1230, 1240, 1340, 1400, 1450, and 1490 kHz mainly for local stations. Nations including the Bahamas, the Dominican Republic, and Cuba later signed on to iterations of the NARBA plan.

The ponderously named Regional Agreement for the Medium Frequency Broadcasting Service in Region 2 superseded NARBA rules in 1981.

But the big shuffle was unequaled in consumer experience until TV's final shift from analog to digital, in 2009. — *HH*

Also March 29:

1927: Sunbeam 1000 HP Breaks 200 MPH Mark at Daytona

Also 1941:

May 12: Fog of War Shrouds Advanced Z3 Computer (see page 134)
October 19: Wind Turbine Feeds First Electricity to Grid

March 30
1848: Niagara Falls Runs Dry

Niagara Falls stops. No water flows over the great cataract for forty hours. People freak out.

An American farmer out for a pre-midnight stroll March 29 noticed the absence of the familiar thundering roar. He went to the river's edge and saw hardly any water. Came the dawn, people awoke to unaccustomed silence. The mighty Niagara was a mere trickle. The bed of the river was exposed. Fish died. Turtles floundered about. Brave—or foolish—people walked on the river bottom, picking up exposed guns, bayonets, and tomahawks as souvenirs.

Was it the end of the world? Thousands filled the churches to pray for the falls to start flowing and the world to continue, or for salvation and forgiveness of their sins if it didn't.

Word eventually arrived from Buffalo, three hours away, that storm winds had pushed huge chunks of ice to the extreme northeastern tip of Lake Erie, blocking its outlet into the Niagara River. The ice jam had become an ice dam. Thousands came from nearby cities and towns to see the spectacle of Niagara without water. People crossed the riverbed on foot, on horseback, and in horse-drawn buggies. Mounted soldiers paraded up and down the empty riverbed.

There was no telling when the rushing waters might return, but the owner of the *Maid of the Mist* sightseeing boat sent workers out with explosives to blast away some dangerous rocks the boat always had to avoid. No water flowed for all of March 30 and the daylight hours of March 31.

But that night, a distant rumble came from upriver. The low-pitched noise drew nearer and louder. The ice jam had cleared, and a wall of water came roaring down the upper Niagara River and over the falls with a giant, sudden thunder.—*RA*

Also March 30:

240 BCE: Chinese Sight Halley's Comet
1842: It's Lights-Out, Thanks to Ether
    (see page 275)

Also 1848:

January 24: Gold Discovered in California

March 31
1901: Wuppertal Monorail Opens

A suspended monorail opens in Germany, whisking passengers on an 8.3-mile loop some forty feet over the Wupper River. Though it is not the world's first single-track hanging-rail system, it's the world's oldest monorail still in operation and Europe's only suspended railway.

The hilly Wupper Valley was an extremely difficult place to build a traditional railway: tight corners, rivers, sharp drops. Plans for a hanging monorail there date back to 1826, but the project lay dormant until engineer Carl Eugen Langen built an

The monorail in 1913.

electric hanging-monorail prototype in Cologne in 1897. Construction of the Wuppertal *Schwebebahn*, or "suspended railway," began in 1898. A test run of the monorail was conducted with German emperor Wilhelm II on board a specially designed *Kaiserwagen*, which is still used for special events more than a century later.

The monorail officially opened to the public March 31, 1901. Construction had cost 16 million marks (about $4 million then, $100 million today).

The Wuppertal *Schwebebahn* has operated with very few incidents and only one accident resulting in fatalities. Human error claimed the lives of five riders in 1999, when a train derailed and crashed into the river after hitting a piece of equipment that had been left hanging from the track. Supervisors were convicted of involuntary manslaughter.

But the rail line's first big accident took place in 1950 and involved both human and pachyderm error. During a promotional stunt for a local circus, an elephant named Tuffi was put on board the monorail. She got spooked and bolted through the side of the train car, landing safely in the Wupper River. Tuffi is now memorialized in a nearby mural. — *KB*

Also March 31:

1932: Ford Introduces Flathead V-8 Engine
1963: LA Streetcars Take Their Last Ride
1999: *The Matrix* Opens in Movie Theaters

Also 1901:

December 5: Birth of Walt Disney and Werner Heisenberg (see page 341)

April 1
1998: World Discovers Disney Is Buying MIT — April Fool!

Massachusetts Institute of Technology students, long known for their pranks, hack the school's home page to announce to the world that the Walt Disney Company would be purchasing MIT for $6.9 billion.

The prestigious school would be renamed the Disney Institute of Technology, according to an April Fools' Day press release linked to the hijacked home page.

The school was accustomed to various engineering pranks from its elite students and left the bogus info up for most of the day. "I knew it was a hack as soon as I saw the price," an MIT spokesman quipped at the time. "Only $6.9 billion? Much too cheap!" The fake press release said, among other things: "As part of the acquisition, the entire MIT campus will be moved brick by brick down to the Walt Disney resort complex in Orlando, Florida." The phony statement also claimed MIT departments would be renamed after Disney characters. The Sloan School of Management would become the Scrooge McDuck School of Management. Its engineering department would become the school of Imagineering.

Patti Richards, an MIT spokeswoman, said "nobody got into trouble" over the prank, adding that the campus expects this kind of stunt on April 1: "It's very much the tradition here." Other MIT shenanigans include the overnight dressing of the huge dome of the central campus building as a giant version of *Star Wars*' R2-D2. On another nighttime excursion, students put what looked like a police cruiser atop the dome, but it turned out to be the considerably lighter metal shell of a car, mounted on a wood frame. It was painted to match campus police vehicles and even had a dummy dressed in uniform. — *DK*

Also April 1:

1875: The *Times* (London) Publishes First Weather Map in a Newspaper

1960: First Weather Satellite Launched

1976: Steve Jobs, Steve Wozniak, Ronald Wayne Found Apple Computer

2004: Google Unveils Innovative Webmail Service, Gmail

Also 1998:

April 27: Koko Goes Ape in AOL Chat

September 7: If the Check Says "Google Inc.," We're "Google Inc."

September 18: ICANN Takes Over Running Internet Domain-Name System

April 2
1922: Rorschach Dies, Leaving a Blot on His Name

Swiss psychiatrist Hermann Rorschach dies. He leaves behind a widely known, but no longer widely used, test for diagnosing mental illness with inkblots.

Rorschach spent so much time drawing that his high school buddies nicknamed him Kleck, the German word for "inkblot." (Hmm, Doktor, und vat do you make of that?)

Rorschach believed that what different people perceived in ambiguous inkblots would reveal differences in basic personality structure. He began showing inkblots to patients and asking, "What might this be?" It seems obvious today that someone who repeatedly sees people fighting in a series of inkblots might have a different mind-set from someone who sees dancing or sexual acts. Or that people who see people in the blots differ from those who always see birds or animals, or who see inanimate objects rather than living things. That general projective test might, however, indicate only a person's momentary mood.

But Rorschach, with the characteristic inclination of a nation of watchmakers, set out to devise a precise system for scoring his test based on whether and how much a patient reported movement, color, and form. When his book *Psychodiagnostics* appeared, in 1921, it attracted little notice. But word spread, and the Rorschach test eventually joined the standard arsenal of shrinky examinations. New and complicated systems of scoring evolved, and competing schools of therapy vigorously disputed one another's methods.

The scoring systems aren't used much these days, but the inkblots became a staple of the popular-culture view of psychiatry and mental illness. They've appeared in countless films and TV shows, and even on the masked face of the character Rorschach in the graphic novel and movie *Watchmen*.

All this didn't come in Rorschach's lifetime. He died of appendicitis in 1922. He was only thirty-seven. —*RA*

Also April 2:

1845: French Physicists Photograph the Sun
1889: Hall's Aluminum Process Foils Steep Prices
1973: Legal-Computer-Info Service Lexis Launches

Also 1922:

January 11: Insulin Makes a Nice Shot (see page 11)
November 26: Archaeologist Howard Carter Enters King Tut's Tomb

1973: Motorola Calls AT&T ... by Cellphone

Martin Cooper of Motorola uses the first portable handset to make the first cellphone call ... to his rival at Bell Labs. Rub it in.

Before that, if you wanted to make a mobile-phone call, you'd have to have a radio-phone in your car. You'd need to spend thousands of dollars, stash about thirty pounds of equipment in your trunk, and install a special antenna.

Bell Laboratories (then the research division of AT&T and now part of Alcatel-Lucent) had developed the concept of cellular communications in 1947. But the company was locked in a competition with Motorola in the '60s and '70s to go truly portable. At Motorola, Cooper and designer Rudy Krolopp worked on the "shoe" phone using many of the company's existing electronics patents. They produced the Motorola DynaTAC (an acronym for dynamic adaptive total area coverage): it was nine inches tall, weighed two and a half pounds, and had thirty circuit boards. You could talk for thirty-five minutes, and the phone took ten hours to recharge.

Cooper set up a cellular base station in New York and made his first call to Joel Engel, Bell Labs' research chief. Ouch. Cooper recalls: "I made numerous calls, including one where I crossed the street while talking to a New York radio reporter—probably one of the more dangerous things I have ever done in my life."

Motorola spent another ten years getting the cellphone over technological and regulatory hurdles. Commercial service started in 1983, with a slimmed-down, sixteen-ounce DynaTAC. First adopters paid $3,500 for the phone ($8,000 in 2012 money). It was 1990 before cellphone service reached a million U.S. subscribers.

The world's lightest cellphones now weigh less than 1.5 ounces each. The world's cheapest are free, if you sign a two-year service plan. —*RA*

Also April 3:

1892: Druggist Invents Ice Cream Sundae
1996: Unabomber Nabbed in His Montana Hideout

Also 1973:

May 22: Ethernet Memo Enables Local Computer
   Networks (see page 144)

April 4
1975: Bill Gates, Paul Allen
Form a Little Partnership

Bill Gates and Paul Allen create a company called Micro-Soft. It will grow into one of the largest U.S. corporations and place them among the world's richest people.

Gates and Allen had been buddies and fellow BASIC programmers at Lakeside School in Seattle. Allen graduated before Gates and enrolled at Washington State University. They built a computer based on an Intel 8008 chip and used it to analyze traffic data for the Washington state highway department, doing business as Traf-O-Data.

Allen went to work for Honeywell in Boston, and Gates enrolled at Harvard. News in late 1974 of the first personal computer kit, the Altair 8800 (see page 355), excited them, but they knew they could improve its performance with the computer language BASIC (see page 123). Allen spoke to Altair manufacturer Micro Instrumentation and Telemetry Systems and sold the firm on the idea. Gates and Allen worked night and day to complete the first microcomputer BASIC. Allen moved to Albuquerque in January 1975 to become director of software for MITS. Gates dropped out of his sophomore year at Harvard and joined Allen in New Mexico.

Allen was twenty-two; Gates was nineteen. Altair BASIC was functioning by March. The "Micro-Soft" partnership was sealed in April but didn't get its name for a few more months. The fledgling company also created versions of BASIC for the hot-selling Apple II and Radio Shack's TRS-80.

Microsoft moved to Washington state in 1979. It incorporated in 1981, a few weeks before IBM introduced its personal computer with Microsoft's 16-bit operating system, MS-DOS 1.0.

Microsoft stock went public in March 1986. Adjusting for splits, a share of that stock was worth about 320 times its original value twenty-five years later. — *RA*

Also April 4:
1581: Circumnavigator Drake Knighted by His
    Grateful Queen

Also 1975:
March 5: Silicon Valley's Pioneering Homebrew
    Computer Club Holds First Meeting
June 7: Before Digital, Before VHS ...There Was
    Betamax (see page 160)

April 5
1951: Birth of Dean Kamen, Inventor of Segway

Dean Kamen, a prolific inventor referred to by *Smithsonian* magazine as "the Pied Piper of Technology," is born in Rockville Centre, New York.

Kamen holds hundreds of patents, many for innovative medical devices that have transformed health care. He was still in college when he invented the wearable infusion pump, which has applications in such diverse medical fields as oncology, neonatology, and endocrinology. At twenty-five, he founded a company to manufacture infusion pumps, and he also devised the first portable insulin pump for diabetics.

He is a tireless booster of technology. In 1989, Kamen founded FIRST (For Inspiration and Recognition of Science and Technology), which sponsors various robotics competitions, among other activities, to expose technology to a younger generation. But Kamen is probably best known to that demographic as the inventor of the Segway transportation system, those little personal scooters that have caught on oh so slowly. When the self-balancing, two-wheeled personal transporter was introduced in 2001, Kamen told *Time* magazine it "will be to the car what the car was to the horse and buggy." A decade on, that hasn't exactly happened. Sales have been well below expectations, although Segways have found niche uses in retirement communities and on urban sightseeing tours.

Happy birthday just the same. — *TL*

Also April 5:
1998: Tamagotchi Virtual Pet Distracts Driver;
   Car Kills Cyclist

Also 1951:
May 11: RAM Is Born — Matrix Core Memory Patent
October 31: First Zebra Crossing for Pedestrians
   (see page 306)

April 6
1938: Teflon Gets Off to a Slippery Start

Fiddling around in the lab one day, Roy Plunkett accidentally discovers polytetrafluoroethylene, soon to be known as Teflon, a slippery substance that will have many practical applications.

Plunkett, a chemist at DuPont's Jackson research lab, in New Jersey, made his discovery in the time-honored scientific way: as the result of a mistake, and with an assistant's help. Plunkett and his assistant, Jack Rebok, were testing the chemical reactions of tetrafluoroethylene, a gas used in refrigeration. The gas was contained in some pressurized canisters, one of which failed to discharge properly when its valve was opened.

Rebok picked up the canister and found it was too heavy to be empty. He suggested cutting the canister open to see what had happened. Despite the risk of blowing the lab to kingdom come, Plunkett agreed.

Of course it was heavy: The gas hadn't accidentally escaped. It had solidified into a smooth, slippery white powder as a result of its molecules bonding, a process known as polymerization. This new polymer was different from similar carbon-based solids, like graphite: it was lubricated better, due to the presence of dense fluorine atoms that shielded the compound's string of carbon atoms.

Setting other work aside, Plunkett began testing the possibilities of polytetrafluoroethylene, eventually figuring out how to reproduce the polymerization process that had occurred accidentally the first time. DuPont patented the polymer in 1941, registering it under the trade name Teflon in 1944. The first products—most having military and industrial applications—came to market after World War II. Teflon didn't became a household word until the early 1960s, when it was used to produce convenient nonstick cookware.

Teflon is used virtually everywhere today, from the aerospace industry to clothing manufacturing to pharmaceuticals. —*TL*

Also April 6:

1903: Birth of Harold Edgerton, Father of High-Speed
   Photography
1909: Robert Peary and Matthew Henson Reach
   North Pole, but Claim Is Disputed

Also 1938:

October 22: Chester Carlson Produces First Xerox Copy
   (see page 297)
October 30: "War of the Worlds" Radio Drama
   Induces Panic

April 7
1969: A Birthday for the Internet

The publication of the first request for comments, or RFC, document paves the way for the birth of the Internet.

April 7 is often cited as the symbolic birth date of the net because RFC memoranda contain research, proposals, and methodologies applicable to Internet technology. RFC documents let engineers and others kick around new ideas in a public forum. Those ideas are sometimes adopted as new standards by the Internet Engineering Task Force. The opener, RFC 1, was titled "Host Software" and written by UCLA's Steve Crocker. But other dates for birth of the net also have their supporters.

Thanks to those early RFCs, ARPANET — sponsored by the Pentagon's Defense Advanced Research Projects Agency — was able to link computers at two remote sites on October 29, 1969. The first two nodes of what would become a global network were UCLA and the Stanford Research Institute, a few hundred miles away in Silicon Valley. The system relied on packet-switching, the same concept that runs the Internet today: Data is transmitted in small, arbitrary packets. That allows huge amounts of data to efficiently utilize the available bandwidth.

Others prefer a much later birth date for the Internet: January 1, 1983, when the National Science Foundation's university network backbone started connecting campus computer networks to one another.

As for the World Wide Web — the multimedia portion of the Internet — you'll find different birthdays there too. Some cite Christmas Day 1990, when Tim Berners-Lee completed the first practical HTML browser, called WorldWideWeb. How about February 26, 1991, when he introduced it to colleagues? Or August 7, 1991, when CERN unveiled it to the world? Or maybe it's April 22, 1993, when the University of Illinois released NCSA Mosaic 1.0, the first web browser to attain public popularity (see page 114).

Take your pick. — *TL*

Also April 7:

1933: *King Kong* Opens in Movie Theaters
1964: IBM Bets Big on System/360 Mainframe
    Computers

Also 1969:

June 22: Umm, the Cuyahoga River's on Fire ... Again
    (see page 175)

April 8
1879: The Milkman Cometh... with Glass Bottles

Milk is sold in glass bottles for the first time in the United States. It's a clear improvement in hygiene and convenience.

Heyday of the milk bottle, c. 1939.

Until that time, people bought milk as a bulk item, with the seller dispensing milk out of a keg or bucket into whatever jugs, pails, or other containers the customers brought. That practice left a lot to be desired on the cleanliness front. Echo Farms Dairy introduced the first purpose-made milk bottles in New York City, delivering the milk from Litchfield, Connecticut. Other dealers initially feared the expense of breakage, and some customers didn't like the drugstore look of the containers.

But the new method of delivery eventually caught on. By the first decade of the twentieth century, some cities required that milk be delivered in glass bottles. The earliest wax containers appeared in the 1890s. Shapes ranged from simple boxes to cylinders to cones to truncated pyramids. Some even imitated the shape of a typical round glass bottle.

What finally prevailed in the 1940s was a rectangular-column design with a small, round pull-up cap on a flat top piece. The containers were lightweight and compact, wasting little space in milk trucks. Flat-top boxes were replaced in the 1950s by square cartons with gable tops that opened out into spouts for easy pouring. This design had actually been patented in 1915.

Milk in glass bottles is a specialty or niche product these days, and home milk delivery is pretty much a thing of the past. The wax gable-top design and the more recent plastic bottles account for nearly all the retail milk sold in the United States, and you have to go to the store to buy them. — *RA*

Also April 8:

1869: Birth of Neurosurgery Pioneer Harvey Cushing
1953: *Man in the Dark*, Hollywood's First 3-D Movie, Released

Also 1879:

October 21: Edison Gets the Bright Light Right (see page 296)
November 4: Ka-ching! Ritty Patents the First Cash Register

1806: I.K. Brunel Is Born, a Giant Among Engineers

Isambard Kingdom Brunel is born. His engineering works will change the face of Britain and British commerce with railroads, bridges, tunnels, docks, and ships.

Brunel was not quite five feet four inches tall, but he was called the Little Giant: he thought big and built big. Some of his best projects are still hard at work today.

Brunel completed his father Marc Brunel's Thames Tunnel in East London. The world's first underwater tunnel opened in 1843 and carried London Underground trains until 2007. He designed the record-setting (702-foot-long) Clifton Suspension Bridge in 1830. It still carries pedestrians, bicycles, and cars.

I. K. Brunel's *Great Eastern* in 1866.

Brunel's bridge at Chepstow wasn't dismantled until 1962. The 1841 Great Western Railway from London to Bristol is still in use, with bridges, tunnels, viaducts, and stations designed by Brunel. Brunel's paddle-wheeled *Great Western* was the first steamship in regular transatlantic service, cutting the one-month (one-way) travel time in half. The *Great Britain,* launched in 1843, was the world's largest ship and the first screw-propeller iron steamship to cross the Atlantic.

Brunel's biggest ship was the 692-foot *Great Eastern*, so big it had to be launched sideways, and even that took three attempts. The *Great Eastern* was the largest ship to sail the seas until the *Lusitania,* in 1906. But it flopped financially as a passenger ship and was converted to lay the first transatlantic telegraph cable in 1866 (see page 210). Before being scrapped, in 1888, the *Great Eastern* laid four Atlantic cables and one across the Arabian Sea. But it ended life as a floating circus and tourist attraction.

The ship also killed its designer: Brunel died of a stroke, probably from overwork, just after the *Great Eastern* suffered a boiler explosion on its maiden voyage. He was fifty-three. — *RA*

Also April 9:

1860: Phonautograph Transcribes Sound into Lines on Paper but Can't Reproduce Sounds

1959: NASA Introduces Seven Original Mercury Astronauts

Also 1806:

October 7: Carbon Paper Patented

April 10
1849: Walter Hunt Patents Safety Pin

A New York inventor receives a patent for the spring safety pin. He invented it because he needed some cash. Parents and babies have been spared blood and screams ever since.

Walter Hunt began inventing when still a teenager and received a patent for a machine to spin flax in 1826. He also invented a fire-engine gong, a stove to burn hard coal, a knife sharpener, a streetcar bell, synthetic stone, road-sweeping machinery, bicycle improvements, ice plows, and paper collars for shirts. He invented an advanced sewing machine but feared it would throw seamstresses out of work. By the time he applied for a patent for it, others had beaten him to the punch — no "stitch in time" here.

But one day in 1849 when Hunt owed a friend fifteen dollars, he decided to live by his wits and invent something. Hunt took an eight-inch length of brass wire, coiled it in the center, and shielded the clasp at one end. He called it a dress pin, and within three hours he had sold the rights to his invention for $400 (about $11,000 today). Clasp pins had existed before, but Hunt's innovation utilized a single piece of metal (for easier fabrication), used a spring instead of a hinge, and combined the spring with a clasp.

The year 1849 was a relatively good one for Hunt: He also invented the first repeating rifle (a precursor of the Winchester). And he beat Charles Rowley, who got a British patent for a safety pin on October 12 that year.

New York's Museum of Modern Art has an 1849 Hunt safety pin in its collection. — *RA*

Also April 10:

1972: 140 Nations Sign Biological Weapons Treaty
1815: Tambora Explosion's "Volcanic Winter" Disrupts
    Global Weather for More Than a Year

Also 1849:

March 10: Abraham Lincoln Files Patent on Riverboat
    Flotation Device (see page 54)
June 1: Birth of Stanley Twins, Steam-Auto Pioneers

April 11
1984: Shuttle Makes House Call, Repairs Satellite

Shuttle astronauts retrieve the malfunctioning Solar Max solar-research satellite, fix it, and send it back into orbit. It's the first time a satellite has been repaired in space.

The Solar Maximum Mission had been launched February 14, 1980. But the satellite was having trouble with its altitude-control system, and the electronics in the coronagraph instrument were wonky.

Space shuttle *Challenger* launched from Cape Canaveral on mission STS-41C on April 6. Five days later, astronauts George Nelson and James van Hoften used the Remote Manipulator System robot arm to retrieve the buggy satellite from orbit and attach it to a workbench in the aft end of the *Challenger*'s open cargo bay. There, they replaced Solar Max's altitude-control and main electronics box, then released the Solar Max back into orbit. The refurbished satellite continued to observe the solar-activity cycle for another five years before reentering Earth's atmosphere and going down in flames, on December 2, 1989.

Challenger's seven-day flight also deployed a school-bus-size science satellite, the Long Duration Exposure Facility, a rack of fifty-seven experiments to test the effects of long-term exposure to radiation and micrometeoroids. In addition, *Challenger* carried a student experiment to see if honeybees could make a honeycomb in the near zero gravity of space. They could, and they did: just like the astronauts, busy as bees.

The shuttle program went on to recover or repair other satellites, including the Hubble Space Telescope several times.

Challenger flew four more successful missions in 1984 and 1985. Its tenth and final flight came on January 28, 1986, when it was blown apart seventy-three seconds after launch, killing its crew of seven. — *RA*

Also April 11:

1755: Birth of James Parkinson, First to Describe Parkinson's Disease
1888: Concertgebouw, Home of Nearly Perfect Acoustics, Opens
1952: First Successful Surgery to Treat Parkinson's Disease

Also 1984:

September 10: DNA Fingerprinting Discovered (see page 255)
November 20: SETI Institute Starts Search for Extraterrestrial Intelligence

April 12
1994: Lawyers Invent Commercial Spam

Members of more than six thousand Usenet discussion groups find themselves the recipients of a message imploring them to use the legal services of Laurence Canter and Martha Siegel to ensure their place in line for a green card from the U.S. government.

It didn't matter that most recipients had no need for such services. They'd just been spammed by a company—for the first time in Internet history (though there'd been some minor kerfuffles on the old ARPANET). Not surprisingly, some lines of the message were in ALL CAPS.

Canter and Siegel achieved notoriety, claiming they'd made $100,000 from their Perl-script spamming. The two remained unrepentant despite the backlash that led them to lose their hosting. Canter even got disbarred. The husband-and-wife duo went on to create a spam-advertising company and write *How to Make a FORTUNE on the Information Superhighway: Everyone's Guerrilla Guide to Marketing on the Internet and Other On-Line Services.*

But an outdated marketing guide is hardly the duo's true legacy. That would be spam, which now accounts for up to 90 percent of e-mail traffic.

And if there's a medium worth communicating in, it's worth spamming. Message boards, blog comments, and social-networking sites all have to fight the scourge unleashed by the infamous green-card spam of 1994.

By the way, you just inherited millions of dollars in Nigeria, so go out there and buy some genuine Viagra Rolexes. —*RS*

Also April 12:
1961: Soviets Orbit Gagarin, First Man in Space

Also 1994:
April 21: First Planet Discovered Outside Our Solar System
July 29: Video-Game Makers Propose Ratings Board to Congress
October 27: Web Gives Birth to Banner Ads
(see page 302)

April 13
1953: CIA Okays Mind-Control Tests; James Bond's First Adventure

U.S. spies embark on secret tests using hallucinogenic drugs on unwitting subjects.

Central Intelligence Agency director Allen Dulles authorized the MK-ULTRA project. The dubious covert program turned unsuspecting citizens into guinea pigs for research into mind-altering drugs. Dulles wanted to close the "brainwashing gap" after the U.S. government learned that American prisoners of war in Korea were subjected to mind-control techniques.

The CIA sought to devise a truth serum to enhance interrogations of POWs and captured spies. It also wanted to develop "amnesia pills" to create CIA super-agents immune to the enemy's mind-control efforts. MK-ULTRA even hoped to create a "Manchurian Candidate," a programmable assassin, and to control the minds of pesky despots like Fidel Castro.

In addition to drug experiments, the program included more than a hundred subprojects that involved radiological implants, hypnosis and subliminal persuasion, electroshock therapy, and isolation techniques.

Several cases of suspicious deaths and lives left in ruin came to light during 1975 investigations, but the CIA had ordered the MK-ULTRA files destroyed in 1973.

In an amazing confluence of fact and fiction, the first James Bond novel was published the same day MK-ULTRA got under way. Ian Fleming's *Casino Royale* introduced Agent 007, literature's most famous spy. *Casino Royale* elevated the espionage genre to the top levels of popular culture. Fleming was the first to combine style and sexiness with the dangerous world of espionage. After *Casino Royale*'s successful British launch, it was released in the United States as *You Asked for It*. The novel (and Fleming's sequels) did not succeed Stateside until President Kennedy included *From Russia with Love* on his favorites list.

Fleming went on to pen twelve Bond novels and the children's classic *Chitty Chitty Bang Bang.—KZ, ST*

Also April 13:

1970: Apollo 13 Tells Houston, "We've Got a Problem" (see page 109)

Also 1953:

April 25: Riddle of DNA's Architecture Finally Solved (see page 117)

April 14
1945: Troublesome Tech Toilet Costs Skipper His Sub

A malfunctioning high-tech toilet forces a German U-boat to the surface off the coast of Scotland, where it is promptly attacked by a British aircraft.

Anyone who's been on a small boat knows how problematic the flush-plumbing can be. Still, it beats the old days of using a bucket, or just letting go over the side.

Over the side is not usually an option on a submarine.

U-1206, sailing out of Kristiansand, Norway, was cruising at a depth of roughly two hundred feet when the commander, Kapitänleutnant Karl-Adolf Schlitt, decided to answer a call of nature. The submarine was a late-war type VIIC, carrying a new kind of toilet designed for use at greater depths. The techno-toilet was just a little buggy. Schlitt (no, we did not make that name up) had trouble operating it. When he called an engineer for help, the man opened the wrong valve, allowing seawater to enter. The water reached the batteries under the toilet, and the submarine began filling with chlorine gas, forcing Schlitt to order *U-1206* surfaced. Unfortunately for the Germans, the vessel was only ten miles off the Scottish coast, and it was quickly spotted by the British.

The crew was still blowing clean air into their U-boat when an aircraft appeared and attacked, killing four men on deck and damaging the U-boat so badly that it couldn't dive. Schlitt, seeing the game was up, gave the order to scuttle and abandon ship.

It was an ignominious end to Schlitt's only combat patrol of the war as a commander. The wreck of the *U-1206* remained undisturbed until the mid-1970s, when workers laying an oil pipeline came across the hulk sprawled on the seabed at 230 feet. — *TL*

Also April 14:

1860: First Pony Express Rider Reaches Sacramento, California

1932: Zounds! We've Split the Atomic Nucleus

1996: Jennifer Ringley Starts Webcasting JenniCam Pictures of Herself Every Three Minutes

Also 1945:

May 5: Balloon-Propelled Japanese Bomb Lands in Oregon, Killing Six

May 25: Arthur C. Clarke Proposes Geostationary Communications Satellites (see page 147)

April 15
1726: Apple Doesn't Fall Far from Physicist

Isaac Newton tells his biographer the story of how an apple falling in his garden prompted him to develop his law of universal gravitation. It will become an enduring origin story in the annals of science, and it may even be true.

Newton was apparently fond of telling the tale, but written sources do not reveal a precise date for the fabled fruit-fall. We do know that on this day in 1726, William Stukeley talked with Newton, who told him how the idea had originally occurred to him:

> It was occasioned by the fall of an apple, as he sat in contemplative mood. Why should that apple always descend perpendicularly to the ground, thought he to himself. Why should it not go sideways or upwards, but constantly to the Earth's centre?

Newton figured out that the force that makes objects fall extends upward infinitely (reduced by the square of the distance), that the force exists between any two masses, and that the same force that makes an apple fall also holds the moon and planets in their courses.

John Conduitt, Newton's nephew-in-law, tells the story thus:

> In the year [1666] he retired again from Cambridge on account of the plague to his mother in Lincolnshire & whilst he was musing in a garden it came into his thought that the same power of gravity (which made an apple fall from the tree to the ground) was not limited to a certain distance from the earth but must extend much farther than was usually thought—Why not as high as the Moon said he to himself & if so that must influence her motion & perhaps retain her in her orbit.

Voltaire also wrote of the event in 1727, the year Newton died: "Sir Isaac Newton walking in his gardens had the first thought of his system of gravitation, upon seeing an apple falling from a tree." Note that no one, from Newton on down, claims the apple bopped him on the bean. Makes a good cartoon, sure, but if it happened, it might have set the guy speculating instead on why— and how—pain hurts. —*RA*

Also April 15:

1452: Birth of Leonardo da Vinci, Visionary Engineer

1912: RMS *Titanic* Sinks on Maiden Voyage

Also 1726:

June 14: Birth of James Hutton, Pioneering Geologist

April 16
1813: Specifying the Interchangeability Standard

Mass production takes a massive step forward as gunmaker Simeon North gets a government contract to make twenty thousand flintlock pistols, all with interchangeable parts.

North built a small factory in Berlin, Connecticut, in 1799 to make pistols and soon landed a government contract for four hundred (some sources say five hundred) pistols. Another contract the following year called for fifteen hundred guns. Soon, he was operating a pistol factory in nearby Middletown. When the War of 1812 broke out, it was North's idea that War Department contracts should call for interchangeable parts. His 1813 contract for twenty thousand pistols is the first evidence anywhere of such a provision. It stipulated that "component parts of the pistols are to correspond so exactly that any limb or part of one pistol may be fitted to any other pistol of the 20,000." The pistols cost the U.S. Army $7 each (about $82 in today's money) and were to be delivered over the course of five years.

North is generally acknowledged as the inventor of the earliest primitive milling machine. The mechanism replaced hand-filing for the shaping of metal parts and made interchangeability a practical goal. Before long, he was turning out ten thousand pistols a year.

The government sent him to the Harpers Ferry Armory in what was then Virginia to standardize production there. North was about to take interchangeability a step further. His Connecticut factory got a contract in 1828 to produce five thousand rifles whose parts would be interchangeable not only with one another but with those of rifles manufactured at Harpers Ferry. He continued making improvements in weapons design and production and remained a War Department contractor for fifty-three years, until his death at eighty-seven in 1852.

North is enshrined in the Machine Tool Hall of Fame. *—RA*

Also April 16:

1943: Albert Hofmann Accidentally Discovers Psychedelic Nature of LSD

1947: Fertilizer Ship Explodes in Texas City, Texas, Killing 600 and Injuring 3,500

Also 1813:

January 4: Birth of Isaac Pitman, Inventor of Shorthand Alphabet

1970: Houston, We No Longer Have a Problem

Apollo 13 splashes down in the Pacific Ocean, recovering from a barely survivable explosion in space.

Apollo 13 was supposed to be the third manned lunar landing (see page 203). But an oxygen tank exploded 200,000 miles from Earth. The astronauts' supplies of air, water, and electricity were imperiled. Astronaut John Swigert radioed Mission Control: "Houston, we've had a problem here" — not the *Apollo 13* movie's "Houston, we have a problem." NASA decided to scrap the lunar landing and have the astronauts swing around the moon and return home, using the lunar module (LM) as a lifeboat.

Oxygen: Plenty was available from LM tanks that would have supplied liftoff from the moon's surface.

Electricity: Noncritical systems were turned off, reducing power consumption to one-fifth normal. But capsule temperature dropped to 38 degrees F.

Water: The crew conserved water by drinking little and eating only wet foods. They became severely dehydrated, losing about ten pounds each.

Carbon dioxide removal: The LM didn't have enough capacity, so Mission Control had the astronauts build a pipeline to the command module (CM) with plastic bags, cardboard, and tape that NASA had wisely placed on board.

Getting home: The navigation system was transferred to the LM, but precise realignment was essential because too steep a reentry angle would cause the CM to burn up in the atmosphere. Too tangential an angle could skip the module out into space forever. Fire or ice. After four days of alternating terror and hope, the three astronauts climbed back to the CM for reentry an hour before splashdown. Everything worked out.

A review board later determined the cause of the explosion: Oxygen-tank heaters had been upgraded, but the heater *switches* hadn't. NASA also found that other warning signs had been ignored, and the oxygen tank was no longer a potential bomb but a real one. — *RA*

Also April 17:

1790: Death of Scientist-Inventor-Publisher-
      Statesman Ben Franklin
1964: Ford Unveils the Mustang

Also 1970:

April 22: First Earth Day Celebrated
May 17: Heyerdahl Sets Sail to Cross Atlantic on *Ra II*
December 23: Construction Workers Place Highest
      Steel on World Trade Center

April 18
1915: Aerial Warfare Is
About to Make a Quantum Leap

Roland Garros is shot down behind German lines and taken prisoner. His plane is recovered intact by the Germans, which results in a technological leap forward for aerial warfare.

Garros was an aspiring concert pianist who gained fame as an aviator prior to World War I when he flew nonstop across the Mediterranean Sea. He's considered the first true fighter pilot in history.

He joined the French army at the outbreak of hostilities and was soon engaged in the new aerial combat. Finding it too difficult to fly and shoot at the same time, Garros mounted a forward-firing machine gun to his plane and fitted metal deflector plates to the propeller to protect it from the bullets. The enhancement allowed Garros to attack head-on, and he shot down five German planes within two weeks, becoming the war's first ace.

After Garros was shot down (ironically, by ground fire), his plane was turned over to Anthony Fokker, a Dutch aircraft engineer building planes for Germany. Improving on Garros's idea, Fokker's team designed the interrupter gear that synchronized the propeller and the gun's firing action, allowing bullets to pass *between* the prop blades without striking them. This became standard on German aircraft and soon was copied in Allied aircraft. The mayhem was on.

Garros himself managed to escape from a POW camp in early 1918. After making his way back to French lines, he returned to combat. He was shot down and killed in October 1918, a month before the war ended.

If Garros's name sounds familiar, you may have heard it in a different context: the tennis center in Paris where the annual French Open is played is named for Roland Garros. — *TL*

Also April 18:
1906: Earthquake Hits San Francisco; Great Fire
    Begins

Also 1915:
February 4: Goldberger Establishes Nutritional
    Deficiency, Not Infection, as Cause of Pellagra
April 22: Germans First Use Poison Chlorine Gas in
    Warfare

1965: How Do You Like It? Moore, Moore, Moore

Gordon Moore publishes a pithy four-page analysis of the integrated-circuit business in which he correctly predicts that chip complexity will regularly double for the foreseeable future.

Moore was chief of research and development for Silicon Valley's Fairchild Semiconductor, and he went on to cofound Intel. His prediction turned out to be basically correct, and it became a rallying cry for the emerging computer industry. By 1970 people were referring to his analysis as Moore's law, shorthand for the inexorable upward march of computing capabilities — an electronic manifest destiny.

The law went through several permutations and is often misunderstood. Moore's 1965 paper in *Electronics* magazine referred to the number of transistors (see page 359) that could be cost-effectively produced on a single integrated circuit (see page 257). He somewhat optimistically predicted that this number would double every year. Moore later revised his estimate to doubling every two years, which fits the long-term trend more closely.

The number of transistors is a rough-and-ready indicator of a chip's processing power, so people have used Moore's law to refer to a biennial doubling of chip power. The flip side of Moore's law is that the cost of computing power gets cut in half roughly every two years. Moore's 1965 paper correctly foresaw that this meant computers — which were then expensive, precious devices available only to a priesthood of computer scientists — would someday be so common they'd be sold in department stores among the sundries.

He also figured that integrated circuits would prompt revolutionary changes in a wide range of fields as they became cheap enough to be embedded in all kinds of things. Now chips are so ubiquitous you can find them in greeting cards, where they do nothing fancier than play annoying, high-pitched music. — *DT*

Also April 19:
1971: Soviets Orbit First Space Station, Salyut 1

Also 1965:
July 16: Mont Blanc Road Tunnel Links France, Italy
July 25: Dylan Plays Electric Guitar, Shocks Folkie
    Audience
August 29: Orbiting Astronaut Chats with Undersea
    Aquanaut (see page 243)

April 20
1940: Zworykin Demonstrates Electron Microscope

Vladimir Zworykin, known as a coinventor of television, demonstrates the first electron microscope in the United States.

While working for Westinghouse, Zworykin developed and patented the iconoscope and kinescope, which used an electronic system to create and reproduce television images. Westinghouse decided not to pursue the new technology, and Zworykin moved to RCA. Besides helping advance TV to a commercial medium, he worked on text readers, electric eyes, missile-guidance systems, and computerized weather prediction.

Ernst Ruska at Berlin Technical University discovered in the late 1920s that if you bombard a tiny object with electrons, you can create a large image with the focused beam. Combining multiple electron lenses increases magnification, just like an optical microscope. Ruska and Max Knott built the world's first electron microscope in 1931. It had a resolution of 400x — less than an optical microscope, but proof of concept. In two years, Ruska built an electron microscope that bettered its optical counterparts, and Siemens built a 1939 model based on Ruska's work.

Zworykin and his team developed their electron microscope at RCA's research labs in 1939. The device they demonstrated in 1940 stood ten feet high and weighed half a ton. It achieved a magnification of 100,000x. That was more than proof of concept. It was fulfillment.

Zworykin shares credit for the television with Philo T. Farnsworth, John Logie Baird, Allen B. DuMont, and others. His efforts earned him scores of awards from associations and institutions around the world. But science's highest honor eluded him. The 1986 Nobel Prize in Physics went to Ruska for his work in electron optics and "design of the first electron microscope," and to Swiss IBM researchers Gerd Binnig and Heinrich Rohrer for the scanning tunneling microscope. — *RA*

Also April 20:

1841: Poe's "Murders in the Rue Morgue" Paves Way
   for Detective Genre
1926: Warner Bros. Announces Vitaphone Talking
   Movies (see page 73)

Also 1940:

April 23: Batteries Included, and They Don't Leak
   (see page 115)
November 7: Newly Completed Tacoma Narrows
   Bridge Collapses in Windstorm

April 21
1878: Thinking Fast, Firefighter Slides Down a Pole

A Chicago firefighter hears the alarm ringing and slides down a pole that just happened to be nearby. It's the birth of the firehouse sliding pole.

It's not easy to get a whole lot of people down a flight of stairs fast, so some fire-houses had added sliding chutes, like the ones in children's playgrounds. But not Engine Company 21, where firefighters were unloading hay for the horses that pulled their fire engines. When the bell rang, firefighter George Reid was up in the hayloft on the third floor. The long pole for securing the hay to the wagon had been stashed vertically in the loading space for the hayloft. Rather than run all the way down two flights of stairs, Reid decided to slide down the pole. Swift thinking, George.

His captain, David Kenyon, liked the idea and arranged with the chief to cut a hole in the second floor of the firehouse and install a pole to get his firefighters quickly from their living quarters to the fire engines. The crew at Engine 21 trimmed a beam of Georgia pine to make a three-inch-diameter pole, sanded it, and varnished it. Then they waxed it with paraffin.

The thing worked. Engine 21, which was staffed entirely by black firefighters, soon got a reputation for being the first responder among first responders. The fire chief, who'd threatened to make Kenyon pay for repairing the hole in the floor if the idea was a bust, wound up ordering poles installed for every firehouse in Chicago.

Boston improved the idea in 1880. Fire pole 2.0 was made of shiny, slippery (and splinter-free) brass.

Nowadays, the poles are sometimes considered safety hazards, and new fire-houses are often built without them. Single-story firehouses are preferred. —*RA*

Also April 21:
1987: U.S. Okays Patents for Genetically Engineered
    Animals
1994: Discovery of First Extrasolar Planet Announced

Also 1878:
June 15: Muybridge Horses Around with Motion
    Pictures (see page 168)

April 22
1993: Mosaic Browser Lights Up Web with Color, Creativity

NCSA Mosaic 1.0, the first web browser that will achieve popularity among the general public, is released.

Mosaic 1.0 screen shot.

The web in the early 1990s was mostly text. People posted images, photos, and audio or video clips on web pages, but these pieces of multimedia were hidden behind links. If you wanted to look at a picture, you had to click on a link, and the picture would open in a new window.

A team of students at the University of Illinois National Center for Supercomputing Applications, or NCSA, decided the web needed something more stimulating and user-friendly, so they started building a better browser. Borrowing design and user-interface cues from some other early browsers, they went through a handful of iterations before release on April 22, 1993. NCSA Mosaic was the first web browser with the ability to display text and images *in-line*, meaning you could put pictures and text on the same page together, in the same window.

It was a radical step forward for the web. It took the boring document layout of your standard web page and transformed it into something much more visually exciting, like a magazine. And, wow, it was easy. If you wanted to go somewhere, you just clicked. Links were blue and underlined, easy to pick out. You could follow your own virtual trail of breadcrumbs backward by clicking the big button up there in the corner.

Windows and Mac versions came out in late 1993, and you could download them for free. Mosaic's color and images made the web a pleasure to use.

Mosaic evolved into the Netscape Navigator in 1994. That browser dominated the fledgling web before ceding the browser throne to Internet Explorer. But Netscape came back as Mozilla, and then it found new life as Firefox. —*MC*

Also April 22:

1915: Germans First Use Poison Chlorine Gas in Warfare

1970: First Earth Day Celebrated

Also 1993:

August 21: Mars Probe Disappears, Never to Be Found (see page 235)

September 2: Space Race Ends as U.S., Russia Sign Pact to Cooperate in Space

April 23
1940: Batteries Included, and They Don't Leak

Engineer Herman Anthony of Ray-O-Vac receives a patent for the leak-proof battery. His invention is about to go to war.

Wisconsin businessmen founded the company in 1906 as the French Battery Company. By the 1920s, it was making Ray-O-Lite flashlights, and it soon produced Ray-O-Spark batteries for your car's spark-plug ignition and Ray-O-Vac batteries for your vacuum-tube (see page 322) portable radio. The company changed its name to Ray-O-Vac in 1934.

World War II–era advertisement.

Standard old-fashioned zinc-carbon batteries were more portable than wet cells (see page 81) but had some bad habits. The zinc can would often swell and burst its seams. Then its innards would leak and render your flashlight or radio an inoperative mess.

Anthony solved the problem by using a better grade of manganese in the battery, to reduce the swelling, and then encasing the battery in steel. Bingo. Case closed... literally. Ray-O-Vac showed off its new wonder in 1939, although the patent wasn't published until 1940. But the leakproofs were not to be had in great number right away.

When the United States entered World War II, batteries were rationed to civilians, and like many other companies, Ray-O-Vac turned its entire production to supplying the military. Troops used the sealed-in-steel batteries to power not just flashlights but everything from walkie-talkies to bazookas. The firm also produced right-angle flashlights that soldiers could hook on their belts and have the light shine forward.

When the war ended, pent-up consumer demand helped the company sell a hundred million of the new batteries in 1946. Ray-O-Vac made a gold-plated flashlight in 1950 to celebrate the production of its billionth leak-proof battery. —*RA*

Also April 23:

1984: AIDS Virus Disclosed; Vaccine Promised

Also 1940:

February 5: Birth of H.R. Giger, Cyborg Surrealist Artist
February 27: Carbon 14 Discovered; Foundation of Radio-Dating
April 20: Zworykin Demonstrates Electron Microscope (see page 112)

April 24
1184 BCE: Trojan Horse Defeats
State-of-the-Art Security

During the Trojan War, the Greeks depart in ships, leaving behind a large wooden horse as a victory offering. It's hauled inside the walls of Troy. Come nighttime, Greek soldiers descend from the horse's belly to slay the guards and commence destruction of the city.

Whether this actually happened, and whether it happened on the traditional date given (based on calendrical hints in ancient accounts), archaeological evidence has established that a Trojan War did occur in Asia Minor around 1200 BCE.

Today, a Trojan horse is software that seems to perform one action but actually performs another, usually with malicious intentions. What cybersecurity lessons might we learn from the original Trojan Horse?

- Persistence: The Greeks had besieged Troy for ten years without result.
- Epistemology: Things are not always what they seem to be.
- Virgil, updated: Beware of strangers bearing gifts.
- Social engineering: The horse flattered the Trojans, who loved horses and were delighted with the gift.
- Engineering: The horse was on wheels, designed to make it easy for the Trojans to pull it inside their defenses.
- Ignoring warning messages: Two prominent Trojans cautioned against accepting the gift. They were both disregarded.
- Delay: Soldiers inside the Trojan Horse did not do their damage immediately but waited for an opportune moment.
- Size: A handful of Greeks unleashed lots of damage.
- Negating security from inside: They killed the guards and opened the gates from within, rendering Troy's strong walls useless against the waiting Greek army.
- Scope of damage: Troy was burned and destroyed.
- Permanent effects: Troy lost the war.

Today, you could lose only your data, your hard drive, your thesis, your job, your money, your business, your identity, or some awful combination of these. — *RA*

Also April 24:

1880: Birth of Gideon Sundbäck, the Man Who Perfected the Zipper
1967: Cosmonaut Vladimir Komarov Killed When Soyuz 1 Parachute Fails
1990: NASA Launches Hubble Telescope

Also BCE:

June 19, 240 BCE: Eratosthenes Calculates Earth's Size (see page 172)
February 29, 45 BCE: Leap Day, Thanks to Julius Caesar (see page 61)

April 25
1953: Riddle of DNA's Architecture Finally Solved

In the journal *Nature*, James Watson and Francis Crick present their research describing the architecture of the double helix that forms the molecular structure of DNA.

Although by then scientists understood that deoxyribonucleic acid was most likely the molecule of life, absolute certainty eluded them, because key components of how it worked were still missing. Chiefly, they didn't really know what the DNA molecule looked like.

Many, among them Linus Pauling, were actively engaged in DNA research, and a number of structural theories were advanced, all of them wrong in varying degrees. When Watson and Crick finally solved the puzzle, the key was provided by an X-ray diffraction photograph of a DNA molecule—the so-called photograph 51—taken by another researcher, Rosalind Franklin (see page 208).

Franklin's photograph revealed a fuzzy X in the center, confirmation that the molecule had a helical structure, allowing it to carry genetic code and pass genetic material through generations. Combining this with their other research, Watson and Crick concluded that the DNA molecule was a double helix and not a triple, as prevailing wisdom then held.

For their work, Watson and Crick shared the 1962 Nobel Prize in Physiology or Medicine with another DNA researcher, Maurice Wilkins.

Rosalind Franklin received...nothing. She'd died of cancer in 1958, at age thirty-seven, and the Nobel Prize is not awarded posthumously.—*TL*

Also April 25:
1859: Big Dig Starts for Suez Canal

Also 1953:
May 15: Primordial Soup Cooks Up Amino Acids
    (see page 137)

April 26
1956: The Containership's Maiden Voyage

The converted tanker _Ideal X_ leaves Newark, New Jersey, carrying fifty-eight cargo-laden truck trailers on its specially fitted deck. Containerization is born. Globalization sets sail.

The first containership was the brainchild of North Carolina trucking magnate Malcom McLean. He'd noted the wasted time of break-bulk cargo handling, with longshoremen laboriously loading individual crates and bales. He thought it made

more sense to lift whole truck trailers on and off the ship. And he wanted to save taxes by avoiding the fees imposed for excess weight as a truck passed through various states.

McLean wanted to separate the truck container from its bed and wheels, and he conceived an angled-corner-post system to allow easy stacking and to hold the containers in place. He bought some old tankers and beta-tested his idea with the _Ideal X_. By the time

Containerships pass in San Francisco Bay.

the ship arrived in Houston, she already had space booked to ship containers north.

The savings began immediately, and they got bigger. Ports needed to retool and install new, jumbo cranes, but more and more did so as they saw other containerized ports increase traffic. Ships were built to contain nothing but containers, above deck and below. Containers were soon standardized to make the system global. In a classic effect, increasing the nodes in the network increased the value of every other part of the network.

About 90 percent of global cargo is now carried by container. The average cost of shipping a product overseas has fallen from 15 percent of retail to less than 1 percent. There's less breakage and theft, but there is a downside. Ports handle more cargo but hire fewer dockworkers. And low-cost goods from overseas have eliminated millions of jobs in developed economies.

McLean was surely a visionary, but that doesn't make him a saint. — _RA_

Also April 26:

1812: Birth of Armament Manufacturer Alfred Krupp, Germany's Cannon King

1986: Chernobyl Nuclear Plant Suffers Cataclysmic Meltdown

Also 1956:

September 26: First Stretch of Interstate Highway Is Paved (see page 271)

October 6: Sabin Polio Vaccine Ready to Test

October 15: FORTRAN Computing Language Unveiled

April 27
1981: A Star Is Born with First Personal Computer Mouse

The first integrated mouse intended for use with a personal computer makes its appearance with the Xerox Star workstation.

The term *mouse* derived from the device's size, rectangular or oval shape, and tail-like cord extending from it, all suggesting the diminutive rodent.

The first mouse, an experimental pointing device, was invented in 1964 by Douglas Engelbart (see page 345), who was then working at the Stanford Research Institute in Menlo Park, California. Other methods of direction were being tried at the time—a head-mounted device, for example—but Engelbart's hand-operated mouse won out. Engelbart's original design underwent a number of changes at the nearby Xerox Palo Alto Research Center. It emerged as part of the Star workstation, a commercial system that was notable for a few other firsts as well: the graphical user interface, and the use of folders, file servers, and e-mail.

The modern mouse is available in a number of variations (mostly optical ones without rollers on the bottom, and many without cords), each designed to be integrated with a specific operating system or to fulfill a specific function. The trackball, the laptop nubbin, and the touchscreen interface may eventually render the mouse obsolete.

Then you won't have to worry anymore whether the plural of *mouse* (the inanimate kind) is *mice* or *mouses.*—*TL*

Also April 27:

1791: Birth of Telegraph Inventor Samuel F.B. Morse (see page 173)
1998: Koko the Gorilla Chats on AOL

Also 1981:

March 19: Two Technicians Killed in Ground Test of Space Shuttle *Columbia*
May 26: Programmer-Attorney Wins First U.S. Software Patent
June 18: Vaccine Announced for Foot-and-Mouth Disease
August 12: IBM Unveils 5150 Personal Computer
November 19: Philippine President Marcos Bans All Video Games as "Socially Destructive"

April 28
1926: Waving Hello to a New Subatomic Theory

Nuclear physicist Erwin Schrödinger writes a letter to Albert Einstein introducing a new term: *wave mechanics*. Things are getting particularly interesting. If they are things. Well, probably.

As an Austrian artillery officer on the Italian front in World War I, Schrödinger might have bombarded ambulance driver Ernest Hemingway. Or not.

After the war, Schrödinger oscillated from job to job before getting a professorship at the University of Zurich. A footnote in a paper by Albert Einstein inspired Schrödinger to model the motion of an electron around a nucleus as a wave rather than an orbiting particle. He wrote to Einstein in Berlin and proposed the term *Wellenmechanik*, or "wave mechanics." It's a fundamental building block of modern quantum and subatomic physics.

The year 1926 was astonishingly fertile for Schrödinger, much like Einstein's 1905 annus mirabilis (see page 327). Schrödinger published four papers: elucidating wave mechanics; expressing the new formulation in a precise equation; showing how it confirmed Niels Bohr's atomic model; and demonstrating how the Schrödinger theory paralleled—rather than contradicted—Werner Heisenberg's matrix mechanics.

Schrödinger moved in 1931 to the University of Berlin to succeed Max Planck. When the Nazis took power in 1933, Schrödinger—a nominal Catholic with an interest in Eastern philosophies—was disturbed by the dismissal and flight of his Jewish colleagues. He left Germany and took a position at Oxford. During his first week there, Schrödinger won the 1933 Nobel Prize in Physics (shared with Paul Dirac). Unfortunately, Oxford did not take to its new laureate: the dons seemed to mind that Schrödinger was living with two women. He returned to Austria to take a position at the University of Graz, but he was dismissed after Germany annexed Austria in 1938. He lived and taught in Ireland until 1956, when he returned to his homeland for his last five years.

Did Schrödinger, at any time, have a cat? He might have.—*RA*

Also April 28:

1940: BMW Sweeps Italy's Mille Miglia Race
1947: *Kon-Tiki* Sets Sail from Peru to Polynesia
2003: Apple Opens iTunes Music Store

Also 1926:

November 15: National Broadcasting Company Takes
   to the Air (see page 321)

April 29
1873: Railroads Lock and Load

A U.S. patent is issued for a new automatic railroad coupler. Within twenty years, it will be the standard car coupler on every American railroad.

Its inventor, Eli Janney of Alexandria, Virginia, was a Confederate army veteran who went into the dry-goods business after the war. He used his lunch hours to refine his design. In the absence of computers and modeling software, Janney whittled a wood prototype of a coupler that joined rolling stock automatically. It featured opposing couplers that, viewed from above, appeared to be two big knuckles shaking hands. Locked into place, the couplers form a viselike grip virtually unbreakable until released.

Janney was not the only guy working on a better railway coupler. When the time came to select a national standard, Janney's design was chosen from among eight thousand patented competitors.

Prior to Janney's invention, railway workers used a link-and-pin device, which was both inefficient and dangerous, because it required hands-on manipulation. Being injured or killed on the job was an occupational hazard for the nineteenth-century railroad man (see page 66). Much of the carnage stemmed from operating the link-and-pin coupler. Between 1877 and 1887, 38 percent of rail-yard injuries and deaths involved coupling accidents. There were eleven thousand of these casualties in 1892 alone.

Once Janney's coupler came into widespread use, yard accidents plummeted. By 1902, a mere 4 percent of all railroad accidents were related to car coupling.

Janney's coupler, later to be known as the Association of American Railroads coupler, was also stronger and more efficient than its predecessors, enabling longer trains that could carry more cargo or passengers. Ka-ching.

Janney's coupler was so well conceived that the basic design, with some modifications, remains in use to this day. — *TL*

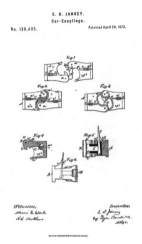

Patent diagram for Janney's knuckle coupler.

Also April 29:

1882: Siemens Tests Electric "Trackless Trolley"
1964: Godzilla, Mothra Clash for First Time

Also 1873:

August 2: San Francisco's First Cable Car Climbs Nob Hill (see page 216)

121

April 30
1939: The Future Arrives at New York World's Fair

The New York World's Fair opens. Its World of Tomorrow theme shapes industrial design, pop culture, and the way the future would envision the future.

The Trylon and Perisphere.

Buildings were inspired by the pages — and covers — of pulp science fiction: huge geometric shapes, sweeping curves, plenty of glass and chrome, and gleaming white walls. The fair was the last great blossoming of the Streamline Moderne style of Art Deco. It also reflected the International Style of architects like Le Corbusier (see page 281).

At the fair's center were the Trylon (a triangular obelisk) and Perisphere, a gigantic sphere accessed by dramatic ramps. Inside was a model city of tomorrow. But the fair's most memorable exhibit was the Futurama ride in designer Norman Bel Geddes's General Motors Pavilion. People stood in line for hours to ride it and experience the exciting possibilities of life in the distant future: the year 1960. Futurama carried visitors past tiny, realistic model landscapes while a narrator described the world of tomorrow. The effect was like glimpsing the future from an airplane window. The GM-sponsored ride focused on what roadways and transportation might look like in twenty years.

The ride presented a utopia forged by urban planning. Two decades before the interstate system (see page 271), model sophisticated highways ran through rural farmland and eventually moved into carefully ordered futuristic cities. Futurama had two factors that compounded the fascination: first, a promise of personal car ownership (and after the Great Depression, that sounded pretty good); and second, a grand overall vision of the future. Until Futurama, manufacturers had exhibited at fairs only to show their *current* products.

The fair left an indelible imprint on popular culture, visible in everything from the art design in sci-fi films to forward-looking marketing campaigns to the widespread use of the suffix *-arama*. — *CB*

Also April 30:

1897: J. J. Thomson Announces the Electron but Names It Corpuscle

1916: Birth of Claude Shannon, Father of Information Theory .

1945: Germany's New-Generation U-Boat Is Too Little, Too Late

Also 1939:

February 11: Physicist Lise Meitner Publishes Work on Uranium Fission

June 19: Dreaded AL S Gets New Name — Lou Gehrig's Disease

September 1: World War II Starts; *Wehrmacht* Puts the *Blitz* in *Blitzkrieg*

May 1
1964: First BASIC Program Runs

In the predawn hours of May Day, two professors at Dartmouth College run the first program in their new language, BASIC.

Mathematicians John G. Kemeny and Thomas E. Kurtz had been trying to make computing more accessible to their undergraduate students. Available computing languages, like FORTRAN and ALGOL, were so complex that you really had to be a professional to use them. So the two professors started writing easy-to-use programming languages in 1956. First came Dartmouth Simplified Code, or Darsimco. Next was the Dartmouth Oversimplified Programming Experiment, or DOPE, which was too simple to be of much use. But Kemeny and Kurtz used what they learned to craft the Beginner's All-Purpose Symbolic Instruction Code, or BASIC, starting in 1963.

The college's General Electric GE-225 mainframe started running a BASIC compiler at 4:00 a.m. on May 1, 1964. The new language was simple enough to use but powerful enough to be useful. Students weren't the only ones who liked BASIC, Kurtz wrote: "It turned out that easy-to-learn-and-use was also a good idea for faculty members, staff members and everyone else." Paul Allen and Bill Gates adapted it for personal computers in 1975 (see page 96), and it's still used today to teach, well, basic programming. (Kemeny and Kurtz formed a company in the 1980s to develop True BASIC, a lean version that meets ANSI and ISO standards.)

Another problem Kemeny and Kurtz attacked was batch-processing, which made for long waits between the successive runs of a debugging process. Building on work by Fernando Corbató, they completed the Dartmouth Time Sharing System, or DTSS, later in 1964. Like BASIC, it revolutionized computing.

In 1972, as president of Dartmouth, Kemeny introduced coeducation to the school after two centuries of all-male enrollment. — *RA*

Also May 1:
1884: First Steel-Frame High-Rise, Chicago's Home Insurance Building
1978: Computer Marketer Spams Four Hundred ARPANET Addresses (see page 104)

Also 1964:
April 7: IBM Unveils System/360, Pioneering Mainframe Computers
April 17: Ford Unveils Mustang
June 19: Last *Twilight Zone* Episode Airs

May 2
1887: Celluloid-Film Patent Ignites Long Legal Battle

Episcopal rector Hannibal Goodwin files a patent application for camera film on celluloid rolls.

Goodwin liked projecting lantern slides of Bible stories to his Sunday school classes, but he found the intricacies of glass-plate photography too daunting for him to make his own. He decided he could invent a better medium for holding the photographic emulsion. Two years of tinkering in his attic laboratory finally produced a flexible film from nitrocellulose, a plastic introduced in 1869. Without a clear understanding of the chemistry involved, he filed a vaguely worded patent application.

George Eastman introduced paper rolls of photographic film in 1888, but they were costly and blurry. Eastman asked chemist Henry Reichenbach to develop a film medium that would be clear, light, flexible, capable of holding the photochemical emulsion, and resistant to folding, shriveling, stretching, wrinkles, blemishes, bubbles, and streaks. Quite a task. Reichenbach developed a formula remarkably similar to Goodwin's, with one additional ingredient: camphor. His tightly worded patent application was approved in 1889. The new Kodak film went on sale the next year and was an immediate success.

Goodwin contested the Eastman-Reichenbach patent. The case wound its way through labyrinthine administrative channels until 1898, when Goodwin finally got his patent. After Goodwin died, his widow sold his company to Ansco, which sued Eastman Kodak. Kodak's improvements had reduced the amount of camphor in the film until its product was virtually indistinguishable from Goodwin's original formula. The U.S. Circuit Court of Appeals ruled for Ansco and Goodwin's heirs in 1914. Kodak had to pay out more than $5 million ($114 million in 2012 money, and 5 percent of George Eastman's net worth then).

Except for one change—substituting acetate for celluloid—Goodwin's technology dominated photography for a century, until the advent of digital cameras. But he's hardly a household name. —*RA*

Also May 2:

1952: First Commercial Jet Flies from London to
    Johannesburg

Also 1887:

March 4: Gottlieb Daimler Tests Benzine Motor
    Carriage
October 6: Birth of Le Corbusier, Architect for the
    Machine Age (see page 281)

May 3
1815: Blown Away by Horn with Valves

A Prussian composer reports a new contrivance: a local chamber musician has modified his brass concert horn by adding valves that allow him to play all the notes in the chromatic scale deftly and with total precision.

Brass instruments were very limited musically at the time, and the invention put the concert horn technically on par with the rest of the orchestra. Gottlob Benedikt Bierey of Breslau (now Wrocław, Poland) was so tickled, he wrote about the valves in a music newspaper: "What a new realm of beautiful effects this has opened up to composers!" Brass instruments, until the fifteenth century or so, were simple, straight tubes, each with a mouthpiece at one end and a fluted bell at the other. They were used mostly as signaling devices or for ceremonial purposes.

The early brass horn was tuned to one note, determined by the length of its tube. By tightening his lips, a player could produce some higher notes, or overtones. The addition of a slide (like a modern trombone's) let players hit the notes in *between* overtones.

Rudimentary valve designs began showing up in the early 1800s, but it is Heinrich Stölzel who is credited with affixing the first valve to a French horn, in 1814. Stölzel's system apparently used two spring-loaded cylindrical piston valves. Holes were bored through the pistons, and depressing the valve redirected air through shorter and longer tubes—each offering a different array of notes.

It let horn players, Bierey wrote, perform any musical passage with "a perfection not hitherto attained." Stölzel's design was widely copied, and variations gave shape to new instruments from manufacturers all over Europe. Soon there were cornets, valve trombones, tubas, euphoniums, and trumpets.—*MC*

Also May 3:
2000: Geocaching Gets Geeks into the Great Outdoors

Also 1815:
April 10: Tambora Explosion's "Volcanic Winter" Disrupts Global Weather for More Than a Year
November 2: Birth of Mathematician-Logician George Boole

May 4
1536: C U @ the Piazza

Francesco Lapi, a Florentine merchant, uses the symbol @ while penning a letter. It's the first recorded use of the at sign outside a monastery.

Ubiquitous in today's Internet culture, the at sign most likely owes its origin to a monk with writer's cramp. Before Gutenberg showed up with his printing press, forever changing human communication, the holy scriptures were considered among the few written works worth copying for wider distribution. The texts were transcribed by hand, a laborious process that encouraged typographical shorthand.

The most plausible theory is that the @ evolved from an *a* with a grave accent (*à*, which means "at" in Italian). The compressed symbol allowed the transcriber to complete the letter in a single stroke. This is just one of a handful of theories, mind. Another hypothesis is that @ evolved directly from the *English* word *at* as a way of allowing the transcribing monk to use one stroke instead of three. That seems less likely. Still another theory posits that the symbol was used as an abbreviation for *amphora,* a unit of measure corresponding to the amount of liquid contained in a vessel of the same name.

Wherever it came from, Signor Lapi glommed onto it. Although Gutenberg's press had existed for nearly a century, it was not yet in widespread use in 1536, and Lapi probably came across the symbol while reading a transcription. He used it in dating his letter, then once again in the text.

Today, of course, @ is all over the place. It has logical uses, as in e-mail addresses, and asinine ones, usually where there's marketing involved.

Among the English-literate, incidentally, a-t remains the preferred spelling of *at.* And th@'s th@! —*TL*

Also May 4:

1904: U.S. Agrees to Complete the Panama Canal
2000: I Love You Virus Infects 55 Million Computers

Also Sixteenth Century:

January 1, 1583: First New Year of Gregorian Calendar
(see page 1)

126

May 5
1992: Wolfenstein 3-D Shoots
First-Person Shooter into Stardom

Id Software releases Wolfenstein 3-D, and it launches a huge computer-game category.

Wolfenstein 3-D may not have been the very first first-person shooter, as the genre came to be known, but it was by far the most successful. Technically, the genre goes back to the '70s, but no one really paid any attention to it. Even Id had released an earlier FPS, called Catacombs 3D, but again, it wasn't nearly as good as Wolfenstein.

Through massive online dissemination of the game's shareware version, Wolfenstein 3D (the hyphen was later dropped from the name) introduced millions to an immersive world in which the action seemed to be happening from the player's perspective. "It was an incredible sensation, really unlike anything gamers had seen before," said Jamie Madigan, who helps operate the GameSpy Network's 3D Action website. "You could move smoothly in 360 degrees. You felt like you were there." A player in the game assumed the role of an American commando battling Nazis and their supernatural servants. It was banned in Germany because of its use of Nazi symbols, like the swastika, and music, like the "Horst-Wessel-Lied."

Developer Id Software leveraged Wolfenstein 3D's success into a franchise of wildly successful first-person shooters, including the seminal Doom and Quake series. These, in turn, begat a slew of sequels, imitators, and adaptations, from Halo to Call of Duty.

Wired.com Game|Life blogger Earnest Cavalli summed it up: "The key to the whole Wolfenstein thing is that its success—which was massive—paved the way for...thousands of games that mimicked them....Plus, who doesn't like killing Nazis?"—*NS*

Also May 5:

1809: Hat-Weaving Tech Gets First U.S. Patent to a Woman

1945: Balloon-Propelled Japanese Bomb Lands in Oregon, Killing Six

Also 1992:

October 9: My Insurance Agent Will Never Believe This (see page 284)

May 6
1840: Queen Victoria Gets Stamped

Britain starts using adhesive postage stamps, with the penny-black stamp allowing prepayment of postage at a fixed, low rate.

The 1840 penny-black stamp.

It used to be the *recipient's* responsibility to pay postage. If delivery was refused, the Royal Mail couldn't collect, and it lost plenty of revenue that way.

Railroads were making a rapid national mail service both possible and, because of the increased freight traffic, necessary. Reformer Rowland Hill campaigned for a cheap, pay-in-advance system that would make the mails available to everyone and help British commerce in the bargain. He proposed the rate of one penny (about forty cents in current U.S. money). Proof of payment would be either pre-paid stationery or a label printed on "a bit of paper just large enough to bear the stamp, and covered at the back with a glutinous wash." *Voilà:* postage stamp.

The first design featured a portrait of young Queen Victoria. All subsequent British stamps have shown the head of the reigning monarch, either as the main subject or as a smaller design motif. The stamps came in sheets of 240, the number of pence in a pound in the pre-1971 system. In 1841, the Royal Mail replaced the penny-blacks with penny-red stamps, largely because the black stamps were too easy to counterfeit. Also, the black ink hid postmarks, and people were reusing canceled stamps. Postal customers had to cut the sheets themselves, because perforations didn't arrive until 1854.

Postal traffic doubled in the first year and quadrupled by 1850. The system spread to more than 150 countries by 1880.

Some say postage stamps originated, at least in concept, in Austria, Sweden, or maybe Greece. But British stamps are still the only ones in the world that need not name the country of origin: the monarch's head suffices. —*RA*

Also May 6:

1937: Passenger Zeppelin *Hindenburg* Explodes, Killing Thirty-Six

1953: Heart-Lung Machine Successfully Aids Human-Heart Surgery for First Time

Also 1840:

June 20: Morse Code Patented (see page 173)

Otto Steiger receives a patent for his Millionaire calculating machine. It may not have been fruitful, but it multiplied.

The history of calculators begins with the invention of the abacus (see page 318) around 2500 BCE and moves through early attempts by mathematicians like Blaise Pascal and on to various nineteenth-century machines, including Charles Babbage's famous difference engine (see page 158). By the late nineteenth century, some of these mechanical wonders could add and subtract but merely simulated multiplication through repeated addition.

The Millionaire calculator.

Steiger advanced the ideas of Ramón Verea's 1878 U.S. patent and León Bollée's 1889 French patent, though neither had put his invention into production.

The Millionaire used a complicated internal clockwork of carriage, cranks, cams, cogs, gears, levers, pins, shafts, and sliders to perform addition, subtraction, multiplication, and division. For multiplication, each turn of the handle read a metallic multiplication table in Braille-like fashion to create a partial product. The device could carry 10s, so you turned the handle a second time for two-digit multipliers, three times for three digits, etc. It was developed for business calculations, but scientists found it useful too, and government agencies became the prime customers. Under the German name Millionär, 4,655 of them were sold over a remarkable forty-year span. Price depended on whether you wanted a hand-operated or electric lever. An upgraded model with keyboard was likewise either hand-operated or electric. U.S. prices ranged from $475 to $1,100 in 1924 ($6,400 to $14,700 in 2012 money).

The inside of the wooden case had extensive printed instructions and a special cleaning brush to keep the works free of dust and grit. At 100 to 120 pounds each, the Millionaire was a far cry from the first pocket calculators of the 1970s or the keychain calculators given out as free promotional swag today. — *RA*

Also May 7:

558: Roof Caves In on Hagia Sophia

1952: Geoffrey Dummer Lectures on Integrated Circuit (see page 257)

1959: C.P. Snow Warns That Science and Humanities Are Diverging into Two Cultures

Also 1895:

November 8: Roentgen Stumbles upon the X-Ray (see page 314)

May 8
1790: *Liberté! Égalité! Métrique!*

In the midst of the French Revolution, the National Assembly decides to create a decimal system of measurement. The metric system is born.

The first meter was based on clock-making: the length of a pendulum with a half-period (one-way swing) of one second. Responding to a proposal by the French Academy of Sciences, the assembly redefined the meter, in 1793, as 1/10,000,000 of the distance from the equator to the North Pole. The system was elegant. All conversions were based on 10, with Greek prefixes (deka-, hecto-, kilo-) for multiples, and Latin (deci-, centi-, milli-) for fractions. The gram unit of weight was defined by the weight of one cubic centimeter (aka one milliliter) of water.

The new Republican measures became legal throughout France in 1795 and were made compulsory in 1799, when definitive platinum meter bars and kilogram weights were constructed. But resistance to the new measures lasted for decades.

France also used a quasi-metric revolutionary calendar with months consisting of three *décades* of ten days each (see page 267).

The current International System of Units—or SI, for Système International—is based on the Treaty of the Meter signed in Paris on May 20, 1875. The United States was a signatory, and the metric system is the legal system in this country, although the legal alternate English system remains more widely used.

The meter was formally redefined in 1960 as 1,650,763.73 wavelengths in a vacuum of the orange-red light radiation of the krypton 86 atom (see page 152). The new standard was a hundred times more precise than the old. The current definition, adopted in 1983, makes the meter the distance traveled by light in a vacuum during 1/299,792,458 of a second.

That's 39.37 inches to counterrevolutionaries.—*RA*

Also May 8:

1886: Pemberton Invents Coca-Cola
1951: DuPont Debuts Dacron Polyester Fiber

Also 1790:

April 17: Death of Scientist-Inventor-Publisher-
    Statesman Ben Franklin
July 31: First U.S. Patent Issued to Potash Process
    (see page 214)

May 9
1960: Easy Birth Control Arrives, but There's a Catch

The birth control pill wins the approval of the Food and Drug Administration. The FDA gives its blessing to the ten-milligram dose of Enovid, which by then had been in clinical trials for four years, and the Searle drug company starts selling the pill a month later.

The first pill contained a synthetic progestin, which is similar to progesterone, a steroid that occurs naturally in the human female fertility cycle. The pill was nearly 100 percent effective but came with some severe side effects, including life-threatening blood clots. Further research found that the approved dose was ten times higher than needed.

Science continued refining the pill until, by the 1980s, safer and effective lower-dose variants were available. Other birth control methods evolved as well, including intrauterine devices, although they fell out of favor after one of them—the Dalkon Shield—was found to cause pelvic inflammatory disease. (The fact that the new techniques prevented conception but not the spread of sexually transmitted infections led to other problems as well.)

Today's woman can still opt for the pill in various forms, although the birth control patch—which slowly releases hormones through the skin—is also proving effective. And *Wired* reported in 2011 that new, safer IUDs are becoming popular.

The pill empowered women to take control of conception in a way that (generally male-controlled) condoms never had. Though the liberated sexual mores of the late twentieth century surely had cultural roots, the technological underpinning of new birth control methods also contributed. — *TL*

Also May 9:

1941: British Remove Secret Enigma Cipher Machine from German U-Boat

Also 1960:

May 16: Researcher Shines a Laser Light (see page 138)
September 24: First Nuclear Carrier, USS *Enterprise*, Launched

May 10
1869: Golden Spike Links Nation by Rail

Four years after the Civil War, the United States is joined from coast to coast by a transcontinental railroad as a ceremonial final spike is driven at Promontory Summit, Utah. Travel time from the Atlantic to the Pacific will soon fall from as much as six months down to one week.

An 1869 poster advertising the transcontinental railroad.

In an early example of a staged media event, two locomotives sat a mere rail tie apart from each other as crowds of people looked on. Railroad financier and former California governor Leland Stanford drove a single golden spike into the final tie with a silver hammer.

The rail lines from east and west were joined. A telegraph operator let the whole country know with a single message: "DONE!"

Congress had ordered the rail line built seven years earlier because westward expansion had been hampered by the dangerous wagon-train journey over the Oregon and California trails. The Central Pacific Railroad built the line eastward from Sacramento, California, and the Union Pacific built westward from Council Bluffs, Iowa. Mountains, rivers, and the Civil War dictated where the rail lines could be built.

The Central Pacific relied on recent Chinese immigrants from California and Mormon laborers from Utah to perform the often deadly work of installing rail ties and blasting through mountains. The Union Pacific hired Civil War veterans and recent Irish immigrants.

After the joining ceremony, the golden spike was removed, to be replaced with a normal steel spike. And it was another year before bridges and extensions created an all-rail link from Atlantic to Pacific. *—KB*

Also May 10:

1960: USS *Triton* Completes First Submerged
Circumnavigation

Also 1869:

April 8: Birth of Neurosurgery Pioneer Harvey Cushing
July 30: First Oil Tanker, the *Charles,* Leaves Port
September 23: Birth of Typhoid Mary Mallon

May 11
868: Signed, Sealed, Delivered

The Diamond Sutra, a sixteen-foot scroll containing one of the most cherished Buddhist texts, is printed. A dated colophon is included, making it the first known block-printed text to carry an explicit date. It appeared six centuries before Gutenberg's press.

The Chinese text, translated from Sanskrit, is a relatively short sutra dealing with the core Buddhist practice of non-abidance (avoiding mental constructs in day-to-day life). The scroll includes six separate sheets that were printed using wood blocks, plus a single woodcut depicting Buddha with his disciples and a couple of cats. Each sheet measures twelve by thirty inches. The printer, Wang Jie, pasted them together to form the scroll.

The Diamond Sutra is not the earliest example of block printing, but it is the earliest to carry a colophon that includes a verifiable date.

Wang Jie's colophon reads:

Reverently made for universal free distribution by Wang Jie on behalf of his two parents on the 13th of the 4th moon of the 9th year of Xiantong.

On the modern calendar, that works out to May 11, 868.

The original scroll eventually disappeared and lay undisturbed until its discovery a thousand years later, unearthed with the Dunhuang manuscripts in one of the Caves of the Thousand Buddhas in Turkestan, along the Silk Road trading route.

The Diamond Sutra now resides at the British Museum in London. —*TL*

Also May 11:
1951: RAM Is Born — Matrix Core Memory Patent
1997: IBM's Deep Blue Wins Match Against Chess
   Champ Kasparov (see page 41)

One Thousand Years Later:
June 23, 1868: Sholes Patents Typewriter
   (see page 176)

May 12
1941: Fog of War Shrouds Computer Advance

German engineer Konrad Zuse unveils the Z3, now generally recognized as the first fully functional, programmable computer.

Because Zuse designed and built his computer inside Nazi Germany, which was already at war, his achievement went unnoticed outside Germany until after the Third Reich's collapse. In the meantime, the Harvard Mark I, a computer produced by an American team, appeared in 1944 (see page 221) and is sometimes cited as the first of its kind.

Complicating Zuse's claim of priority, an air raid destroyed his computer, as well as all accompanying photographs and documentation. Zuse rebuilt the Z3 fifteen years after the war ended to demonstrate its capabilities and to establish his claim to the patents associated with the machine. The Z3, Zuse's third computer in a series of four, used the simple binary system for performing complicated mathematical computations — its outstanding feature.

Zuse is also remembered for devising Plankalkül (calculation plan), an early programming language designed, although never implemented, for engineering purposes. Additionally, he's credited with founding the world's first computer startup company, Zuse-Ingenieurbüro Hopferau, or Zuse Engineering Office of Hopferau (Bavaria), in 1946.

Zuse's achievement, according to his son, was even more remarkable considering he worked independently, even in isolation, and remained unaware of contemporary developments in computer science. And unlike computer pioneers in the Allied countries, Zuse received precious little support from his government. The Nazis saw little military value in his computers and provided only very minimal funding.

Years later, Zuse was generously funded by Siemens and some other German companies when he rebuilt his Z1 computer as part of a retro computing project. Replicas of the Z3 and Z4 are on display at the Deutsches Museum in Munich. — *TL*

Also May 12:

1820: Birth of Florence Nightingale, Nurse and Public-Health Statistician

1936: August Dvorak Patents New Keyboard Arrangement, but QWERTY Still Reigns

1967: Pink Floyd Performs World's First Quadrophonic Surround-Sound Concert

Also 1941:

September 27: First Liberty Ship Launched, More to Follow (see page 272)

Carl C. Magee of Oklahoma City files to patent a device that will elicit curses and contempt from generations of motorists: the parking meter.

Magee, a lawyer and newspaper editor, served on the chamber of commerce traffic committee. Even in 1935, U.S. cities were having thoroughly modern problems: workers parked on downtown streets and stayed all day. That left few spaces for shoppers and people with downtown appointments. Magee's brainstorm was a device that had a coin acceptor and a dial to engage a timing mechanism. The patent application described "meters for measur-

Metered parking in Omaha, Nebraska, 1938.

ing the time of occupancy or use of parking or other space, for the use of which it is desirous an incidental charge be made upon a time basis."

The world's first parking meters were put into nickel-gulping service right there in Oklahoma City in July 1935. Your 5 cents (84 cents in 2012 money) got you fifteen minutes' to an hour's worth of parking, depending on location.

A visible pointer and flag indicated the expiration of the paid period, meaning you had to either move, put in more money, or face the wrath of the local constabulary. Magee's meter not only solved Oklahoma City's parking problems but also got fresh cash flowing into city coffers, from both meter money *and* parking fines.

The idea spread around the world. Meter design remained largely unchanged, so to speak, for more than forty years. Magee managed to make money manufacturing meters and marketing myriads to many municipalities, starting at $23 a pop ($380 in current cash).

So, if you've ever realized a minute too late that you forgot to feed the meter and run back to your car only to see a ticket being slapped on your windshield...blame Carl Magee. — *TB*

Also May 13:

1637: Cardinal Richelieu Invents Dinner Knife
1884: American Institute of Electrical Engineers
Founded

Also 1935:

May 24: Night Baseball Comes to the Major Leagues
(see page 146)

May 14
1944: Birth of George Lucas, Film Tech Innovator

George Lucas is born in Modesto, California. His imagination and drive will make him one of the most successful independent directors and producers in film history.

Lucas met Steven Spielberg and Francis Ford Coppola at the USC film school. Lucas and Coppola formed the indie studio American Zoetrope in 1969.

Lucas's first film for that fledgling enterprise was sci-fi dystopia *THX 1138*. Its fearsome vision of a futurist society governed by dispassionate machines intent on eradicating freedom would reappear in Lucas's world-beating franchise: Star Wars.

But *THX 1138* didn't make money. That was left to Lucas's nostalgic '60s comedy *American Graffiti*, which was nominated for Academy Awards for Best Picture and Best Director.

His firm Lucasfilm spawned special-effects pioneers Industrial Light and Magic, audio innovators Skywalker Sound, the high-fidelity sound reproduction standard THX, game peddler LucasArts, and Pixar—the award-winning animation studio bought in 1986 by Apple guru Steve Jobs (see page 298). But it was his spiritual space opera *Star Wars* that took Lucas from cinema standout to cultural and business influential. Lucas waived his up-front director's fee for *Star Wars* in favor of 40 percent of the box-office receipts and full ownership of the merchandising rights. Smart move: the epochal success of *Star Wars* netted Lucas hundreds of millions in revenue, setting the stage for the franchising and licensing deals we now take for granted.

Sequels and prequels followed. Forbes has estimated that Star Wars films, toys, games, comics, paraphernalia, and ephemera have earned around $20 billion. Lucas himself is worth about $3 billion. Lucas has championed a national educational wireless-broadband network, supported the Martin Luther King Memorial in Washington, and lavishly funded the film school that gave him his break.

Happy birthday, O Master Jedi. —*ST*

Also May 14:

1771: Birth of Utopian Industrialist Robert Owen
1796: Jenner Tests Cowpox Vaccination on Human Subject to Protect Against Smallpox

Also 1944:

June 13: German V-1 Missile Ushers in a New Kind of Warfare
August 7: Harvard Mark I Computer Dedicated (see page 221)

May 15
1953: Cookin' Up Some Primordial Soup

Stanley Miller, just twenty-three years old, publishes his landmark work reporting that he's created amino acids — a necessary component of life — in a jar.

The slight, eight-hundred-word paper in *Science* ignited decades of debate about how life could arise spontaneously. The experiment opened the first peephole into the series of complex processes that appear to have turned inert matter into life, without the divine intervention earlier generations had assumed.

The experiment was simple. Miller and his adviser at the University of Chicago, Harold Urey, combined ingredients they believed were part of Earth's primordial soup — water, ammonia, methane, and hydrogen — and zapped the concoction with electricity as a stand-in for lightning flashes.

"During the run the water in the flask became noticeably pink after the first day, and by the end of the week the solution was deep red and turbid," Miller wrote. When he took the water out and analyzed it, half of the amino acids used to make proteins in living cells appeared.

Though organic molecules — chains of atoms containing hydrogen-carbon bonds that are critical for organisms — had been generated by other experiments, Miller's brought new significance to the idea that life could arise from the primordial soup. Our cellular machinery is just long strings of amino acids.

Since the Miller experiment, creating a replica of the first complex, self-replicating molecule, colorfully termed the Adam molecule by *Time* magazine, has proven incredibly difficult. Scientists now find the Urey-Miller version of the primordial-soup hypothesis a little less convincing than earlier scientists did but still plausible and powerful: The "Miller experiment... almost overnight transformed the study of the origin of life into a respectable field of inquiry," two fellow chemists rhapsodized in *Science* on the paper's fiftieth anniversary. — *AM, BM*

Also May 15:

1859: Birth of Pierre Curie, Radium's Codiscoverer
1930: Stewardesses Make the Skies a Little Bit
      Friendlier

Also 1953:

May 18: Jackie Cochran, First Woman to Break the
      Sound Barrier (see page 140)

May 16
1960: Researcher Shines a Laser Light

Physicist Theodore Maiman uses a synthetic-ruby crystal to create the first laser.

Maiman tinkered with electronics in his teens and earned college money repairing appliances and radios. He was working at Hughes Aircraft when he built the first functional laser. The laser produces monochromatic (all one wavelength), coherent (all waves in phase) light. Lasers are now used in eye surgery, dentistry, range-finding, astronomical measurement, and manufacturing. You'll find them at the heart of scientific instruments, communications networks, weapons, music systems, and supermarket scanners (see page 179). They're everywhere.

The concept was already bouncing around: Arthur Schawlow of Bell Labs and Charles Townes of Columbia University wrote a 1958 paper and submitted a patent application proposing an optical version of the maser (microwave amplification by stimulated emission of radiation). Columbia grad student Gordon Gould jotted the idea in his notebook in 1957 and applied for a patent in 1959. Gould coined the word *laser*.

Maiman altered the Schawlow-Townes concept. He coated the ends of a ruby with silver mirrors, one end with a thinner mirror to let a light beam escape. He used a flash tube to energize the crystal's atoms and enclosed it all in a polished aluminum tube.

The Bell researchers heard of Maiman's realization of their concept with mixed emotions, but they soon bested him by using an arc lamp to produce a continuous, rather than pulse, laser. Bell got its patent in 1960. Maiman applied in 1961 but didn't receive a patent until 1967. Gould spent decades mired in lawsuits before winning some patents in 1977.

The 1964 Nobel Prize in Physics went to Townes for the laser and to Soviets Nicolay Basov and Aleksandr Prokhorov for their earlier work on the maser. Schawlow shared the 1981 Nobel Prize in Physics for his contribution to laser spectroscopy. Maiman was nominated twice but never won. — *RA*

Also May 16:
1988: Nicotine Declared as Addictive as Heroin, Cocaine

Also 1960:
July 12: Etch A Sketch? Let Us Draw You a Picture (see page 195)
September 26: JFK, Nixon Open the Era of TV Debates

May 17
1902: Ancient Antikythera Calculating Mechanism Discovered

A diver exploring a shipwreck off the coast of Antikythera, an island between the Greek mainland and Crete, brings up a heavily encrusted mechanism that turns out to be the world's first known scientific instrument.

The Antikythera mechanism plotted the positions of celestial bodies nineteen years into the future. A dictionary-size assemblage of thirty-seven interlocking dials crafted with the precision and complexity of a nineteenth-century Swiss clock, the machine has been dated to approximately 150 BCE.

Scientists painstakingly reverse-engineered the mechanism, deciphered the script etched on its housing—the world's first instruction manual—and pieced the fragments into physical and later digital models. Most recently, they've made a working replica.

They determined that the mechanism predicted future positions of the moon and sun, and perhaps other planets. But that's not all: Tony Freeth and his Antikythera Mechanism Research Project colleagues found a tiny dial labeled with the locations of Olympic competitions. The feature was probably not integral to its function, said Freeth, but a stylish demonstration of the machine's power, like a watch that displays stock prices. The Olympics were of paramount importance to ancient Greeks, who labeled years in relation to ongoing Olympiads and suspended wars for the games' duration.

"We haven't found anything on the instrument that suggests it was used for astrology, which was suggested in the past," he said. "I think the maker was showing off a huge amount of knowledge and skill. They demonstrated that you could take these theories about how astronomical bodies move, and make a machine that would calculate them. That was a completely revolutionary idea."

The mechanism is a forerunner of all scientific instrumentation. Though its functions are understood, said Freeth, its application remains unknown: "We can only look at the result, and the result is dazzling." —*BK*

Also May 17:
1970: Heyerdahl Sets Sail to Cross Atlantic on *Ra II* Boat Made of Reeds

Also 1902:
June 9: First Automat Restaurant Opens (see page 162)

May 18
1953: Jackie Cochran, First Woman to Break Sound Barrier

Jackie Cochran becomes the first woman to break the sound barrier.

Cochran was already famous (as an aviatrix and racing pilot) and wealthy (through marriage) when she broke the sound barrier over Rogers Dry Lake, California, flying a Royal Canadian Air Force F-86 Sabrejet. In moving from subsonic to supersonic speed, Cochran averaged 652 mph.

Everything in Cochran's life pointed to her being the logical woman to accomplish this feat. Born into poverty, she was nevertheless introduced to flying at an early age. She proved a natural, learning to fly with only three weeks' training and earning a commercial pilot's license before she was thirty. She flew in her first major race in 1934, and she was the only woman to compete in (and win) the Bendix race, a transcontinental, point-to-point sprint. During World War II, Cochran helped deliver American-built planes to Britain and was instrumental in recruiting qualified women pilots into the Air Transport Command, the air-transport service of the U.S. Army Air Corps (predecessor of the Air Force).

Among her other aviation firsts: she was the first woman to take off from an aircraft carrier (see page 18), the first woman to reach Mach 2 (see page 49), the first *pilot* to make a blind instrument landing, and the first woman inducted into the Aviation Hall of Fame.

Cochran died in 1980 at age seventy-four. — *TL*

Also May 18:

1952: Carbon14 Sets Stonehenge Date at 1848 BCE, More or Less
1980: Mount St. Helens Blows Its Top

Also 1953:

June 2: Elizabeth II's Coronation Shown on Global Kluge TV (see page 155)

May 19
1780: New England Discovers Itself Dark at Noon

In the midst of the Revolutionary War, darkness descends on New England at midday. Many people think Judgment Day is at hand. It will take two centuries before modern technology pins down the scientific cause of the strange event.

Diaries of preceding days mention smoky air and a red sun at morning and evening. Around noon this day, darkness fell. Birds sang their evening songs. Farm animals returned to their barns. Humans were bewildered. Some went to church, many sought the solace of the tavern, and some near the edges of the darkened area noticed the strangely beautiful half-light: clean silver had the color of brass.

It was darkest on the New Hampshire coast and nearby Maine and Massachusetts. People ate midday meals by candlelight. "A sheet of white paper held within a few inches of the eyes," someone wrote, "was equally invisible with the blackest velvet." That night's full moon could not be seen. Some feared the sun would never rise again. But rise it did on May 20.

Harvard professor Samuel Williams gathered reports to seek an explanation: A "black scum like ashes" collected in tubs. The air smelled like a "malt-house or coal-kiln." Rain fell "thick and dark and sooty" and tasted and smelled like burned leaves.

As if from a forest fire? Various accounts reveal the darkness moved southwest at about 25 miles per hour. We now know that Canadian forest fires in 1881, 1950, and 2002 each cast a pall of smoke over the northeastern United States.

A definitive answer finally came in 2007 in the *International Journal of Wildland Fire*. Erin McMurry of the University of Missouri and coauthors found old fire scars in the annual rings of many trees around Algonquin Provincial Park in southeastern Ontario.

Counting the rings dated that massive wildfire to...the spring of 1780.—*RA*

Also May 19:

1910: Halley's Comet Brushes Earth with Its Tail

Also 1780:

May 4: Charter Granted to American Academy of Arts and Sciences

October 27: Harvard Astronomers Miscalculate, Miss Solar Eclipse Totality

May 20
1747: A Limey Ship, and Proud of It

Aboard one of His Majesty's ships, a British doctor begins clinical testing that will uncover the cause of scurvy and lead to its cure.

Scurvy was the sailor's scourge: fatigue, anemia, swollen and bleeding gums, loose teeth, slow-healing wounds, and subcutaneous hemorrhaging. It often proved fatal to sailors a long way from shore.

Though the disease had been documented since the Crusades, its precise cause (a vitamin deficiency) would remain a mystery until vitamin C was discovered. Sailing ships had fruits and vegetables in their stores, but without refrigeration (see page 197), they had to be eaten early in the voyage. As European empires expanded, voyages lengthened, and scurvy cases increased.

Royal Navy surgeon James Lind had a hunch that diet was involved, and he put his theory to the test when he shipped aboard the *Salisbury*. Taking a dozen men stricken with scurvy, Lind divided them into six groups of two and administered specific dietetic supplements to each group. The two lucky sailors who were fed lemons and oranges for six days recovered, and one was even declared fit for duty before the *Salisbury* reached port.

The Royal Navy was slow to react to Lind's evidence, though Captain James Cook is credited with mitigating scurvy by careful management of his crews' diets. It would take nearly half a century before the Admiralty accepted Lind's findings and began issuing lemon or lime juice to its sailors as a standard ration.

When that happened, scurvy all but vanished from the fleet. And British sailors (and Brits in general) came to be known as limeys. — *TL*

Also May 20:

1873: Riveted-Pocket Blue Jeans Patented
1990: Hubble Space Telescope Sends First Image
　　　Back to Earth

Also 1747:

February 14: James Bradley Announces Wobble of the
　　　Earth to Britain's Royal Society
March 28: Benjamin Franklin Writes Letter on His
　　　Electrical Experiments

May 21
1906: Research on Motion Creates Tire Rim

Louis Henry Perlman applies for the "demountable tire-carrying wheel rim." A flat tire is still a hassle, but thanks to Perlman, you can deal with it.

In the early days of motoring, radiators were brass, headlamps had wicks, cars were made of wood as much as metal, and wheels and tires were a single unit. The tires were solid rubber, and the wheels were wooden hub-and-spoke setups not unlike what you'd find on a horse-drawn wagon. Each one was affixed to the car by a single nut.

The tires were about the width of a business card and provided roughly the same level of grip. They were pretty darn tough but (like everything else on a car) would eventually wear out. At that point, you had to replace the tire *and* wheel, even if the wheel was just fine.

Perlman found a better way. His demountable tires worked pretty much like the ones on your car right now. A bead—that's the inner rim of the tire—held the tire against a groove machined into the wheel. The friction of shallow notches kept the tire from rotating on the wheel, though some early versions used a cumbersome screw-clamp system. The tire-and-wheel assembly had to be balanced to prevent vibrations and ensure a smooth ride, but that wasn't a big problem. Today people don't think twice about it (except when they forget to get it done).

Perlman's invention led to the adoption of inflated pneumatic tires, which provided much better performance. It also allowed automobile owners to choose their own wheels.

That, in turn, has led to absurdly oversize wheels and silly accessories like wheel spinners. Don't blame Perlman for that. He made cars *more* practical. —*TB*

Also May 21:

1860: Birth of William Einthoven, Inventor of
   Electrocardiogram
1901: Connecticut Sets First Speed Limit at 12 MPH
1927: Lindbergh Flies Solo Nonstop New York to Paris
1956: U.S. Tests Hydrogen Bomb on Bikini Atoll

Also 1906:

April 18: Earthquake Hits San Francisco; Great Fire
   Begins
December 24: First Public Voice-over-Radio Broadcast
   (see page 360)

May 22
1973: Enter Ethernet

Bob Metcalfe of the Xerox Palo Alto Research Center writes a memo outlining how to connect the think tank's new personal computers to a shared printer. The memo puts forth the basic properties of—and gives a name to—Ethernet.

PARC was installing the Xerox Alto, the first personal computer, and EARS, the first laser printer. It needed a system that could add additional PCs and printers without someone having to reconfigure or shut down the network. It was the first time computers were small enough for hundreds to be in the same building, and the network had to be fast to drive the printer.

Metcalfe circulated his plan in a memo titled "Alto Ethernet." It contained a rough schematic drawing and suggested using coaxial cable for the connections and using data packets like Hawaii's AlohaNet and the Defense Department's ARPANET (see page 99). The system was up and running on November 11, 1973.

Metcalfe didn't base the name Ethernet on an anesthetic (see page 275). It refers instead to the discredited scientific theory of luminiferous aether, an undifferentiated universal medium that some eighteenth- and nineteenth-century scientists thought necessary for the propagation of light. Metcalfe saw it as an apt metaphor for a medium that would propagate information.

Metcalfe shares four patents for Ethernet. He and PARC colleague David Boggs published the concept in a 1976 paper, "Ethernet: Distributed Packet-Switching for LANs." Metcalfe convinced funders Xerox, DEC, and Intel to let Ethernet become an open networking standard. It eventually supplanted competing technologies like IBM's Token Ring and General Motors' Token Bus as the predominant standard for local area networks.

Metcalfe is also known for Metcalfe's law: The value of a network grows as the square of the number of its users. —*RA*

Also May 22:
1990: Microsoft Releases Breakthrough Windows 3.0

Also 1973:
August 25: CT Scan Goes into Use in U.S. (see page 239)

May 23
1962: Give That Kid a Hand!

Boston doctors reattach the severed arm of an injured boy. It is the first successful reattachment of a human limb.

Red Knowles had been trying to hop a freight train and was thrown against a stone wall that ripped his right arm off cleanly at the shoulder. Knowles walked away from the tracks, using his left hand to hold his right arm inside a bloody sleeve. A police ambulance rushed the twelve-year-old to Massachusetts General Hospital, where emergency-room staff discovered the extent of his injury.

Surgeons had attached *partly* severed limbs before but never had the ideal candidate for a complete reimplantation, or replantation. Mass General's thirty-year-old chief surgical resident, Dr. Ronald Malt, ordered Knowles's arm put on ice, and he assembled the team of experts he needed.

In hours of surgery, twelve doctors reconnected the blood vessels, pinned the arm bone, and grafted skin and muscle together. All the techniques had been used before, but never all at once to save an entire limb. To everyone's delight, Knowles's hand turned pink, and a pulse returned to the wrist.

Doctors waited until September to reattach four major nerve trunks. Within weeks, Knowles was complaining of severe pain in the arm, which in the unusual circumstances was a good sign. A year later, Knowles's arm and fingers were sensitive to heat, cold, and touch, and he could move his fingers and bend his wrist. He could also play first base, but only with his left hand. The next year he was playing tennis and more baseball. After four years of recovery, Knowles had the same use of his right arm and hand as a natural lefty. He eventually drove a six-wheel truck and lifted sides of beef at his job. — *RA*

Also May 23:

1985: Spy Gets Life Sentence for Selling Stealth
   Bomber Secrets to Soviets

Also 1962:

June 14: Western Europe Joins the Space Race
   (see page 167)

May 24
1935: Reds Nip Phils as Night Baseball Comes to the Major Leagues

The first night game in major-league baseball history is played at Crosley Field in Cincinnati.

Crosley Field, built in the same era as Boston's Fenway Park and Chicago's Wrigley Field, was smaller than either. In fact, with a seating capacity of around thirty thousand, it was the smallest stadium in the majors. But in the mid-1930s, the Cincinnati Reds were struggling to fill even those seats. Team president Larry MacPhail proposed night baseball as a way of keeping major-league baseball in the Queen City. Minor leagues (all white in those segregated days) and the Negro League had been playing night baseball since 1930. But major-league team owners, notoriously skeptical of any kind of change, were initially cool to the idea. They relented only after MacPhail made it clear the Reds were in danger of folding.

Eight light standards were erected around Crosley Field, housing a total of 632 individual lamps. When the lights were switched on, the hometown Redlegs, with Paul Derringer toeing the slab, beat the visiting Philadelphia Phillies 2–1 in front of 20,422 fans.

Night baseball caught on quickly around the majors, except in Chicago. The Cubs kept it pure for decades but eventually bowed to the economic pressures of prime-time TV. After playing 5,687 consecutive day games at Wrigley Field, the Cubs finally switched on the lights in 1988. Coincidentally, the Philadelphia Phillies were the opponents for that milestone as well.

The Reds and Cubs also figure in another of baseball's technology firsts: the season before lights came to Crosley Field, the Reds became the first major-league club to travel by plane, flying from Cincinnati to Chicago on June 8, 1934. — *TL*

Also May 24:
1883: Brooklyn Bridge Opens (see page 2)

Also 1935:
May 29: Hoover Dam Set in Concrete (see page 151)

May 25
1945: Sci-Fi Author Predicts Future by Inventing It

Arthur C. Clarke distributes a paper that proposes using space satellites for global communications.

It was a bold suggestion for 1945, but Clarke, a physicist and budding science-fiction author, had his head firmly in the future. "The Space-Station: Its Radio Applications" suggested that space stations could be used for broadcasting television signals. TV, of course, was barely a commercial reality at this point (see page 302).

Clarke followed his private paper with an October 1945 article in *Wireless World* titled "Extra-Terrestrial Relays: Can Rocket Stations Give World-wide Radio Coverage?" The paper discussed how rocket technology like Germany's V-2s could be turned to peaceful ends by launching artificial satellites into orbit.

Satellites in the smallest orbits would circle the earth in about ninety minutes. But a satellite 22,500 miles high would orbit in twenty-four hours, matching the earth's rotation. Clarke wrote:

> A body in such an orbit, if its plane coincided with that of the earth's equator, would revolve with the earth and would thus be stationary above the same spot on the planet. It would remain fixed in the sky of a whole hemisphere and unlike all other heavenly bodies would neither rise nor set.

Clarke wasn't the first to propose such a geostationary orbit, but his essay did popularize the idea and its communication possibilities. It was less than twelve years before Sputnik (see page 279), only seventeen years before the first TV broadcast satellite, Telstar (see page 206), and nineteen years before the first geostationary satellite.

There are more than three hundred communications satellites in Clarke orbits today. Communications evolved much as Clarke predicted, and he regarded his satellite proposal as definitely more significant than his science-fiction classic *2001* (see page 12). —*DT*

Also May 25:

1961: JFK Vows to Put American on Moon by Decade's End

2001: First Towel Day Honors *Hitchhiker's Guide* Author Douglas Adams

Also 1945:

January 25: Grand Rapids, Michigan, Becomes First U.S. City to Fluoridate Drinking Water

August 6: U.S. Drops Atomic Bomb on Hiroshima (see page 220)

May 26
1676: Leeuwenhoek Points Microscope at Bacteria

Antoni van Leeuwenhoek turns his microscope on rainwater from his roof and finds some "very little animalcules."

Leeuwenhoek's microscopes.

Leeuwenhoek's varied career included stints as a fabric merchant, a wine assayer, and trustee of the bankrupt estate of painter Jan Vermeer. But neither Leeuwenhoek nor Robert Hooke invented the microscope. The compound microscope (using an ocular and an objective lens in series) was invented in the 1590s, four decades before their births. Leeuwenhoek didn't even use a compound microscope. Back then, it couldn't produce a clear image beyond 20x or 30x magnification.

Inspired by Hooke's illustrated book *Micrographia*, Leeuwenhoek learned to grind lenses. He built simple microscopes using one lens mounted in a brass plate. A sharp point held the specimen. One screw positioned the specimen laterally; another moved it into focus. You had to hold the four-inch instrument close to your eye. It required good lighting and sharp eyesight.

Leeuwenhoek built the best microscopes of his day, achieving magnifications above 200x. He wrote to England's Royal Society, describing dental plaque:

> [T]here were many very little living animalcules, very prettily a-moving. The biggest sort...had a very strong and swift motion, and shot through the water (or spittle) like a pike does through the water. The second sort...oft-times spun round like a top...and these were far more in number.

The "unbelievably great company of living animalcules...were in such enormous numbers," he wrote, "that all the water...seemed to be alive." These are among the first recorded observations of living bacteria. Leeuwenhoek also discovered microscopic fossils, nematodes, rotifers, sperm cells, and blood cells—confirming William Harvey's work (see page 156).

Some of Leeuwenhoek's original samples were rediscovered in 1981 in the Royal Society's vaults. They had been so well prepared that they could still be examined under modern microscopes. —*RA*

Also May 26:

1908: Drilling in Persia Discovers First Big Mideast Oil Field

Also Seventeenth Century:

October 2, 1608: Up Close and Personal with Hans Lippershey (see page 277)

June 8, 1637: Descartes Builds Rationalism, Scientific Method (see page 161)

1931: Wind Tunnel Lets Airplanes "Fly" on Ground

The world's first full-scale wind tunnel opens at Langley Field near Hampton, Virginia, with a test area sixty feet wide and thirty feet high.

Small versions already existed. The Wright brothers had built a small, six-foot-long wind tunnel to test scale models of wing sections before their historic 1903 flight (see page 353). But using models has limitations, so the government built a wind tunnel large enough for engineers to examine actual airplanes in flight without the planes' leaving the ground.

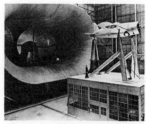

A Brewster Buffalo airplane mounted for wind-tunnel testing.

In a massive building covering more than two acres, the wind tunnel used a pair of thirty-five-foot propellers connected to 4,000-horsepower electric motors. Air was sucked through large funnel-like structures that directed a smooth flow of air past the staging area. Wind tunnels let engineers test their calculations by examining real-world aerodynamics around objects. Even today, wind tunnels reveal performance variables that the most advanced computational fluid-dynamics software doesn't predict. In the age of slide rules and drafting tables, wind tunnels were even more critical for discovering bugs in aircraft design.

Orville Wright, Charles Lindbergh, and Howard Hughes (see page 308) watched together in 1934 when Boeing's P-26 Peashooter fighter plane was tested at Langley. Every World War II American fighter aircraft was tested in the facility. It remained the world's largest wind tunnel until 1944. Langley tested airplanes, helicopters, race cars, a Navy SEAL submarine, space capsules—even semi trucks—hoping to reduce drag and increase fuel efficiency.

Bigger and faster aircraft eventually made the original wind tunnel obsolete. It shut down in 2009, after seventy-eight years of research. But Langley is still home to nine other wind tunnels. —*JP*

Also May 27:

1937: Golden Gate Bridge Opens — World's Longest for a Generation

1941: Royal Navy Sinks German Battleship *Bismarck*

Also 1931:

October 5: First Nonstop Transpacific Flight Completed (see page 280)

December 8: Coaxial Cable Patented

May 28
1959: Inventing a New Language for Business

A meeting at the Pentagon lays the foundations for the computer language COBOL, which will become a mainstay of business computing for four decades.

COBOL, for Common Business-Oriented Language, was one of the earliest computer languages. Along with FORTRAN, it was also one of the first to be based on English words. It owes its existence to pioneer programmer Grace Hopper, who cut her programming teeth in the U.S. Naval Reserve writing machine code for the Harvard Mark I computer (see page 221) during World War II. In the late 1950s, she proposed computer languages that would resemble human language, making them far more understandable than the assembly language and machine code then being used.

The 1959 Pentagon confab included various computer manufacturers so the language would be machine-independent. A fast-working subgroup finished specifications by December and named the language COBOL. The first COBOL compilers were built in 1960.

COBOL's appeal to business programmers was its readability, its accessibility, and the ease with which it could be used to compute business functions. By 1997, the Gartner Group estimated, 80 percent of the world's businesses ran on COBOL, with a cumulative total of 200 billion lines of code in existence. That legacy turned into an enormous burden with the belated discovery that COBOL's language constructs had encouraged programmers to store year data with just two digits. Fears of potential system crashes when the year 2000 arrived drew thousands of COBOL programmers out of retirement to comb through old code and update programs.

COBOL is no longer a field of active research and programming, but it spurred development of many other high-level computer languages that use quasi-English syntax, like BASIC (see page 123), and helped put computer programming within reach of many more people than before. — *DT*

Also May 28:

585 BCE: Predicted Solar Eclipse in Asia Minor Ends a Battle

1987: German Teen Flies Small Plane Right Through Massive Soviet Air Defenses to Moscow

1999: Scientifically Restored *Last Supper* Returns to the Public Eye

Also 1959:

September 14: Soviet Luna 2 Probe Lands on Moon (see page 259)

May 29
1935: Hoover Dam Set in Concrete, but Not Its Name

The last concrete is poured at the Hoover Dam site, four months before President Franklin Roosevelt dedicates one of the largest hydroelectric projects in U.S. history.

Hoover Dam was conceived in the early 1920s to irrigate desert, improve water supply to seven states, and generate electric power for burgeoning Southern California. The site (about thirty miles southeast of Las Vegas) was adjacent to Boulder Canyon, and the undertaking was christened the Boulder Canyon Dam Project. It was then the largest public-works program in U.S. history.

The upstream face under construction, May 1935.

Construction began in 1932, and the first concrete was poured in June 1933. To prevent uneven cooling and contraction (which could cause cracks), concrete was poured in five-foot increments rather than continuously. A system of cold-water pipes sped the cooling.

The dam was built at considerable human cost: 112 workers died from accidents, heatstroke, or heart failure. A brief workers' strike in 1931 failed, although working conditions improved in its wake. The Six Companies, which ran the project, began providing water to employees on a regular basis; probably a good idea, because temperatures routinely reached 120 degrees.

Hoover Dam was the world's largest concrete structure until 1942, when the Grand Coulee Dam opened. The dam stands 726 feet high. Its seventeen turbines generate up to 2,074 megawatts of hydroelectric power. Damming the Colorado River also created Lake Mead, named for the dam's project manager, Elwood Mead.

President Hoover's interior secretary dubbed it Hoover Dam while visiting the site. The name stuck until FDR swept Hoover out of the White House in 1932. Roosevelt's interior secretary changed the name back to Boulder Dam. President Truman, under pressure from Congress, restored Hoover's name in 1947. — *TL*

Also May 29:

1919: Solar Eclipse Confirms Einstein's General Theory of Relativity

1953: Hillary, Tenzing First to Climb Mount Everest and Return

Also 1935:

January 24: First Canned Beer Sold (see page 24)

May 30
1898: Krypton Discovered
Decades Before Superman Arrives

British researchers discover the element krypton. It's real but will inspire fantastic fiction.

William Ramsay.

William Ramsay and his student Morris Travers were searching for gases in the helium family. They boiled liquefied air until they got rid of the water, oxygen, nitrogen, helium, and argon. Then they placed the residue in a vacuum tube connected to an induction coil. It produced a spectrum with bright yellow and green lines. Because they had to look for it by removing all that other stuff, Ramsay and Travers gave element number 36 the name krypton, from the Greek *kryptos,* for "hidden" (think *cryptography* or *encryption*).

Within weeks, the dynamic duo detected a duet of other noble gases: neon and xenon. Ramsay was already responsible for discovering helium (with Lord Rayleigh) in 1894 and argon in 1895, giving him ownership of nearly an entire column of the periodic table. He received the 1904 Nobel Prize in Chemistry.

Krypton is used today in high-speed photographic flashes, fluorescent lights, and so-called neon signs (see page 347) that are greenish yellow. Between 1960 and 1983, krypton wavelengths defined the length of the meter (see page 130).

When Jerry Siegel and Joe Shuster created Superman in *Action Comics* no. 1, in 1938, they named their superhero's home planet after the element discovered forty years earlier. Superman and his legion of fans have made the fictional planet Krypton far better known than the real element. The fictional mineral kryptonite, which threatens Superman's strength and vitality, even has a real-life counterpart, almost.

Mineralogists in Jadar, Serbia, in 2007 unearthed some sodium lithium boron silicate hydroxide and learned that's what's written on a case containing kryptonite in the film *Superman Returns.* Naming the find kryptonite would violate international nomenclature because it contains no krypton. So they called it jadarite.

Isn't Jadar Superman's cousin or something? — *RA*

Also May 30:

1911: Gentlemen, Start Your Engines — First Indy 500

Also 1898:

July 30: First Automobile Advertisement Appears (see page 213)

May 31
1977: Trans-Alaska Pipeline a Source of Oil...
and Worry

The eight-hundred-mile-long Trans-Alaska Pipeline System is completed.

The Alaskan pipeline, built to help slake America's insatiable thirst for oil, was designed to move oil from the fertile fields of the North Slope to Valdez, Alaska's northernmost ice-free port.

The pipeline was an engineering marvel when you consider the terrain that had to be negotiated: three mountain ranges and numerous rivers and streams stood between all those thirsty gas-guzzlers and their sustenance. The project, which was privately funded, cost $8 billion ($30 billion in 2012 dollars).

Since the spigot was opened, on June 20, 1977, more than sixteen billion barrels of oil have flowed to the storage tanks at Valdez. The *ARCO Juneau* was the first tanker to carry crude through Prince William Sound. Around twenty thousand ships have made the trip since, most notoriously the *Exxon Valdez*, which ran aground in 1989, spilling roughly eleven million gallons of oil and causing one of the worst ecological disasters in U.S. history.

The effects of that catastrophe are still being felt, which hasn't stopped energy companies from pushing to open more oil fields at the North Slope for exploration and exploitation. — *TL*

Also May 31:

2006: Swedish Police Shut Down Pirate Bay File-Sharing Website

Also 1977:

March 27: Canary Island Runway Collision Kills 583 in Worst Airline Accident

July 13: Massive Power Blackout Plunges New York into Rioting

August 3: The TRS-80 Is Bad, and That Ain't Trash Talk (see page 217)

August 4: U.S. Department of Energy Created

September 10: France Performs Last Execution by Guillotine

June 1
1495: King James Will Have a Scotch, Good Sir

The Scottish government records it has commissioned Friar Jon Cor to make Scotch whisky—the first mention in print of an elixir that has since brought down many a government, made friends of enemies and enemies of friends, and lubricated good and bad writing.

Today is the anniversary of some momentous writing from the royal records:

> To Friar John Cor, by order of the King, to make aqua vitae VIII bolls of malt.—Exchequer Rolls 1494–95, Vol. x, p. 487.

We don't know if the order was actually fulfilled (eight bolls is enough for fifteen hundred bottles of whisky). But not giving the king his due was not a good career (or life) move. The king of Scots at the time was James IV, known in military history for his disastrous rout by the English at Flodden in 1513. It has also been said of James: "[A]s his gallantries were numerous, he had many illegitimate children." So the man knew a thing or two about partying. Nor do we know the kind of Scotch whisky Friar Cor whipped up or if it would have pleased our modern palates some five hundred years later. *Malt* is a clue—by today's definition, it could have been either a single or a blended malt, as opposed to grain. The original Celtic alchemy of fermentation was already perhaps a thousand years old, for this was by no means the first Scotch, just the first *recorded* Scotch.

And that *aqua vitae* business? The word *whisky* is derived from the Gaelic *uisge beatha,* or *usquebaugh,* which means "water" (*aqua*) "of life" (*vitae*). *Gaelic* in this case refers to that branch of Celtic spoken in the Highlands of Scotland.—*JCA*

Also June 1:
1849: Birth of Stanley Twins, Steam-Auto Pioneers
1890: Census Bureau Uses Hollerith Punch-Cards,
    Forerunner of Eighty-Column IBM Cards

Also Fifteenth Century:
March 9, 1454: Birth of Amerigo Vespucci (see page 70)

June 2
1953: Coronation Shown on Global Kluge TV

Elizabeth II is crowned as queen of the United Kingdom. Television cameras show the ceremony live in Europe, but viewers everywhere else have to wait a few hours.

The 1937 coronation of Elizabeth II's father, George VI, was the first one broadcast on radio, and about ten thousand people with early televisions also saw the outdoor processions.

Conservatives in 1953, including prime minister Winston Churchill, opposed televising from inside Westminster Abbey, calling it an intrusion on a mystical moment. They lost the argument to those, including the young queen, who wanted the ceremony democratized. The only compromise: the sacred anointing before the actual crowning was out of view of TV cameras.

People throughout Britain watched the rest on small black-and-white screens, often newly bought for the occasion, and often in the company of many neighbors who could not yet afford the expensive entertainment novelty. Elsewhere, people could only *listen* to the ceremony on a live global radio hookup. Communications satellites were a decade in the future (see page 206). Even videotape was still a few years off.

Instead, networks filmed the BBC television signal received at Heathrow Airport. They also rushed newsreel film by motorcycle relays from Westminster to the airport. They then loaded the undeveloped film in batches onto air force fighter jets. The films were processed en route to North America. The freshly developed films were then rushed to TV studios to be broadcast, sight unseen, to a waiting public. CBS lagged ten minutes behind the other networks, but NBC took criticism for running too many commercials and for its coronation "interview" with *Today* show chimpanzee J. Fred Muggs.

The instantaneous global village did not arrive until the state funerals of John F. Kennedy in 1963...and erstwhile coronation-on-TV-opponent Churchill in 1965. —*RA*

Also June 2:

1883: The L Comes to Chicago...Indoors
1954: Convair XFY-1 Pogo Aircraft Takes Off and Lands Vertically

Also 1953:

June 30: Chevrolet Introduces Corvette, First Fiberglass Car (see page 183)

June 3
1657: William Harvey Taken Out of Circulation

The blood stops circulating in the body of the scientist who definitively established that blood indeed circulates.

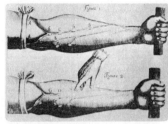

An illustration from William Harvey's 1628 *De Motu Cordis* shows how valves in the veins of a human arm allow blood to flow toward the heart but not away from it.

Most scientists and physicians in William Harvey's time still blindly followed the classical Greek physician Galen, who thought the liver converted food to blood and that arteries and veins were distinct systems. In the century before Harvey, Belgian anatomist Andreas Vesalius literally resurrected the dissection of human cadavers. The widely forbidden practice often had to be performed in secret on newly dead bodies stolen from cemeteries by illicit "resurrection men."

Cairo physician Ibn al-Nafis had established the "lesser circulation" between heart and lungs in the eleventh century. Italian Hieronymus Fabricius published a work on the valves in the veins in 1603, but he saw them as imposing a speed limit on blood flowing *from* the heart. However, al-Nafis's work was not widely known in Christian Europe, and no one had applied it to Fabricius's research.

Harvey experimented on animals and even on surface veins in the limbs of living humans. In 1628, he published his magnum opus, *Anatomical Disquisition on the Motion of the Heart and Blood in Animals*—often called *De Motu Cordis* for the literal heart of its Latin title.

It demonstrated conclusively that the heart pumps blood to the rest of the body and that veins return blood to the heart. With good microscopes still decades off (see page 148), Harvey correctly hypothesized the existence of capillaries.

Harvey also served as a royal physician to King Charles I. (The man was talented but also had the good judgment to marry the daughter of another royal physician.) Harvey left the court before the king was beheaded in a public execution in 1649. Plenty to learn about the motion of blood there.

Harvey died of a stroke, a circulatory disease. —*RA*

Also June 3:

1889: First Long-Distance Transmission of Electricity, Fourteen Miles to Portland, Oregon

1948: Fly-In Movie Theater Opens in New Jersey (see page 159)

1979: Ixtoc 1 Oil Well Explodes, Polluting Gulf of Mexico

Also 1657:

June 16: Astronomer Christiaan Huygens Patents First Practical Pendulum Clock

1937: Humpty Dumpty and the Shopping Cart

The Humpty Dumpty supermarket chain in Oklahoma City places a newspaper ad showing a woman exhausted by the weight of her shopping basket: "It's new — It's sensational. No more baskets to carry."

Owner Sylvan Goldman had invented the shopping cart. He was looking for an easier way for his customers to carry their purchases. If they weren't weighed down by handheld baskets, he figured, they would buy more.

The solution came when he adapted a folding chair. With the help of employee Fred Young, Goldman devised a prototype shopping cart: wheels at the bottom of the chair legs and two metal baskets on top of each other in place of the chair seat.

The launch turned out to be a flop: Customers didn't want to use the new invention. Men feared looking weak if they used the carts. Women thought they were unfashionable and looked too much like baby carriages.

Instead of giving up, Goldman hired male and female models to push things around in his store, pretending to be shopping. It worked. Soon the carts in Goldman's stores were a success. By 1940, the shopping cart's popularity had grown so much that buyers faced a seven-year waiting list.

Though Goldman is considered the cart's inventor, research by Catherine Grandclément found other frame-on-wheel solutions emerging in the mid-1930s. The telescoping frame, to let carts fit into one another for compact storage, was devised by Orla Watson of Kansas City in 1946.

Goldman continued to modify his design. The baskets grew bigger as stores realized that customers purchased more when cart size increased. Today, warehouse stores have warehouse-size carts.

Another innovation: kiddie seats. Alas, more than twenty thousand children a year are injured in shopping-cart accidents. Take care, folks. — *KS*

Also June 4:

1783: Montgolfier Hot-Air Balloon Demo Reaches Six Thousand Feet

1942: Battle of Midway Heralds Dominance of Aircraft Carriers

1977: VHS Comes to America

Also 1937:

March 6: Birth of Valentina Tereshkova, First Woman in Space

July 2: Aviator Earhart Vanishes over the Pacific

July 13: Gibson Plugs In the Electric Guitar (see page 196)

June 5
1833: Ms. Software, Meet Mr. Hardware

Ada Byron meets Charles Babbage. He designed an early computer, and she would write the first computer program.

Ada's father was the poet Lord Byron, but her parents separated when she was a month old. She never knew her poetically wild father. Ada was fifteen when she met Cambridge mathematics professor Babbage, who was working on a "difference engine" that could do mathematical calculations. He nurtured an even bigger idea: an "analytical engine" that "could not only foresee but could act on that foresight."

After Babbage gave a seminar on the analytical engine in Italy in 1841, Ada (who had married and become Countess of Lovelace) translated an article about the presentation and showed it to him. Babbage, apparently better at conceiving things than explaining them (unheard-of in a mathematician, eh?), suggested she expand the article with her own notes.

Countess Ada's "notes" tripled the original article. She predicted that a computing machine could compose music, draw graphics, and find application, so to speak, in business and science.

She also wrote a plan for the analytical engine to calculate Bernoulli numbers. It's now considered the first computer program. The countess originated the idea of a loop in a program, which she likened to a "snake biting its tail."

Countess Ada was a friend of novelist Charles Dickens and scientist Michael Faraday. She was also an opium addict who had numerous affairs and gambled away much of her family fortune, dying at age thirty-six.

The Countess of Lovelace has attained recent fame through Betty Toole's 1992 edition of her correspondence *Ada, the Enchantress of Numbers,* and Lynn Hershman-Leeson's 1997 film *Conceiving Ada*, starring Tilda Swinton. The U.S. Department of Defense named a computer language Ada in her honor. — *RA*

Also June 5:

1977: Apple II Personal Computer Goes on Sale

1981: Centers for Disease Control Publish First Report of AIDS

2002: Mozilla Open-Source Web Browser Released (see page 114)

Also 1833:

September 13: India Initiates Ice Imports from New England (see page 258)

June 6
1933: A Car, a Movie, Some Popcorn, and Thou

The world's first drive-in movie theater opens in Camden, New Jersey.

The concept was developed by Richard Hollingshead Jr., who experimented with projection and sound techniques in his driveway. Using a 1928 Kodak projector mounted on the hood of his car and aimed at a screen pinned to some trees, Hollingshead worked out the spacing logistics to make sure that all cars had an unobstructed view of the screen.

He received a patent for his idea in May 1933 and opened his first drive-in only three weeks later. They quickly fanned out across the country. Their popularity soared after World War II. They offered cheap family entertainment: a place where parents could take the kids without shelling out for a babysitter or worrying about the little ones' behavior. In fact, that was Hollingshead's original hook: "The whole family is welcome, regardless of how noisy the children are." Drive-ins always included a snack stand (profit center) and a play area where the kids could go when they got bored. For teens, drive-ins offered a sort of private place to make out in the car.

Another feature of the early drive-in was the tinny sound—delivered to the car through a single, monaural speaker. But the technology improved over time, sometimes using the car's FM radio. The drive-in's heyday lasted from the late 1950s until the mid-'60s, when nearly five thousand theaters were operating in the United States.

The rising cost of real estate contributed to the decline of the drive-in. So did video rentals. Fewer than four hundred remain today. The survivors often rely on additional sources of income to pay the rent: flea markets, swap meets, motorcycle schools, and even outdoor churches. —*TL*

Also June 6:

1944: Prefabricated Artificial Harbors Assist D-Day
Invasion of Normandy

Also 1933:

March 27: Just One Word: *Plastics* (see page 88)
April 7: *King Kong* Opens in Movie Theaters
September 12: Physicist Leo Szilard Conceives Nuclear
Chain Reaction

June 7
1975: Before Digital, Before VHS...
There Was Betamax

Sony introduces the Betamax video recorder.

Revolutionary for its day, the Betamax format was on its way to becoming the consumer standard until the appearance of JVC's VHS a year later. Betamax was sharper and crisper, but VHS let you record an entire movie at maximum speed (and quality) on a single tape of 120 or 160 minutes. Standard Beta tapes maxed out at an hour.

Other theories have been floated as to why VHS emerged victorious despite the superior quality of Betamax. Besides the longer tapes, VHS machines were cheaper and easier to use. Another possibility is that if you wanted to watch prerecorded X-rated movies, you had to buy a VHS, because Sony wouldn't license its technology to the porn industry.

There's also a tipping-point issue. People borrowed tapes (both prerecorded and home-recorded) from friends. Once VHS established a slight edge, it had the additional advantage of offering you a larger network of friends to swap tapes with, and video-rental stores started concentrating on the larger market too. So the Betamax sector shrank quickly.

As the earlier system, Betamax was sued by the entertainment industry (with Disney and Universal taking the point). Hollywood felt threatened by consumers recording TV shows or movies. The court ruled in Sony's favor, agreeing that consumer recording represented fair use.

Betamax enjoyed a long run as the standard for professional television and video production, and it still enjoys a connoisseur's niche. But DVDs, DVRs, and digital downloads have rendered both Beta and VHS passé.

Sony built its last Betamax recorder in 2002. — *TL*

Also June 7:
1968: Legoland Opens in Denmark

Also 1975:
October 20: Atari Patents Sit-Down Cockpit Arcade
    Game (see page 295)

June 8
1637: Descartes Builds Rationalism, Scientific Method

René Descartes publishes his *Discourse on the Method for Guiding One's Reason* and *Searching for Truth in the Sciences*. He outlines his rules for understanding the natural world through reason and skepticism, forming the foundation of the scientific method still in use today.

After brief legal and military careers, Descartes decided to devote his life to tearing down everything he had been taught was true and building it back up on a firm foundation, a project he said came to him in a dream.

In the *Discourse*, Descartes described four rules he established to make sure he always came to true conclusions.

1. Doubt everything.
2. Break every problem down into smaller parts.
3. Solve the simplest problems first, and build from there.
4. Be thorough.

By following these simple guidelines, he said, "there cannot be anything so remote that it cannot eventually be reached nor anything so hidden that it cannot be uncovered." The senses sometimes lie, Descartes reasoned, but the fact that he wondered about the truth of his thoughts and sensations meant that something must be doing the wondering. He expressed this thought concisely and memorably as *cogito ergo sum*: I think, therefore I am.

Descartes also expressed his preference for mathematics as the basis and language of his new method. The essay titled "Geometry" introduced the Cartesian coordinate system, which influenced the development of calculus (see page 304) and is still widely used.

The *Discourse on Method* is widely regarded as one of the most influential works in the history of science and as marking the beginning of the scientific revolution. "For it is not enough to have a good mind; it is more important to use it well." —*LG*

Also June 8:

1949: George Orwell's *1984* Is Published

1959: Ballistic Missile from U.S. Sub Carries "Rocket Mail" in Publicity Stunt

Also 1637:

May 13: Cardinal Richelieu Invents Dinner Knife

June 9
1902: Put a Nickel In, Take Your Food Out

Joe Horn and Frank Hardart open the Automat at 818 Chestnut Street in Philadelphia. It's America's first coin-operated cafeteria.

A customer put a nickel (or several) into a slot, turned a knob, and opened a little glass door to withdraw the food. Horn and Hardart used Swedish-patented equipment they'd imported from Berlin, which already sported a successful "waiterless restaurant."

The company branched out to New York in 1912 and continued to expand. The place was a bargain. A cup of coffee still cost a nickel in 1950 (about 50¢ in today's money) before it finally rose to two nickels.

Employee "nickel throwers" at the head of the line exchanged currency or large coins for the nickels needed for the coin slots. One nickel for coffee, five for the turkey and gravy, another nickel for pie. You could also have a famous macaroni and cheese, chicken potpie, Salisbury steak, mashed potatoes, creamed spinach, or baked beans. Desserts were also renowned: huckleberry, pumpkin, coconut cream, and custard pies; vanilla ice cream with real vanilla beans; and rice pudding with plump raisins.

It was all prepared in centralized, assembly-line kitchens using standardized recipes that called for quality ingredients. This, plus eighty-five locations in Philadelphia and New York, made it America's first fast-food chain.

Irving Berlin composed "Let's Have Another Cup of Coffee" about the Automat. Edward Hopper painted it. The Broadway set for *The Producers* included it. But the chain succumbed to the ever-rising price of the ingredients in its original recipes and to the growing popularity of fast-food chains and pizza parlors. Philly's last Automat closed in 1990, and New York's a year later.

The Automat lives on in the Smithsonian's National Museum of American History: an elaborately decorated, thirty-five-foot section of Philadelphia's original 1902 Horn & Hardart, complete with mirrors and marble. —*RA*

Also June 9:
1928: Four-Man Crew Completes First Transpacific Airplane Flight — in Eight Days

Also 1902:
July 17: Carrier Invents Air-Conditioning (see page 200)

1943: The Ballpoint Pen — Ink Dry for Me, Argentina

Brothers László and Georg Bíró, Hungarian refugees living in Argentina, patent the ballpoint pen.

Lewis Waterman's invention of a practical fountain pen, patented in 1884, had already solved the problem of portability. American banker John L. Loud patented a ballpoint pen in 1888. It used a ball-and-socket mechanism to deliver sticky, quick-drying ink. Too sticky: it didn't really write well on paper. (Figuratively a good idea on paper, but literally not.)

László Bíró was a journalist who saw an idea in quick-drying newspaper inks. His chemist brother Georg helped him with technical aspects. They used a tiny, precisely ground ball bearing to distribute ink evenly from cartridge to paper and to hold the rest of the ink inside the cartridge.

They applied for a patent and sought financial backing. Englishman Harry Martin realized that the ballpoint solved a problem faced by Britain's Royal Air Force: conventional pens were unsuitable for writing aircraft logs because they leaked, were sensitive to altitude, and couldn't write uphill.

Martin flew to Washington and London and convinced both the United States and Britain to adopt the new technology. When the Allies won World War II, the ballpoint shared the luster of victory. Ballpoints went into commercial production in 1945 and were a sensation. The Reynolds Pen sold for $12.50 (about $150 in today's money). Yet people swarmed a New York department store to buy eight thousand the first day. (Sounds like modern times; see page 182.) Some of the earliest commercial ballpoints leaked and smudged, but manufacturers eventually worked out the bugs. (That sounds familiar too.)

A ballpoint is now what most people mean when they say *pen*. And in much of the world, the generic word is *biro*, after its inventors. In Argentina, land of its birth, it's a *birome*. — *RA*

Also June 10:

1952: DuPont Registers *Mylar* as Trademark for Its Polyester Film

2000: London's Millennium Bridge Has Wobbly Opening

Also 1943:

July 5: Defeat at Kursk Heralds Twilight of the Panzers (see page 188)

June 11
1910: Champion of the Wine-Dark Sea

Jacques Cousteau is born. He will become a scientist, naval officer, author, photographer, filmmaker, researcher, explorer, innovator, and ecologist.

But he is remembered mainly for developing the Aqua-Lung and raising the world's consciousness to both the splendor and the plight of the oceans.

During his service in the French navy he began to seriously study the sea under his ships. He borrowed a pair of diving goggles from a friend and had an epiphany, recounted in his book *The Silent World:*

> I was astonished by what I saw in the shallow shingle at Le Mourillon, rocks covered with green, brown, and silver forests of algae and fishes unknown to me, swimming in crystalline water.... Sometimes we are lucky enough to know that our lives have been changed, to discard the old, embrace the new, and run headlong down an immutable course. It happened to me at Le Mourillon on that summer's day, when my eyes were opened on the sea.

But the only way to remain submerged a long time was to don a cumbersome diving suit with air pumped through an umbilical cord from the surface. The suit restricted movement, and the cord restricted distance.

With engineer Emile Gagnan, Cousteau developed a regulator to supply compressed air from tanks on a diver's back: a self-contained underwater breathing apparatus, or scuba. They patented it as Aqua-Lung in 1943.

Cousteau bought an old American minesweeper in 1950, rechristened it *Calypso,* and converted it into an oceanographic research vessel. Cousteau became a household name in 1966 with the airing of his first hour-long television special, *The World of Jacques-Yves Cousteau.* That started three decades of TV work during which Cousteau alerted the public to the growing ecological threat to the world's oceans.

He died in 1997, steeped in honors. —*TL*

Also June 11:

1644: Torricelli Demonstrates Mercury Barometer to Measure Air Pressure
1985: Karen Ann Quinlan Dies, but Debate over Brain Death Continues

Also 1910:

March 28: First Successful Seaplane Flight
December 11: Claude Shows Off Neon Light in Paris
(see page 347)

June 12
1897: The Swiss Army Gets Its Own Knife

Karl Elsener registers his "soldiers' knife" for use by the Swiss army.

Elsener, a manufacturer of surgical instruments and cutlery, was disturbed to learn that the Swiss army was importing Solingen blades from neighboring Germany. Elsener set out to develop a homegrown multifunctional tool worthy of the mountain nation.

His 1891 prototype included a blade, screwdriver, and can opener. But Elsener was not happy with it and tinkered endlessly, adding a second blade with a revolutionary spring mechanism and strengthening the housing. The extra space needed for the spring opened up room for other tools.

Elsener created a company and named it Victoria, after his mother. After switching to stainless-steel (see page 227) blades in 1921, the company was renamed Victorinox, from the French *inoxydable,* "rust-resistant."

More than 34,000 Swiss army knives are manufactured every day. The remarkably versatile tool is standard equipment for everyone from Boy Scouts to building contractors to mountaineers. It was even carried into space aboard the space shuttle.

Oh, and every new recruit in the Swiss army is still issued a knife. Victorinox supplies the army with about 50,000 knives each year. Many variations are available, from relatively simple to jaw-droppingly complex. The SwissChamp boasts large blade, small blade, can opener with small screwdriver, bottle opener with large screwdriver and wire stripper, scissors, pliers with wire cutter, wood saw, fish scaler with hook disgorger and ruler, metal saw with metal file and nail file, magnifying glass, reamer with sewing eye, Phillips screwdriver (see page 190), corkscrew, hook, wood chisel, fine screwdriver, mini-screwdriver, ballpoint pen, straight pin, tweezers, key ring, and...toothpick.

A commemorative knife known as the Giant (nine inches thick) contains eighty-five devices for 110 functions. Other models reflect advances in technology with features like a laser pointer, a USB flash drive, and even an MP3 player. — *TL*

Also June 12:

1957: Monsanto House of the Future Opens at Disneyland, Glimpses World of Our Today
1997: U.S. Introduces High-Security $50 Bills to Foil Counterfeiters

Also 1897:

March 7: First Morning of the Cornflake (see page 68)
April 30: J.J. Thomson Announces the Electron but Names It Corpuscle

June 13
1983: Pioneer 10 Reaches an End... and a Beginning

Pioneer 10 becomes the first human-made object to pass outside Pluto's orbit.

Pioneer 10 is one of the most successful spacecraft of all time. Designed for deep-space exploration, which at the time of its launch, in 1972, meant pretty much anything beyond the moon, Pioneer 10 achieved a number of firsts while sending back valuable data. Among the milestones:

Artist's rendition of the Pioneer 10 spacecraft.

- A breakaway speed of 32,400 mph, making it the fastest human-made object to leave the earth. It shot past the moon in just eleven hours and crossed Mars's orbit in twelve weeks.
- First spacecraft to pass through the asteroid belt
- First direct observations and close-up images of Jupiter, which confirmed that the solar system's largest planet is mostly liquid
- Clearing Pluto's orbit, considered the boundary of the planetary solar system before astronomers deplanetized Pluto (see page 238)
- Continuing to send back data regarding solar wind until its scientific mission ended, in 1997

Attempts to contact Pioneer 10 were terminated following the spacecraft's last transmission of telemetry data, on April 27, 2002. Nevertheless, NASA's Deep Space Network received a final, faint signal on January 22, 2003. It's been silence ever since.

Although lost to contact forever, Pioneer 10 continues its journey through interstellar space. It's headed in the general direction of Aldebaran, the brightest star in the constellation Taurus, forming the bull's eye. According to NASA, it will take about two million years for Pioneer 10 to reach Aldebaran.

So Pioneer 10's mission, originally intended to go twenty-one months, lasted twenty-five years and change. As project manager Larry Lasher said, "I guess you could say we got our money's worth." — *TL*

Also June 13:

1944: German V-1 Missile Ushers in a New Kind of Warfare

Also 1983:

January 19: Apple's Lisa Debuts, First Commercial Computer with a Graphical User Interface
November 10: First Computer Virus Is Born
(see page 316)

June 14
1962: Western Europe Officially Joins the Space Race

An agreement establishing the European Space Research Organisation, forerunner to the European Space Agency, is signed.

The Western Europeans were as shocked as the Americans by the Soviet Union's successful launch of the Sputnik satellite in 1957 (see page 279). Their gloom only deepened as the United States and the USSR embarked on a superpower space race that left the historic European powers in the dust.

The 1962 agreement actually set up two agencies: the European Launcher Development Organisation, or ELDO, and the European Space Research Organisation, or ESRO. The first was tasked with developing a launch system, while the latter handled research and administration. The first seven satellites sent up by the Europeans used U.S. launch systems.

ELDO and ESRO merged in 1974 to form the European Space Agency. The ESA, based in Paris, stepped up the tempo just as the two superpowers began to cut funding for their own space programs.

ESA points to many accomplishments:

1975: Cos-B satellite monitoring gamma-ray emissions in the universe.

1978: Joins NASA and U.K. in launching IUE, the world's first high-orbit telescope, which operates successfully for eighteen years.

1986: Giotto probe studies comets Halley and Grigg-Skejllerup.

1988-2006: Ariane rockets establish ESA as world leader in commercial space launches.

1990s: SOHO, Ulysses, and Hubble Space Telescope together with NASA.

2003: Mars Express orbiter and its lander, *Beagle 2,* launched.

2005: ESA Huygens probe lands on Saturn's moon Titan (see page 361).

2008: Columbus laboratory launched on space shuttle *Atlantis* to the International Space Station.

2009: Herschel and Planck radiation observatories launched.

2010: European Node-3 and Cupola installed on ISS; CryoSat-2 launched to study Earth's ice cover.

Although it continues to collaborate with NASA, the ESA now considers the Russian Federation an important partner. —*TL*

Also June 14:

1948: *TV Guide* Prototype Hits NY Newsstands

Also 1962:

July 10: Three-Point Seat Belt Patented (see page 193)

June 15
1878: Muybridge Horses Around with Motion Pictures

Photographer Eadweard Muybridge uses high-speed stop-motion photography to capture a horse's motion. The photos prove the horse has all four feet in the air during parts of its stride. The shots settle an argument and start a new medium.

Former California governor Leland Stanford financed Muybridge's experiments. Legend has it Stanford wanted to settle a $25,000 bet by proving that horses "flew," but most historians doubt that colorful bit.

Eadweard Muybridge's photos of a horse's gait.

Photographs then usually required exposures of fifteen seconds to one minute. Glass plates had a speed equivalent to about ISO 1. Muybridge devised faster emulsions and rigged a trip wire across a racetrack. The "automatic electro-photograph" he made on July 1, 1877, showed Stanford's racehorse Occident with all four feet off the ground. The press and the public wouldn't accept it as proof because what *they* saw was a retouched reproduction.

Muybridge devised a more elaborate system. On June 15, 1878, before assembled gentlemen of the press, Stanford's top trotter raced across the trip wires at about forty feet per second, setting off twelve cameras in rapid succession in less than half a second. About twenty minutes later, Muybridge showed the freshly developed photographic plates. The horse, indeed, lifted all four legs off the ground during its stride.

Stanford was vindicated, and the press astounded. Muybridge refined his invention and studied the motion of other four-footed animals, human athletes, a nude descending a staircase, and even birds.

He also adapted the zoetrope, a children's toy that produced the illusion of motion by spinning animation-style drawings behind a viewing slit. He fitted one with glass disks to project his trotting sequences onto a screen. This "zoopraxiscope" created the first photographic motion pictures. Thomas Edison and the Lumière brothers soon advanced Muybridge's concept to create movies as a commercial art form. —*RA*

Also June 15:

1667: First Human Blood Transfusion Performed (see page 320)

1919: Alcock, Brown Complete First Nonstop Transatlantic Flight

Also 1878:

September 25: Physician Calls Tobacco a Health Hazard (see page 270)

June 16
1884: A Technology with Plenty of Ups and Downs

The first gravity roller coaster designed and built specifically as an amusement ride opens at Coney Island, New York.

LaMarcus Adna Thompson's Coney Island coaster, for a nickel ($1.25 in 2012 money), hurtled passengers down an undulating six-hundred-foot-long track at speeds up to a blistering 6 mph.

Passengers faced sideways, and the track was not laid out in a continuous loop. The ride began atop a fifty-foot-high platform, and when it reached the other end, passengers had to disembark so the cars could be switched over to the return track for the ride back to the starting point. Within a year, the original tracks were replaced by an oval course that allowed riders to remain seated from start to finish. The seats on this new coaster, the Serpentine Railway, faced forward.

Thompson reportedly developed his idea from the Mauch Chunk switchback gravity railroad in Pennsylvania, a coal-hauling device that provided thrill rides to the locals when not delivering coal. But the coaster's origins go back even further, to seventeenth-century Russia and slides carved from specially constructed ice hills outside St. Petersburg. The first man-made coaster using structural support is believed to have been built on the orders of Catherine the Great.

The world-famous Matterhorn Bobsleds at Disneyland, which opened in 1959, became the first roller coaster to use tubular steel track. This innovation allows designers to incorporate maneuvers like loops and corkscrews into the course.

Built in 1912, the Luna Park Scenic Railway in Melbourne, Australia, is the oldest continuously operating roller coaster in the world. Leap-the-Dips in Altoona, Pennsylvania, is ten years older but stood idle from 1985 to 1989. — *TL*

Also June 16:

1922: Henry Berliner Makes First Controlled Helicopter Flight

1959: George Reeves, Superman, Felled by Speeding Bullet

Also 1884:

August 16: Birth of Pioneering Sci-Fi Publisher Hugo Gernsback

September 15: Ophthalmologists Learn of Cocaine as Local Anesthetic

October 13: Greenwich Adopted as Prime Meridian (see page 288)

June 17
1862: Worst Mash-Up Ever Has Farmers Tillin' 'n' Killin'

In the midst of the American Civil War, inventors W.H. Fancher and C.M. French of Waterloo, New York, receive a patent for the New and Improved Ordnance Plow, a horse-drawn plow outfitted with a firearm.

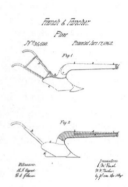

Patent drawings for the combined plow and gun.

It was to be "used in border localities, subject to savage feuds and guerrilla warfare." Many farmers living in Union border states endured Southern cavalry carrying out raids on their farmlands. The patent didn't specify the caliber of gun the plow was to be outfitted with. It just described it as capable of firing "grapeshot or a piece of light ordnance ... its capacity may vary from projectile of 1 to 3 pounds." Civil War cannon typically fired balls weighing six to twelve pounds, so this is no mere rifle. More like an agricultural bazooka.

In the event of trouble, the farmer would theoretically unhitch his horses, thrust the iron plowshare into the soil (for stability against recoil), and fire at the marauders. The patent had several practical problems that prevented the machine from ever being built. One was visibility: raiders would eventually get wise to the ordnance plow and shoot the farmer-gunner before he could use it. And we're also talking about a plow here, not a Ferrari. It's cumbersome and not so easy to turn and plant. All an attacker had to do was flank his plow-packin' target in order to knock him dead.

No, this wasn't the first poorly devised combination in history, and it wouldn't be the last. But it was surely one of the absolute worst. — *DD*

Also June 17:

1867: Joseph Lister Performs First Antiseptic Surgery
1947: Pan Am Launches Round-the-World Air Service

Also 1862:

March 9: USS *Monitor* Fights CSS *Virginia* in First Naval Battle of Ironclads
July 16: Swift Sights Comet Swift-Tuttle Three Days Before Tuttle

1908: Prescient Letter Creates Concept of TV

A Scottish electrical engineer publishes a brief letter in the journal *Nature* describing the essentials of making and receiving television images. But it will take forty years before the well-delineated concept finally achieves commercial success.

Think twice before you throw away the silly idea you scribbled on a bar napkin: many an invention was born on a piece of paper.

Edinburgh-born Alan Archibald Campbell-Swinton made his concepts public in the 1908 letter and again three years later in a lecture to the Roentgen (see page 314) Society of London. He described using cathode-ray tubes to capture and display imagery. An electron gun in the neck of the cathode-ray tube would shoot electrons toward the flat end of the tube, which was coated with light-emitting phosphor. Sweeping the electron stream back and forth in rows from top to bottom, Campbell-Swinton proposed, would display a moving image on a flat screen.

Many inventors in the early twentieth century filed patents for technologies that would eventually make their way into the television, but Campbell-Swinton's concept was central because he proposed a modification of the cathode-ray tube that let it be used as both a transmitter and receiver of light. Campbell-Swinton didn't know how to actually make his television work. Inventors Kalman Tihanyi, Philo T. Farnsworth, John Logie Baird, Vladimir Zworykin (see page 112), and Allen DuMont all built on Campbell-Swinton's ideas to devise the first working televisions.

Progress was slowed by World War I and the Great Depression. The first practical consumer TV receivers were a hit at the New York World's Fair in 1939 (see page 122). Then World War II intervened, and it was only after the war that Campbell-Swinton's vision blossomed into a mass medium.

He was not around to see it. Campbell-Swinton died in 1930. — *BXC*

Also June 18:

1981: Vaccine Announced for Foot-and-Mouth Disease
1983: Sally Ride Becomes First American Woman in Space

Also 1908:

July 8: Movies Get Color (see page 191)
October 1: First Production Model T Rolls Out of Ford Plant

June 19
240 BCE: The Earth Is Round, and It's This Big

Greek astronomer, geographer, mathematician, and librarian Eratosthenes calculates Earth's circumference.

Eratosthenes was the librarian of the great Library of Alexandria. His fans called him Pentathalos, a champion of multiple skills. His detractors called him Beta, because he came in second in every category.

He invented the Sieve of Eratosthenes, an algorithm for finding prime numbers that's still used in modified form today. He sketched the course of the Nile from the sea to Khartoum, and he correctly predicted the river's source would be found in great upland lakes.

Eratosthenes knew that at noon on the summer solstice, the sun was observed to be directly overhead at Syene (modern-day Aswan): you could see it from the bottom of a deep well, and sundials cast no shadows. But there were shadows at Alexandria, to the north, where the sun wasn't directly overhead. Therefore, the earth must be round—already conventionally believed by the astronomers of his day.

Eratosthenes computed that the angle of the shadow in Alexandria was one-fiftieth of a full circle. He then estimated the distance between the two locations and multiplied by 50 to derive the circumference. We don't know the exact length of the measurement unit Eratosthenes was using when he calculated the figure of 252,000 stades. Depending on which classical source you trust, that's somewhere between 24,663 and 27,967 miles. The accepted figure for equatorial circumference today is 24,902 miles. Pretty darn good for a guy without modern measurement tools.

Eratosthenes also computed the tilt of Earth's axis to within a degree and suggested that calendars should have a leap day every fourth year, an idea taken up by Julius Caesar two centuries later (see page 61).—*RA*

Also June 19:
1939: Dreaded ALS Gets New Name — Lou Gehrig's Disease
1964: Last *Twilight Zone* Episode Airs

Also 240 BCE:
March 30: Chinese Astronomers Spot Halley's Comet

June 20
1840: A Simple Matter of Dots and Dashes

Samuel F.B. Morse patents his dot-dash telegraphy signals, known to the world as Morse code.

The code went into practical use in 1844, after Morse and Alfred Vail produced a working electromagnetic telegraph transmitter. Vail worked on various refinements to the transmitter before leaving the business altogether in 1848, feeling he was being lowballed on his salary.

Some scholars argue that Vail, not Morse, actually came up with the dot-dash system. Vail did hold a small piece of Morse's patent but didn't get rich from it. When Morse received a patent for the telegraph itself, it came from the Ottoman Empire. Others, notably Englishmen Charles Wheatstone and William Cooke, had patents on similar (and, some say, superior) hardware, but Morse eventually triumphed in the legal battle. His adept promotion, one-wire transmission system, and simple software—the Morse code—won the day.

International Morse code was introduced at a conference in Berlin in 1851. Original Morse code contained not only dots and dashes but also spaces in five letters: *C, O, R, Y,* and *Z.* (*C,* for example, was rendered like this: • • •) Numerals 0 through 9 were also different. But American Morse remained in widespread use until the 1920s, when everyone finally lined up behind the international version.

Morse code has been around for more than 170 years. It still has practical applications in the modern world because almost anything can be used to tap out or flash a message—from a telegraph key to a flashlight to a pencil to a fingertip. Severely disabled people use Morse to communicate, sending out the code by eye movement or puffing and blowing.—*TL*

Also June 20:

1963: Cuban Missile Crisis Spurs Moscow-DC Hot Line

Also 1840:

February 5: Birth of Hiram Maxim, Inventor of
   Machine Gun (see page 36)

June 21
1948: Columbia's Microgroove LP
Makes Albums Sound Good

Columbia Records releases the first workable twelve-inch, 33⅓ rpm LP microgroove record: Nathan Milstein playing Mendelssohn's Violin Concerto in E Minor, with Bruno Walter conducting the New York Philharmonic.

Columbia engineer Peter Carl Goldmark and his staff had worked since 1939 to extend the 78 rpm record's playback time from less than five to more than twenty minutes per side, shrinking vinyl grooves to an accessible, acceptable millimeter size.

RCA Victor debuted the first commercially available long-playing records in 1931, but these 33s crumbled under the heavy tonearms and were quickly withdrawn. Technical difficulties also ended Columbia's own ten-inch-LP experiment in 1932.

The Great Depression slowed everything down, including Goldmark's team and its microgroove innovations. But once it ended, the record business boomed, pulling in more than $10 million in sales by 1945. Cue the applause track.

When World War II ended, Columbia was free to resolve the LP's technical difficulties—heavy pickups, wide grooves, short playback times, and terrible audio fidelity. "Goldmark assigned individual researchers to individual problems: cutting-motor and stylus design, pickup design, turntable design, amplifier, radius equalization," Martin Mayer wrote in *High Fidelity* magazine in 1958. "The 33-1/3 speed had been established before work began, and it already had become clear that a very narrow groove, something like the .003 inch groove finally adopted, would be necessary to record 22 minutes of music to a side."

Columbia president Goddard Lieberson introduced the LP at the Waldorf-Astoria Hotel in 1948, and the commercially available long-playing record went supernova. Columbia owned the brand name LP outright, so other labels couldn't use the buzzword to market their own releases. But the public called all of them LPs. —*ST*

Also June 21:

2004: SpaceShipOne — First Privately Financed
   Manned Craft Reaches Edge of Space
   (see page 254)

Also 1948:

September 7: Where the Rubber *Is* the Road
   (see page 252)

1969: Umm, the Cuyahoga River's on Fire...Again

The Cuyahoga River catches fire near Cleveland, Ohio. Unrestricted dumping of waste by local industries, which leaves the river clotted with oil and other combustible effluents, is blamed.

Actually, the famous fire (famous mainly because *Time* magazine gave it big play nationwide) was simply the latest in a series of conflagrations that plagued the river between 1936 and 1969. But it was an effective symbol of the ravages of industrial pollution at a time when the nation's consciousness was being raised in this area (see page 34).

In the article accompanying the pictures, *Time* described the Cuyahoga as a river that "oozes rather than flows" and as a place where a person "does not drown but decays." Hyperbole, perhaps, but effective.

The fact is the Cuyahoga River was badly polluted, and the factories and industries around Cleveland were the chief culprits. It's very telling of the mind-set of the time that Cleveland fire chief William Barry described the burning river as "a run-of-the-mill fire." Neither Cleveland daily treated the fire as big news.

But *Time* magazine did, and the image of a burning river next to a major American city was alarming. This incident played a significant role in the establishment of the Environmental Protection Agency in 1970 and the expansion of the Federal Water Pollution Control Act in 1972. — *TL*

Also June 22:
1675: King Charles II Creates Greenwich Observatory
1783: Iceland's Laki Volcano Disrupts Europe's
    Economy

Also 1969:
January 16: Two Soviet Spacecraft Dock in Orbit,
    Transfer Cosmonauts
July 20: Humans Land on Moon (see page 203)

June 23
1868: Tap, Tap, Tap, Tap, Tap...Ding!

Christopher Latham Sholes receives U.S. patent 79,265 for a type-writing machine.

It wasn't the first typewriter, or even the first to be patented. But it was the first to have actual practical value, so it became the first typewriter to be mass-produced. Sholes, a Milwaukee printer, perfected his typewriter in 1867 with the help of two partners. Patent in hand, Sholes licensed production to Remington and Sons, the famous gunmaker. The first commercial typewriter, the Remington Model 1, hit the shelves in 1873.

It couldn't produce multiple copies of an entire page, like a printing press, but it simplified—and democratized—the making of a single printed copy, using keys that inked the paper by striking a ribbon.

The notion of a machine for the individual writer had been around long before Sholes. The first known typewriter patent went to Englishman Henry Mill in 1714. Alas, no example of Mill's machine exists, and the blueprints, if there were any, have been lost too. American William Burt patented a typographer machine in 1829, but it was cumbersome to use and ultimately didn't go anywhere either. Sholes's patent was the decisive one.

You'll find the fingerprints of Thomas Edison, whose name seems to appear on practically everything invented during the latter part of the nineteenth century, on the typewriter too. Edison is credited with building the first electric typewriter in 1872. The idea was not popular, and electric typewriters didn't come into widespread use until the 1950s.

Sholes's other great contribution to mass communications? In the 1870s, he developed the QWERTY keyboard to minimize tangling of the rapidly moving typebars. That problem is long gone, but, despite alternatives, this keyboard arrangement is likely the one you use regularly. — *TL*

Also June 23:
1912: Birth of Computer Pioneer Alan Turing
1983: Domain-Name Test Sets Stage for Internet Growth

Also 1868:
July 14: Spring Tape Measure Patented
August 23: Birth of Paul Otlet, Inventor of Mundaneum (see page 237)

June 24
1812: Coal-Powered Locomotive Hauls Coal

John Blenkinsop shows off the world's first rack-and-pinion steam locomotive.

Blenkinsop, the manager of the Middleton Colliery in West Yorkshire, England, was looking for a cheaper way to move coal from Middleton to Leeds. Cavalry fighting in the Peninsular War had driven up the cost of horse feed (see page 48), making it expensive to transport coal in horse-drawn wagons.

Blenkinsop was inspired by earlier experiments with motive steam power: Nicolas-Joseph Cugnot tested his *fardier à vapeur* in 1769, and Richard Trevithick began building prototype locomotives in 1801. All these vehicles failed partly because of excess weight, but Blenkinsop feared a lighter locomotive's wheels would slip and spin rather than move along the rails when the train tried to haul a heavy load.

Blenkinsop thought a rack-and-pinion system would carry heavy loads better than adhesion. His design featured a cogwheel, or pinion, on the locomotive that engaged with a toothed rack rail. The two main rails guided the locomotive's wheels, while the turning cogwheel meshed with the rack rail to provide traction.

Blenkinsop turned to engineer Matthew Murray to execute the design. Murray built a steam engine with two vertical cylinders in the top of the boiler and pistons cranking the cogwheel. The locomotive weighed five tons and could haul ninety tons of coal at nearly 4 miles an hour—equal to the work of fifty horses and two hundred men. Blenkinsop christened the engine *Salamanca*, after the English victory on July 22, 1812, in the war that had forced Blenkinsop to build a locomotive in the first place.

William Hedley, manager of the competing Wylam Colliery, demonstrated that a railway didn't need rack and pinion. Hedley's adhesion locomotive was hauling coal at 5 miles an hour by 1814. — *KB*

Also June 24:

1947: Pilot Kenneth Arnold Sights UFOs Near Washington's Mount Rainier

1967: First Consumer Electronics Show Opens in New York

1993: Geek Band Concert Goes Live on Net

2000: President Clinton Goes Live on Net

Also 1812:

February 27: Lord Byron Pleads for the Luddites (see page 59)

June 25
1867: Barbed Wire — the Beta Version

Lucien Smith patents barbed wire, an artificial thorn hedge.

Smith's design called for spools of four short, sharp metal spikes at right angles. The spools would revolve loosely and be set every two to three feet along the fence wire. William Hunt patented a similar design that year, and Michael Kelly did so the next. A patent battle followed, but none of these guys would win.

The great need was in the Great Plains. As American settlement moved west, the spaces to enclose got bigger, while wood and stones for building fences got scarcer. Growing hedgerows took time, and water was also scarce. The farther you shipped fence materials, the more expensive it was.

Fencing wire available at the time was brittle, and cattle could rub against the smooth wire with impunity until it broke or the fence posts loosened. Then the critters could wander into your kitchen garden, your cash crops, your neighbor's ranch, or the wide-open spaces where the deer and the antelope roamed.

Joseph Glidden got *his* idea for barbed wire when he saw Henry Rose's invention at a county fair: boards with sharp nails hanging from a smooth-wire fence. Glidden thought the board unnecessary and expensive: Why not put the barbs directly in the wire? He rigged the crank of a household coffee-bean grinder — his wife's suggestion, the legend goes — to twist the wire into loops that were then clipped off into sharp points. Irritating.

Glidden's wire caught on, as it were. There were soon 570 different patents for various types of wire, twists, and barbs. A three-year legal battle ensued, and Glidden triumphed. By the time of his death, in 1906, he was one of America's richest men. — *RA*

Also June 25:

1876: Indians May Have Had Better Guns at Custer's Last Stand
1997: Unmanned Soviet Spacecraft Dents Mir Space Station

Also 1867:

June 17: Joseph Lister Performs First Antiseptic Surgery
July 16: Concrete Gets Some Positive Reinforcement (see page 199)

June 26
1974: By Gum! UPC Is New Way to Buy Gum

A supermarket cashier scans some chewing gum across a bar-code reader in Troy, Ohio. It's the first product ever checked out by Universal Product Code.

The UPC started out not to save labor for checkout clerks but to keep track of inventory. Bernard Silver and Norman Woodland began working in 1948 with ink patterns that glowed in ultraviolet light. Too expensive. They considered Morse code (see page 173), but optical readers would require the clerk to line up the code at a precise angle. Not practical.

On the beach one day, Woodland punched some dots and dashes into the sand, then idly lengthened them into vertical lines and bars. *Voilà!* Elongated marks would be readable from nearly any angle.

Woodland and Silver's invention went through many changes before the supermarket industry finally selected IBM's laser-reader system, in 1973.

But at 8:01 that fateful morning, Clyde Dawson grabbed a fifty-stick pack of Wrigley's Juicy Fruit gum from his shopping cart (see page 157), and cashier Sharon Buchanan made the first UPC scan. The register rang up sixty-seven cents (three bucks in 2012 dollars). Retail history was made: the pack of gum is now in the Smithsonian.

The scanner cost $4,000, but prices came down as more stores used the system. Adjusting for inflation, a scanner now costs just ¼ percent of what it did in 1974.

Retailers today use UPCs not only for checkout and inventory but to track individual preferences, spit out coupons, and provide information for future sales campaigns. Car-rental companies use bar codes to track their fleets; airlines track luggage; shippers track packages; and researchers track animals. Fashion houses even stamp bar codes on their models to make sure the right model wears the right parts of the right outfit at the right time in the fashion show.

A far cry from drawing lines in the sand. —*RA*

Also June 26:

1498: Emperor of China Patents Toothbrush of Hogback Bristles Set in Bone or Bamboo

1978: First Dedicated Oceanographic Satellite Launched

Also 1974:

November 24: Lucy Skeleton Discovered, First Recognizably Human Primate

December 19: Altair 8800 Kits Go on Sale (see page 355)

June 27
1954: Soviets Open World's First Nuclear Power Plant

The world's first nuclear power plant, in Obninsk, outside of Moscow, starts delivering electrical power.

The nuclear reactor heralded Obninsk's new role as a major Soviet scientific city, a status it retains in the Russian Federation, where it carries the sobriquet of First Russian City of Science. Obninsk, population of a bit more than 100,000, currently houses no fewer than twelve scientific research institutions and a technical university. Research is focused on nuclear power engineering, nuclear physics, radiation technology, the technology of nonmetallic materials, medical radiology, meteorology, and environmental protection.

The Obninsk reactor didn't produce the very first nuclear-generated electricity. An experimental breeder reactor in Idaho, devised by Chicago Pile veteran Walter Zinn, had done that on December 20, 1951. But the Soviet plant was no experiment: it sent electricity to the grid.

Since 1954, most of the industrialized West, along with India and China, have tried—and even embraced—nuclear power (see page 338). But the backlash against this energy source continues in the wake of accidents such as those that occurred at Three Mile Island in 1979 (see page 89) and Chernobyl in 1986.

Italy, for example, has decommissioned its nuclear plants. And after Japan's giant 2011 tsunami triggered a meltdown at the Fukushima Daiichi nuclear plant, Germany decided to transition away from nuclear.

The Obninsk reactor wasn't decommissioned until 2002. The city, by the way, claims Oak Ridge, Tennessee, as a sister city—another town that has more than a passing relationship with things nuclear.—*TL*

Also June 27:

1898: Joshua Slocum Finishes First Solo Sail Around World

Also 1954:

January 21: First Lady Mamie Eisenhower Christens USS *Nautilus*, First Nuclear Sub
August 30: President Eisenhower Signs Atomic Energy Act
October 7: IBM Builds First Transistorized Calculating Machine (see page 282)

June 28
1846: Belgian Inventor Patents Saxophone

Emerging from his Paris workshop, musician-inventor Adolphe Sax files fourteen patents for an instrument destined to revolutionize American music nearly a century later. His new invention: the saxophone.

Initially crafted from wood, Sax's new instrument flared at the tip to form a music-amplifying bell. Designed in seven sizes, from sopranino to contrabass, the saxophone combined the easy fingering of large woodwinds with the single-reed mouthpiece of a clarinet.

Adolphe Sax.

The saxophone became popular with French army bands, but Sax spent decades in court trying to fend off knockoffs and made only meager profits before his patents expired, in 1866. Myriad modifications followed, improving ease of play. U.S. production began in 1888 when Charles Gerard Conn of Elkhart, Indiana, started manufacturing the instruments for military bands. By the early 1900s, the saxophone was a comedy fixture on the vaudeville circuit, where musicians used it to mimic chicken sounds.

Eventually produced in baritone, tenor, alto, and soprano models, the saxophone became a creative tool of the first magnitude only in the early 1920s, when New Orleans clarinet player Sidney Bechet grew weary of being drowned out by his bandmate's much louder cornet. When the jazz musician switched to soprano saxophone and began projecting a stronger "voice" within the ensemble, other players took note.

Coleman Hawkins, inspired by Louis Armstrong's syncopated trumpet solos, became the first jazz virtuoso to exploit the deep, throaty tones of the tenor sax. Ben Webster and Lester Young soon followed, showcasing soulful tenor solos with the Duke Ellington and Count Basie bands during the 1930s.

In the mid-1940s, alto sax player Charlie Parker pioneered bebop music by producing rapid-fire flights of improvisational fancy, light-years removed from the genteel European ditties imagined in the 1840s by Monsieur Sax. — *HH*

Also June 28:

2005: Design Unveiled for World Trade Center Replacement

Also 1846:

September 10: Elias Howe Patents First Practical Sewing Machine

September 30: Ether He Was the First or He Wasn't (see page 275)

June 29
2007: iPhone, You Phone, They All Wanna iPhone

Apple puts the iPhone on sale. It sells...fast.

CEO Steve Jobs had announced the iPhone in January, so everybody knew it was coming. But nobody predicted how the iPhone would change the way we look at phones. It was not the first smartphone, just dramatically better than its multiuse predecessors, like the BlackBerry.

Days before the iPhone landed, hundreds of fanatical consumers camped outside Apple and AT&T stores waiting for the six-hundred-dollar gadget. The price tag was a turnoff, but the price soon dropped to four hundred dollars, and then—with the iPhone 3G—to two hundred. Apple sold over ten million iPhone 3G units in five months.

Just as 1977's Apple II was the first computer made for consumers, the iPhone was the first phone whose software was designed with the user in mind. It was the first phone to make listening to music, checking voicemail, and browsing the web as easy as swiping, pinching, and tapping a sleek and sexy touchscreen. Apple put a miniature computer in the consumer's pocket.

Operating on an Apple-controlled platform, the iPhone was limited to the few apps that Apple offered. That gave birth to an underground world of hackers seeking to add third-party applications. Repeated attempts to fend them off finally made Apple realize apps were in high demand.

So the company launched the App Store and shook up the mobile industry again by reinventing software distribution. Apple keeps 30 percent of sales and gives developers 70 percent. Apple's App Store now features over 50,000 applications, and over 1 billion apps have been downloaded. The App Store spurred other smartphone companies to launch their own application stores.

Apple flipped the cellphone industry on its head and transformed mobile software into a viable business. What's next?—*BXC*

Also June 29:

1888: Handel Oratorio Becomes First Commercial Music Recording

1956: Eisenhower Signs Interstate Highway Act (see page 271)

Also 2007:

March 4: Estonia Becomes First Country with Internet Voting in National Election

October 6: Jason Lewis Completes First Human-Powered Circumnavigation, by Pedal Boat

June 30
1953: Corvette Adds Some Fiber, Flair to American Road

Chevrolet introduces the Corvette, the first production car with a body made entirely of a new wonder material called fiberglass.

The Corvette was born of the boom years following World War II and was a response to the growing popularity of the small, nimble two-seat sports cars American GIs brought home from Europe. Legendary GM designer Harley Earl designed the car, and GM adman Myron Scott named it (after a class of fast, compact, powerful warships). A fiberglass car was lighter — and more futuristic — than one made of steel.

The gorgeous hand-built '53s were all painted polo white with sportsman-red interiors, black tops, whitewall tires, and analog instruments — including a 5,000 rpm tachometer. Suggested retail price was $3,513 (about $30,000 in today's money). The first-generation Corvette looked like a sports car but didn't drive like one. The six-cylinder truck engine was sluggish, the drum brakes were weak, and the car had a two-speed automatic transmission.

Two things saved it: Chevy's introduction in 1955 of the small-block V-8 engine, and the arrival of engineer Zora Arkus-Duntov. As director of high-performance-vehicle design, he gave the Corvette its performance pedigree and earned the name "Father of the Corvette."

The 'Vette got more powerful engines. A fully independent rear suspension and disc brakes on all four corners made the car a real runner. Race-trimmed Corvettes driven by Roger Penske, A. J. Foyt, Jim Hall, and Dick Guldstrand were outright terrors on the track.

More than that, the Corvette became a cultural landmark, immortalized in countless movies and songs. It's been restyled six times over the years, but the car has stayed true to its sports-car heritage. The Corvette remains the pinnacle of American sports-car design, and it's still made of fiberglass. — *TB*

Also June 30:
1908: Mysterious Tunguska Explosion Rocks Siberia

Also 1953:
March 17: Airplane Black Box Is Born (see page 78)

July 1
1858: Darwin and Wallace Shift the Paradigm

Charles Darwin's theory of evolution gets its first public airing, but it's not his theory alone.

Scientists knew evolution occurred. The question was, How? The answer, Darwin realized back in the 1830s, was natural selection. It accounted for the biodiversity he'd observed while traveling on the *Beagle* (see page 363). He was writing a multivolume treatise to prove his radical new idea.

In June 1858, he received a short paper from naturalist Alfred Russel Wallace in Malaysia, proposing the very same theory in slightly different language. Darwin was crestfallen, fearing he would lose credit for two decades of work.

Two scientists who had seen early drafts of Darwin's work helped arrange for a joint paper to be read at the meeting of the Linnaean Society of London on July 1.

As a reader of this book, you will not be surprised by simultaneous discovery. But this instance is extraordinary in multiple ways:

1. Publication was simultaneous: same place, same day, same presentation. (Darwin's *Origin of Species* was not published until 1859.)
2. Both scientists got the idea from the same place: Malthus's 1798 essay on population. Malthus observed that in each generation, not every individual would survive to reproduce. After decades of no one else picking up on that, both Darwin and Wallace realized that *who did* survive would determine evolution.
3. Both authors acknowledged they didn't know *how* successful traits were passed to the next generation. And although Gregor Mendel published his literally seminal work on genetics in 1865 (see page 39), no one made this connection until *three different botanists* published on it in 1900.

4. Finally, it involved not a simple invention or discovery but a complete shift, inventing the reigning paradigm that organizes modern biology—and much of modern science.—*RA*

Also July 1:

1910: Automated Bread Factory Opens in Chicago

1934: Federal Communications Commission Begins Operation

1941: TV Goes Commercial — Bulova Runs First Ad (see page 302)

Also 1858:

October 14: Big Ben Bell Hoisted into Parliament's Clock Tower (see page 289)

December 4: Birth of Earmuff Inventor Chester Greenwood

1928: America's First TV Station Goes on the Air

W3XK, the first American TV station, begins broadcasting from suburban Washington, DC.

The station was the doing of Charles Francis Jenkins, who devised a way to transmit pictures over the airwaves, a process he called radiovision. He sold several thousand receivers, mostly to hobbyists, and, after obtaining permission to start an experimental TV transmitting station, aired "radiomovies" five nights a week until he closed the station in 1932.

Jenkins essentially brought the wrong technology to the field: his receiving sets relied on a forty-eight-line image projected onto a six-inch-square mirror to create the picture, rather than on electronics, the technology that ultimately determined the future of television.

A wholly electronic scanning technology was outlined by A.A. Campbell-Swinton as early as 1908 (see page 171), but it wasn't ready for prime time (or almost any other time) until the public finally got a good look at it, during the New York World's Fair in 1939 (see page 122).

An interesting aside: Jenkins was also the first to air a television commercial. He was fined by the government for doing so. Unfortunately those fines stopped when the medium matured.

The first legal TV commercials didn't appear until 1941, and the first show with its own sponsor didn't air until five years after that (see page 302). — *TL*

Also July 2:

1937: Aviator Earhart Vanishes Over the Pacific
1982: Weather Balloons Carry Lawn-Chair Pilot Three
      Miles High

Also 1928:

March 12: St. Francis Dam Disaster Kills More Than
      Five Hundred in California
October 12: Iron Lung, Savior to a Generation
      (see page 287)

July 3
1886: Front-Page Days Get Rolling, Thanks to Linotype

The *New York Tribune* becomes the first newspaper to use Linotype, a complex but highly efficient typesetting machine that revolutionizes the printing process.

The Linotype machine was the brainchild of Ottmar Mergenthaler, a German-born inventor who became a naturalized U.S. citizen. To see one of these machines in action is to see the age of mechanization in excelsis.

Employing a ninety-character keyboard, the Linotype operator punched out individual characters to form a line of type (or line o' type, hence Linotype) that went to the page compositor in a "stick." Those sticks of type were arranged by hand inside a metal frame, called a chase, to correspond to a page layout supplied by the editorial department.

The Linotype keyboard did not resemble a standard typewriter keyboard. Letters were arranged in columns by their frequency in English (the first two columns were e-t-a-o-i-n and s-h-r-d-l-u). Lowercase letters were on the left, caps on the right, with five rows of special characters and numerals dividing the two. Type could be set either justified or ragged through the use of space bands to fill out an individual line.

Mergenthaler's invention had a profound effect on the newspaper business. Before Linotype, typesetting was done by hand, a laborious process that necessarily limited the size of newspapers. Before 1886, no daily paper was longer than eight pages.

Because of the Linotype's use of molten lead, printers called the process hot type. It remained the dominant newspaper-production method until the 1970s, when it was replaced, first by cold type (photo-typesetting and pasteup), and then by computerized pagination and desktop publishing. — *TL*

Also July 3:
1999: Thirty-Three-Year-Old Scores Perfect 3,333,360 Pac-Man Points (see page 285)

1999: Thirty-Three-Year-Old Scores Perfect 3,333,360 Pac-Man Points (see page 285)

Also 1886:
May 8: Druggist John Pemberton Concocts Coca-Cola

July 4
1776: Preserving the Declaration

The Continental Congress passes the Declaration of Independence. But it will take 127 years before someone gets around to saying, "Hey, maybe we should preserve this thing."

During the Revolutionary War, the official parchment copy of the Declaration of Independence was rolled up and toted around like a Thomas Bros. map. Given the vicissitudes of war, that's perhaps understandable.

Less understandable is what came later: Water was spilled on it while it was being copied in 1823. Then it was tacked up on the wall at the U.S. Patent Office for about forty years, subjected to strong light. Finally, the suggestion was made in 1903 that maybe it shouldn't be exposed to sunlight, and maybe it should be kept dry too. The latter turned out to be a bad idea, because parchment actually needs a bit of moisture to keep from cracking.

It wasn't until 1951 that the document was sealed inside a bronze-and-bulletproof-glass case at the National Archives in Washington. Humidified helium replaced oxygen to prevent further erosion, and the glass was filtered to cut light exposure.

Beginning in 1987, using camera equipment developed for the Hubble Space Telescope, preservationists were able to monitor the Declaration for the tiniest signs of fading or flaking ink. After undergoing careful inspection for additional erosion in 2003, the document was resealed in a titanium casement filled with non-reactive argon gas (see page 152). Similar preservation techniques protect the Bill of Rights and the Constitution.

The Declaration of Independence remains on display in the rotunda of the National Archives, where roughly six thousand tourists see it every day. At night, when the crowds have all gone home, the case is lowered twenty-two feet into a vault.

That's almost as much protection as the French give to Napoléon's bones. — *TL*

Also July 4:

1054: Chinese Astronomers Spot Supernova, Origin of Crab Nebula

Also 1776:

March 24: Chronometer Inventor John Harrison Dies, Like Clockwork, on His Birthday
December 5: Phi Beta Kappa Founded

July 5
1943: Defeat at Kursk Heralds Twilight of the Panzers

The Battle of Kursk begins. It features the largest tank engagement in history.

Soldiers with a Tiger I of Germany's SS-Panzergrenadier-Division.

Following their catastrophic defeat at Stalingrad five months earlier, the Germans lacked the strength to attack the entire eastern front. The high command focused instead on a salient close to Kursk in Russia, near Ukraine. There was an opportunity there to encircle and destroy a dozen Soviet armies. After that, the Germans planned to strike northward at Moscow. The operation, about which even Hitler expressed misgivings, was code-named Zitadelle, or Citadel.

The Germans committed nearly a million men, 2,700 tanks, and 2,000 aircraft to this attack. They would face a Soviet force of 1.3 million men, 3,600 tanks, 2,400 planes, and more than 20,000 artillery pieces.

The Battle of Kursk was the apogee of tank warfare (see page 260). The Soviets were equipped with probably the war's most consistently efficient tank, the T-34, which had given the Germans a nasty shock early in the eastern campaign. Germany's tanks (called *Panzer* in German) were also excellent—at least on paper. Their Panther tank had been designed specifically to take on the T-34.

The decisive engagement was fought July 12 at Prokhorovka, with German and Soviet tanks blasting away at point-blank range. Although the Soviets suffered considerably heavier losses, overall German strength was ebbing fast. With the Allied invasion of Sicily on July 11 and a new Soviet offensive beginning in the north, Hitler called off Zitadelle on July 13. Kursk was where the operational initiative on the eastern front passed to the Red Army. Germany could no longer dictate the course of war to an enemy that was only growing in strength. With the country's dwindling forces being siphoned off to meet threats in other theaters, Germany's fate was sealed.—*TL*

Also July 5:

1687: Isaac Newton Publishes *Principia Mathematica*, Foundation of Modern Physics

1937: Hormel Introduces Spam, Canned Spiced Pork Shoulder

1946: Paris Pool Sees First Bikini Bathing Suit Worn in Public

1997: Birth of Dolly the Sheep, First Cloned Mammal

Also 1943:

July 26: LA Gets First Big Smog (see page 209)

1947: The AK-47, an All-Purpose Killer

The AK-47, one of the world's first operational assault rifles and probably the most durable and enduring small-arms weapon ever made, goes into production in the Soviet Union. More than sixty years later, it remains the standard infantry weapon in numerous armies, and a mainstay in the arsenals of rebels, drug traffickers, and terrorists worldwide.

The AK-47 was the brainchild of self-taught inventor Mikhail Kalashnikov, who was inspired by his fellow Russian soldiers' complaining about the poor quality of Soviet-made small arms. Elements of Nazi Germany's Sturmgewehr 44 are plainly evident in Kalashnikov's original AK-47. The Red Army adopted the Avtomatni Kalashnikova model 1947 as its standard infantry weapon in 1949, and it was standard issue for the major Warsaw Pact armies as well.

Characteristic of assault rifles, a selector level lets the shooter choose between automatic (continuous firing) and semiautomatic (single-shot firing). The weapon is gas-operated, chambering a 7.62mm round. The genius of the AK-47 lies in its sheer simplicity, which makes it cheap to manufacture, reliable to use, and durable in the field.

Revolutionaries from Cuba to Angola to Vietnam clamored for the weapon. Because many of these movements were directly supported by the Soviet Union, obtaining AK-47s (or, more colloquially, Kalashnikovs) in bulk was not difficult. American soldiers in Vietnam, toting M16s prone to jamming, gained a grudging respect for the AK-47 too.

About 100 million of them, in their many variations, are believed to be in circulation. Kalashnikov's original 1947 model, however, is a rare find.

Kalashnikov expressed regret that his AK-47 became the weapon of choice for terrorists, but he suffers no pangs of guilt: "My conscience is clean. I constructed arms to defend my country." — *TL*

Also July 6:

1886: Rabies Vaccine Saves Boy — and Pasteur
1920: U.S. Navy Pilots Start Using Radio Compass

Also 1947:

August 18: Hewlett-Packard Incorporates (see page 232)

July 7
1936: Get a Grip—Phillips Screws Up the Toolbox

Henry Phillips patents a new kind of screw and the new screwdriver needed to make it work. It's enough to make you cross.

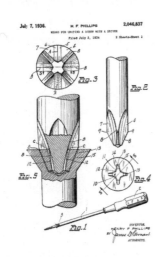

Patent diagrams for the Phillips screw and screwdriver.

Phillips wasn't focused on hand tools; he was trying to solve an industrial problem. To drive a slot screw, you need hand-eye coordination to line up the screwdriver and the slot. If you're a machine—especially a 1930s machine—you don't have eyes, and your coordination may depend on humans.

The Phillips head screw was designed for power tools, especially on assembly lines. The shallow, cruciform slot in the screw allows the tapering cruciform shape of the screwdriver to seat itself automatically. That saves a second or two, and if you've got hundreds of screws in thousands of units, you're talking big time. A power Phillips driver also stays engaged, and it's hard to overscrew: the screwdriver just pops out when the screw is completely fastened.

After years of rejection, Phillips got the American Screw Company to develop a manufacturing process. American Screw convinced GM to try the newfangled fasteners on the 1936 Cadillac. Nearly all American automakers switched to Phillips screws by 1940. American jeeps and tanks and aircraft of World War II were assembled with speed and efficiency, thanks in part to Henry Phillips.

If you're a home handyperson, the variety of screw types may be annoying. The Phillips cam-out—when you've gone far enough and the tool pops out of the screw—has led to plenty of workshop profanity.

Still, remember Henry Phillips gently. His screws are holding your life together. —*RA*

Kinemacolor, the first successful color-motion-picture process, is demonstrated at a scientific meeting in Paris.

British inventor Edward Turner patented a three-color motion-picture process in 1899, but it didn't work well. American expatriate Charles Urban and Albert Smith tried fixing Turner's process but then moved to a simpler two-color system using black-and-white film. Instead of exposing the then-standard sixteen frames a second, the new process exposed thirty-two frames.

The Kinemacolor projector.

A spinning wheel of transparent filters exposed alternate frames in red and green. A similar wheel was used to project the film, and just as we perceive a succession of still movie frames as continuous motion, so also do two partial-color images merge into full color.

Urban previewed the system for the press in London before giving it a scientific debut in Paris, which film pioneers Auguste and Louis Lumière attended. Kinemacolor got its name in 1909 and was used to film George V's coronation as emperor of India in 1912.

The process was more economical than the frame-by-frame hand tinting employed by some producers, which sometimes used stencils to create several hundred color prints for commercial distribution. But Kinemacolor was deficient in presenting blues and getting a true white. And rapid motion caused color fringing, because red frame and green frame were shot one-thirty-second of a second apart. Kinemacolor was also expensive for theater owners and difficult for projectionists.

Kinemacolor spawned offshoots, including side-by-side, rather than alternating, red and green images, and Urban had to fight patent battles. After World War I, Kinemacolor also faced a superior technology, one that used stationary prisms instead of moving wheels to film and project color separations. Devised by MIT-trained engineers in Boston, it was called Technicolor. —*RA*

Also July 8:

1947: Roswell Incident Launches UFO Controversy
1967: Buck Rogers Stops Here; Comic Strip Ends

Also 1908:

September 17: First Airplane-Passenger Death
(see page 262)

July 9–10
1856: Visionary Tesla Born at Midnight

Scientific genius and visionary inventor Nikola Tesla is born at the stroke of midnight in Smiljan, Croatia. He wastes little time in revolutionizing the world through foundational developments in electromagnetism, electrical current, wireless power and communications, weaponry, robotics, computer science, and more.

"Tesla is like a character out of a science-fiction novel, the quintessential mad genius," says author Tom McNichol. "Whether he was more mad than genius depends on who you're talking to."

Tesla job-hopped from Budapest to Paris to New York, where he joined his scientific contemporary and lifelong nemesis Thomas Edison. Tesla had already privately built a successful prototype of the alternating-current induction motor. He soon upgraded Edison's inefficient direct-current motors and generators, and he was perfecting the polyphase system for distributing AC power when he left Edison's employ after a salary dispute, in 1886.

During Tesla's AC/DC war with Edison (see page 4), the eccentric inventor was bankrolled by engineer-entrepreneur George Westinghouse (see page 66). Westinghouse won the current war with hydroelectric power generated from Niagara Falls, leading to more than a century of AC supremacy. Tesla's subsequent innovations in wireless communications and power gave birth to everything from the radio to the Wi-Fi network and our probably inevitable cordless future.

Tesla's personal quirks, like his obsessive love of pigeons and a physical revulsion to jewelry, didn't help his career. He died in 1943, penniless and mostly alone, in New York. His personal papers were quickly impounded and eventually declared top secret by the FBI.

Also July 9:

1955: Russell-Einstein Manifesto Calls for Nuclear Disarmament

1958: Surf's Up, as 1,700-Foot Wave Scours Alaskan Bay

1993: DNA Tests Confirm Identity of Czar's Family's Bones

Also 1856:

August 19: Gail Borden Patents Condensed Milk (see page 315)

"The modern industrial world, powered by alternating current, is very much a child of Tesla," explains McNichol. "Tesla was truly a visionary in the sense that he saw things no one else did." —*ST*

July 10
1962: Swedes Set to Belt Us All...Safely

Swedish engineer Nils Bohlin receives a U.S. patent for the three-point lap-and-shoulder vehicle safety belt. It's considered one of the most important and widespread safety innovations of all time.

Bohlin designed pilot-ejection systems for Saab Aircraft before Volvo hired him as its *first* safety engineer, in 1958. The automobile seat belts of the time were two-point lap belts that didn't restrain the upper body. The buckle itself often caused internal injuries in high-speed crashes.

Bohlin took just a year to devise, engineer, and test a double-strap, triple-anchor design that restrained the upper body, could be buckled securely with one hand, and kept the buckle away from the passenger's soft abdomen. It was simple and efficient.

Volvo introduced the new belt in 1959. It started saving lives almost immediately. Volvo made the design "freely available" to other car manufacturers and sent Bohlin abroad to promote seat-belt adoption and legislation.

Bohlin received letters from all over the world from thankful car-crash survivors. He delighted in hearing of lives saved by his invention, and there have been plenty of them. Volvo estimated in 2002 that three-point seat belts had saved more than a million lives. They prevent an estimated 100,000 injuries a year just in the United States.

West Germany cited Bohlin's invention in 1985 as one of the eight most important patents of the century. Bohlin received a Swedish Gold Medal and was enshrined in the Automotive Hall of Fame in 1999. He died in 2002, on the very day he was inducted into the National Inventors Hall of Fame.

He was eighty-two and died from complications of a stroke and a heart attack. Bohlin's family assured the world that he buckled up every time. —*RA*

Also July 10:

1997: Neanderthal DNA Suggests a Separate, Unequal Being

1999: Inventor of Reddi-wip Sputters Out

Also 1962:

July 23: Telstar Provides First Transatlantic TV Link (see page 206)

July 11
1979: Look Out Below! Here Comes Skylab!

The Skylab space station reenters Earth's atmosphere after six years in orbit.

The Skylab space station.

The first U.S. space station, launched as a science and engineering laboratory, was not a success. Originally intended to remain in orbit as a shelter for crews from the new space shuttle program, Skylab was damaged during liftoff and then plagued by a power deficit.

NASA wanted Skylab to remain in low orbit until a space shuttle equipped with a re-boost module could reach it and send it into a higher orbit. Subsequent shuttle missions were supposed to overhaul Skylab, making repairs and replacing various components.

Three crews traveled to Skylab aboard Apollo spacecraft—spending a total of 171 days there and returning by splashdown—and some repairs were made. Skylab scored a few achievements, especially in solar research and astronauts' adaptation to longer periods in space. The space station was placed in a parking orbit after the third Apollo crew departed. But the shuttle program was enduring delays, and Skylab's orbit was deteriorating. NASA had to bring it down.

As Skylab neared Earth's upper atmosphere, news coverage was amped up and sensationalized, souvenirs were hawked, and bookmakers took bets on when the 77.5-ton space station would burn up and where debris would land. The *San Francisco Examiner* offered $10,000 (about $30,000 in today's money) to the first person who could deliver a chunk of Skylab debris to the paper's newsroom.

With only limited control, NASA engineers struggled to coax Skylab to break up over the Indian Ocean. Most of it did, but parts landed in western Australia. Stan Thornton grabbed the small piece that landed on his roof, and the plucky seventeen-year-old flew off to San Francisco to collect his ten grand. —*TL*

Also July 11:

1975: Archaeologists Unearth China's Terra-Cotta Army

Also 1979:

September 24: CompuServe Debuts, First Online Service
   for Consumers (see page 269)

1960: Etch A Sketch? Let Us Draw You a Picture

The Etch A Sketch goes on sale.

The technology behind this children's toy is both simple and complex. Simple, in that an internal stylus is manipulated by turning horizontal and vertical knobs to etch a sketch onto a glass window coated with aluminum powder. Complex, because the Etch A Sketch employs a fairly sophisticated pulley system that operates the orthogonal rails that move the stylus around when the knobs are turned. (Many a user has painstakingly cleared a large section of the screen to examine the apparatus inside.) The stylus etches a black line into the powder-coated window to create the drawing.

Along with the aluminum powder, the guts of the toy include a lot of tiny styrene beads that help the powder flow evenly when the sketch is being erased (by shaking), recoating the screen for the next drawing. As for how the aluminum powder sticks to the window, well, it pretty much sticks to everything.

Arthur Granjean, a Frenchman, was the Etch A Sketch's inventor. He called it L'Ecran Magique, or "the Magic Screen." After failing to get some of the bigger toy companies to bite, he sold his invention to the Ohio Art Company, which has manufactured it ever since.

Although the traditional Etch A Sketch comes in a red plastic housing, it is now available in several colors. You can also get it as an app for your smartphone or tablet. The more things change... —*TL*

Also July 12:

1854: Birth of George Eastman, Father of
    Mass-Market Photography (see page 124)
1895: Birth of Buckminster Fuller, Popularizer of
    Geodesic Domes

Also 1960:

April 1: First Weather Satellite Launched
August 16: Geronimo-o-o-o-o-o-o! Record Parachute
    Jump (see page 230)

July 13
1937: Gibson Plugs In the Electric Guitar

Guy Hart, general manager of the Gibson guitar company, gets the first patent for an electric guitar pickup. Gibson's is not the first electric guitar to market, but its pickup design is superior to competing models'.

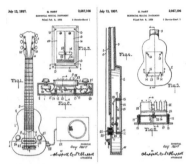

Patent drawing for the original Gibson "lap-steel" electric guitar.

The first viable electric guitar was designed by guitarist George Beauchamp to boost the instrument's volume over the drums, brass, and audience chatter. He began manufacturing them with engineer Adolph Rickenbacker (see page 224). Their guitars were "lap steel": the player held the instrument flat in his lap and slid a metal bar up and down the strings. All guitar pickups employ essentially the same design: one or more magnets wrapped in a thin coil of wire. Beauchamp's pickup used two horseshoe-shaped magnets over the strings.

Guy Hart saw the market for lap-steel guitars growing and decided it was time Gibson got in the game. Hart and Gibson's in-house engineers developed a new design: a fat steel blade positioned vertically underneath the strings. The blade was then sandwiched by two heavy magnets at the bottom, and the wire was coiled around the blade, above the magnets.

Gibson rolled out the E-150, its first electric, Hawaiian-style lap-steel guitar, in 1935. It came with an amplifier (like all electric guitars then), and the whole package sold for $150 (about $2,500 in today's leaf). Unlike Rickenbacker's "frying pan" design, Gibson's guitar actually looked like a guitar, complete with round feminine curves, shoulders, and scooped waist. Hart's next design won the patent race: not a traditional hollow-bodied guitar getting a boost from a pickup, but rather a guitar that makes noise *only* when it's plugged in.

As Hart predicted, the future wasn't in acoustic boxes that got louder with the help of his device. He knew the secret: If you really want to rock, you've got to plug in! —*MC*

Also July 13:

1977: Massive Power Blackout Plunges New York into Rioting

Also 1937:

May 27: Golden Gate Bridge Opens
June 4: Humpty Dumpty and the Shopping Cart
(see page 157)

1850: What a Cool Idea, Dr. Gorrie

Florida physician John Gorrie uses his mechanical ice-maker to astonish guests at a party.

Cooling back then relied on ice blocks carved from frozen lakes and rivers in the North; they were kept in shaded sheds and cellars, sometimes delivered at great expense by specially fitted ice ships (see page 258). Dr. Gorrie improved the survival rate of his feverish patients by cooling them down with pans of ice water placed high in their sickrooms. But ice was expensive in the Florida summer and often completely unavailable.

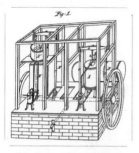

John Gorrie's ice-making machine.

Gorrie wanted to make ice mechanically by compressing air, cooling it partially with water, and then letting it expand again, to create freezing temperatures. He had a working model by the mid-1840s. It could be driven by wind, water, steam, or the brute force of an animal.

He arranged a dramatic demonstration for a social, rather than medical, occasion. It was a muggy July in Florida. Ice from the North had been exhausted. Gorrie attended an afternoon reception given by the French consul to honor Bastille Day (in 1847, 1848, or 1850, depending on the source). He complained about drinking warm wine, then signaled waiters to enter with bottles of sparkling wine on trays of ice. It was a sensation. *Smithsonian* magazine dubbed it the "chilly reception."

Gorrie failed at business. His inefficient, leaky machines were mocked in the press by the ice-shipping establishment. He died in poverty and ill health in 1855, in his early fifties. It took Ferdinand P.E. Carré's closed, ammonia-absorption system (patented in 1860) to make mechanical refrigeration practical.

So, have a happy Bastille Day (or *une joyeuse fête nationale*), chill out, and lift a cold one to the Father of Refrigeration with the very words spoken that sweltering day: "Let us drink to the man who made the ice: Dr. Gorrie." —*RA*

Also July 14:

1868: Spring-Click Tape Measure Patented
1965: Mariner 4 Photo Brings Mars Up Close and Cardinal

Also 1850:

December 6: The Eyes Have It, Thanks to Ophthalmoscope (see page 342)

July 15
1783: Marquis Invents Steamboat, Misses Esteem Boat

A young French nobleman demonstrates the first successful steamboat on the river Saône at Lyon.

Artist's rendering of the fashionable crowds watching the Marquis d'Abbans's steamboat demo.

In 1776, on his first steamboat, Claude-François-Dorothée de Jouffroy, Marquis d'Abbans, used a Newcomen engine (see page 56). The engine of the forty-two-foot vessel moved oars equipped with rotating, hinged flaps modeled on the webbed feet of waterfowl. The marquis called it the Palmipède, or Webfoot, but the flaps were a flop.

Undaunted, de Jouffroy adapted James Watt's designs to build a parallel-motion, double-acting steam engine for steamboat 2.0, the Pyroscaphe (Greek for "fire-boat"). It was equipped with a large paddle wheel (like those that ran water mills) on each side of the hull. The Pyroscaphe easily tripled the size of his earlier attempt: 148 feet long. The horizontal engine moved a reciprocating double rack, which geared to ratchet wheels on the shaft that carried the paddle wheels.

Thousands of people lined the banks of the Saône when de Jouffroy showed his pride and joy. The Pyroscaphe steamed upstream at 6 mph without a sail. The crowds cheered. But after fifteen minutes, the boat began to break up under the pounding of the engine. De Jouffroy quickly and cannily steered the boat ashore and then bowed to the cheering multitudes.

The marquis continued experimenting, but the French Academy of Sciences refused to recognize his achievement, ostensibly because the demonstration was not done in Paris but perhaps because of jealous rival inventors. The French Revolution soon followed, and though the nobleman kept his head, he never got his patent. De Jouffroy ended life discouraged and poor, dying of cholera in 1832 at age eighty.

American steamboat pioneer Robert Fulton, whose own experiments began not on the Hudson but the Seine, acknowledged de Jouffroy as the inventor who deserved the glory. —*RA*

Also July 15:

1954: Boeing 707 Makes First Flight
1999: U.S. Admits That Thousands of Aerospace Workers Were Poisoned by Beryllium

Also 1783:

June 4: Montgolfier Hot-Air Balloon Demo Reaches 6,000 Feet
June 22: Iceland's Laki Volcano Disrupts Europe's Economy

1867: Concrete Gets Some Positive Reinforcement

F. Joseph Monier patents a new construction material: reinforced concrete. It combines the compressive strength of ordinary concrete with the tensile strength of iron.

Ancient Romans perfected concrete, but the technology was lost with the fall of Rome. It did not reappear in the West until the seventeenth and eighteenth centuries. Builders soon found a variety of uses for concrete. It was relatively inexpensive—hey, it's mostly *sand*—and would assume the shape of whatever you poured it into. Solid concrete is not bad for holding things up: it's great under pressure. Pour a thick-walled foundation, and build a wood house on top of it. When the wood is gone, the concrete will still be there.

The problem was that it didn't stand up well to tension or shearing. Thin concrete walls crumbled, pillars buckled, and horizontal beams collapsed when loaded.

Enter Monier, a French gardener looking for a better flowerpot. Clay pots broke easily, and wood weathered poorly. He tried concrete, but the weight of the plant and soil stretched concrete pots and broke them. Why not embed an iron mesh in the concrete to strengthen it? *Et voilà!* Monier was soon using this reinforced concrete, or ferroconcrete, for beams and posts as well as pots.

Monier exhibited reinforced concrete at the 1867 Paris World's Fair. It was a hit. By 1875, a reinforced-concrete castle was built on the New York–Connecticut state line. Monier himself fabricated ferroconcrete wall panels for buildings, beams, and even complete bridges.

François Hennébique adapted the new material he'd seen at the Paris Exposition. He patented a construction method using hooked connections on the reinforcing bars. Hennébique built Europe's first multistory reinforced concrete building in 1898: a flour mill. Reinforced concrete would become ubiquitous in the twentieth century.—*RA*

Also July 16:

1862: Swift Sights Comet Swift-Tuttle Three Days Before Tuttle

1945: Trinity Test Blast Opens Atomic Age

1965: Mont Blanc Road Tunnel Links France, Italy

Also 1867:

June 25: Barbed Wire — the Beta Version (see page 178)

July 17
1902: An Invention to Beat the Heat, Humidity

With human comfort the last thing on his mind, a young mechanical engineer completes the schematic drawings for what will be the first successful air-conditioning system.

Willis Haviland Carrier was trying to help a Brooklyn printing company. Its paper was expanding and contracting with changing heat and humidity. Because ink was applied one color at a time, small changes made for misaligned, muddy illustrations.

Carrier devised the air conditioner (or, more accurately, humidity controller): Air was nozzle-sprayed over coils that were chilled with refrigeration coolant. The cold air was then expelled into a closed space using a fan, cooling the room and stabilizing the humidity.

It was just what the Brooklyn printer needed. The humidity problem vanished, and other companies began clamoring for Carrier's machine. Paper victories aside, the salubrious effect on *humans* was also recognized early on. Carrier Engineering went on to supply cooling systems to hotels, department stores, theaters, and, eventually, private homes. Early big-ticket customers included the U.S. Congress, the White House, and New York's Madison Square Garden.

The impact of air-conditioning can't be overstated. Sun Belt cities and other places where stifling-hot weather was a factor enjoyed an economic boom as people settled in large numbers, protected from the elements by Carrier's invention. That population shift in turn changed the political balance of the nation.

The chlorofluorocarbons used in air-conditioning have in recent years been blamed for the hole in Earth's ozone layer. And some people believe that sealed air-conditioned buildings with circulated forced air are breeding grounds for communicable diseases.

Carrier died in 1950, but his company lives on. There's nothing like a good air conditioner to beat the heat. — *TL*

Also July 17:

1938: Wrong Way Corrigan Flies from New York to Ireland Instead of California

1975: Apollo and Soyuz Spacecraft Dock in Orbit

Also 1902:

September 1: Pioneering Sci-Fi Movie *A Trip to the Moon* Premieres (see page 246)

1942: World's First Operational Jet Takes Wing

The third prototype of the Messerschmitt 262 fighter becomes the first nonexperimental jet plane when it takes to the skies over Bavaria at the height of World War II.

Engine problems, other teething difficulties, and political bungling delayed its debut as a combat aircraft until 1944, but when it arrived, the twin-jet Me-262 showed that with an experienced pilot at the controls, it was more than a match for the best Allied fighters, including Britain's own jet fighter, the Gloster Meteor.

In truth, the Me-262 should have been ready for front-line service much earlier. The original design—which, in the end, looked a lot like the finished product—existed as early as April 1939. But high costs and the belief of many high-ranking Luftwaffe officers that conventional aircraft could win the war prevented Germany from making the Me-262 a priority.

The first prototype flew in 1941, but the BMW-made turbojets weren't ready yet, so the first Me-262 went aloft equipped with 700-horsepower Jumo 210G piston engines.

Like the type XXI U-boat, the Me-262 appeared too late in the war to help Germany stave off defeat. History will remember it as the world's first operational jet plane, but the Me-262's true legacy is the influence it had on the design of a new generation of warplanes. — *TL*

Also July 18:

1876: Britain Appoints Royal Commission on Pollution; Environmental Regulations Ensue

Also 1942:

January 20: Nazi Wannsee Conference Unleashes "Final Solution," Genocide by Technology

June 4: Battle of Midway Heralds Dominance of Aircraft Carriers

August 11: Actress Hedy Lamarr, Pianist George Antheil Patent Stealthy Electronic Torpedo

October 29: 1,500-Mile Alaska Highway Completed in Eight Months as Wartime Necessity

December 1: Mandatory Gas Rationing, Lots of Whining

December 2: Nuclear Pile Gets Going Under Football Stadium (see page 338)

July 19
1961: Fasten Your Seat Belts, *By Love Possessed* Will Begin Shortly

Trans World Airlines becomes the first airline to offer regular in-flight movies.

The first film shown, on a flight from New York to Los Angeles, was *By Love Possessed,* starring Lana Turner. The screening was in the first-class cabin only and resulted from a eureka moment for the technology's mastermind, David Flexer.

Flexer, who owned a small chain of movie houses, told *Life* magazine he was on a transcontinental flight in 1956 when he realized, "Air travel is the most advanced form of transportation and the most boring."

There'd been some experimental in-flight movie screenings. In 1921, for example, attendees at a Chicago fair watched a promotional film while flying over the city in a small plane. But no airline regularly offered the service.

Flexer had to overcome some technical problems. He had to find a way to put full films onto single reels that wouldn't need switching. He also needed to build a projector tough enough to withstand turbulence but light enough to mount on a plane. His engineers told him it wasn't doable.

But after three years of development and $1 million of Flexer's own money, his team developed a horizontal design that projected large 16mm reels and weighed less than a hundred pounds. Flexer's new company, Inflight Motion Pictures, went looking for a partner.

TWA agreed to give Inflight a shot. Flexer and his team took a Boeing 707, fitted it with their equipment, and spent early 1961 flying around fixing bugs. When the airline started offering films, the response was extraordinary: fliers began paying huge fees to get into first class just to catch the show.

Before long, in-flight movies were everywhere, and in-flight entertainment now includes satellite radio, video on demand, and even wireless Internet.—*AW*

Also July 19:

1963: Cracking the 100-Kilometer Altitude Barrier ... in a Plane
1989: Human Heroics Overcome Aircraft Failure in Sioux City

Also 1961:

February 10: New Niagara Falls Hydroelectric Power Plant Starts
April 12: Soviets Orbit Gagarin, First Human in Space
May 25: JFK Vows to Put American on Moon by Decade's End
October 30: Soviets Detonate 100-Megaton Hydrogen Bomb

July 20
1969: One Small Step... One Giant Leap

NASA's Neil Armstrong becomes the first human to set foot on the lunar surface, realizing humanity's age-old dream. And effectively winning the space race for the United States, even though the Soviet Union had been first to land a spacecraft on the moon, in 1959.

Armstrong and fellow astronaut Buzz Aldrin left the Apollo 11 command module (piloted by Michael Collins) in orbit and performed a landing in the lunar module *Eagle*. At 4:18 p.m. EDT, Armstrong announced to a watching and waiting world, "The *Eagle* has landed."

Six and a half hours later, he stepped onto the powdery surface with the words that were heard as "That's one small step for man, one giant leap for mankind." In fact, he'd spoken the more logical "That's one small step for *a* man, one giant leap for mankind," but the transmission clipped an all-important article. Aldrin soon followed Armstrong down the ladder to become the second man on the moon.

The mission was by no means a slam dunk. There was real fear that once on the lunar surface, the astronauts might end up marooned and beyond rescue. In fact, President Richard Nixon had a condolence speech ready to go in the event things turned out badly.

But things went as planned, and Armstrong and Aldrin returned to the command module, leaving behind a plaque inscribed with the words: "Here men from the planet Earth first set foot upon the Moon, July 1969, A.D. We came in peace for all mankind."

Five more Apollo missions carried astronauts to the moon before the program ended, in 1972. (There would have been six, but Apollo 13 was a near disaster; see page 109.) The last human to leave his footprint on the moon was Apollo 17 commander Eugene Cernan, on December 14, 1972. — *TL*

Also July 20:

1960: First Polaris Missile Launched from Submerged Submarine

Also 1969:

February 9: Boeing 747 Jumbo Jet Makes First Test Flight
September 2: First U.S. ATM Starts Doling Out Dollars (see page 247)

July 21
1904: All Aboard for Siberia

The Trans-Siberian Railway is officially completed.

As you'd expect with a project of this size, complexity, and scope, *officially* is a relative term. Trains were already operating on parts of the line, while other sections weren't up and running for years.

The idea of rails connecting European Russia to Siberia and the remote Far East had been kicked around for years. Despite the pleas of Siberian provincial governors, the powers that be in St. Petersburg (the Russian capital at the time) were not enthusiastic. Czar Alexander II green-lighted the project in 1881 but was assassinated soon thereafter. Alexander III restarted it in 1886, but construction didn't begin until 1891.

It was one of the most ambitious engineering projects of all time. As with the transcontinental railroad in the United States (see page 132), the Trans-Siberian built from both ends toward an eventual meeting point. Thousands of workers hacked through dense forests, bridged fast-moving rivers, negotiated vast swamplands, dynamited the permafrost layer, and blasted tunnels through mountains.

Most construction lacked the benefit of heavy machinery. Shovels and picks, along with a little dynamite, made up the railroader's basic tool kit. Besides extreme cold, workers faced floods and landslides, armed bandits, cholera, anthrax... even an occasional tiger.

The Trans-Siberian hauled under steam. Conversion to electricity began in 1927, but the last steam engine wasn't retired until 1987. The railway opened Siberia to the rest of the country. People flowed east, and Siberian agricultural products flowed west.

The modern Trans-Siberian Railway is actually four different routes, although the name is usually associated with the Moscow–Vladivostok run: 6,000 miles, seven time zones, six days. The train to Pyongyang, North Korea, runs 6,400 miles and is the world's longest single continuous rail service. —*TL*

Also July 21:

1911: Birth of Marshall McLuhan, Mass
     Communications Theorist
1925: John Scopes Found Guilty of Teaching Evolution

Also 1904:

January 7: Wireless Distress Call CQD Precedes SOS
March 5: Physicist Nikola Tesla Expounds on Ball
     Lightning (see page 192)
November 16: Vacuum Tube Heralds Birth of Modern
     Electronics (see page 322)

July 22
1952: Genuine Crop-Circle Maker Patented

Frank Zybach patents the center-pivot irrigator. You've seen the results if you've flown over farmland: big green circles of irrigated land, making repeated dot patterns.

Zybach thought it inefficient to have workers connect pipes and sprinklers, disconnect them, move them to a different field, and then do it all over again. Why not have the equipment move itself? He built his first prototype within a year. It rotated around a center wellhead. Guy wires and support towers held sprinkler-fitted water pipes above the ground. The first towers moved on skids, but Zybach soon replaced those with wheels propelled by the irrigation water itself.

Center-pivot sprinkler irrigation in Nebraska.

Sales of the Zybach Self-Propelled Sprinkling Apparatus didn't take off, partly because Zybach kept making improvements before partner A. E. Trowbridge could sell the models they'd already manufactured. They sold the patent rights to farm-equipment manufacturer Valley Manufacturing in 1954. Valley improved the device to deliver variable amounts of water or apply fertilizer and pesticides automatically. End guns let water reach the dry corner patches. Business took off in the 1960s. The amount of land tended by each irrigation worker quadrupled from 400 acres to 1,600 acres.

More than a quarter million center-pivot irrigation systems now water fields around the world. Modern setups run in forward or reverse on rubber wheels driven by electric motors. Some systems use GPS and wireless to control water flow. They take directions from laptops and cellphones.

And those empty corners between the circles? New systems bury radio wires in the corners to signal the pivot arm to telescope the pipe outward toward the corner, then retract it again, following the border of the field. As that technology spreads, the circles you see from your jet window seat may become a thing of the past. — *RA*

Also July 22:

1933: Wiley Post Completes First Solo Flight Around the World

1962: $18.5 Million Mariner 1 Venus Probe Blown Up Due to Software Punctuation Error

Also 1952:

November 4: Univac Gets Election Right, but CBS Balks (see page 310)

July 23
1962: Telstar Provides First Transatlantic TV Link
1996: Stand By—High-Definition TV Is on the Air

A signal day in television history, so to speak.

1962: The Telstar 1 communications satellite relays the first transatlantic television signal in history.

Telstar was the product of an international collaboration to develop satellite communications. NASA, AT&T, and the French and British post offices were all involved, with AT&T's Bell Labs actually building it.

A spherical satellite festooned with solar panels and packed with transistors, Telstar used a helical antenna to receive microwave signals from one ground station and then amplified and rebroadcast them to another.

The satellite was launched July 10 aboard a NASA Delta rocket and placed in an orbit that provided a narrow twenty-minute transmission period every two and a half hours.

President John F. Kennedy was supposed to inaugurate this historic moment, but Telstar was ready before JFK was. To fill time, producers ran a Phillies-Cubs baseball game from Wrigley Field, so Ernie Banks may have been the first human image beamed across the Atlantic. — *TL*

1996: WRAL-HD becomes the first U.S. television station to broadcast a high-definition signal.

This milestone was the culmination of a twenty-year global initiative to improve the over-the-air TV signal that hadn't changed in four decades.

After working for a few years with an industry consortium called the Grand Alliance, the FCC granted the first HD license, just five weeks before the broadcast, to WRAL-TV in North Carolina's Raleigh-Durham-Fayetteville market. Technicians worked at a fever pitch to upgrade the station for the nation's first HD broadcast.

The only viewers were the two hundred–plus invited members of the media and the television industry at the WRAL studios and at an experimental station in Washington, DC.

HDTV sets didn't get into stores until 1998. And they cost a lot, considering that there was precious little HD programming to receive. — *JCA*

Also July 23:
1956: Bell X-2 Rocket Plane Sets Aircraft Speed Mark
1995: Inventors Hall of Fame Opens Doors

Also 1962:
August 5: First Quasar Discovered (see page 219)

Also 1996:
November 19: Canadian Bridge Crosses Eight Miles of Icy Ocean (see page 325)

July 24
1950: America Gets a Spaceport

Cape Canaveral, Florida, launches its first rocket.

Cape Canaveral was just an obscure spit of land when a U.S. Air Force committee recommended it for missile testing in 1948. The cape was actually the committee's second choice. A California site was rejected after Mexico refused to let rockets cross over Baja California. (That decision was probably influenced by a near miss when a wayward rocket from White Sands, New Mexico, crashed into a Juarez cemetery.)

The first rocket launch from Cape Canaveral.

The fact that the downrange trajectory of a rocket launched eastward was over the ocean was desirable. And Canaveral was closer than the California site to the equator, which made it easier to launch rockets. So, President Harry Truman inked legislation in 1949 establishing the Joint Long Range Proving Ground at Cape Canaveral.

The first rocket to lift off there was a Bumper V-2, modified from the World War II–era German V-2s that pounded London. The two-stage rocket—a V-2 booster topped by a WAC-Corporal second stage—was launched mainly to conduct atmospheric tests. A second Bumper went up three days later.

The Florida installation, which originally included four launchpads and a few buildings, expanded rapidly during the early 1950s. More land was acquired; Patrick Air Force Base was established nearby, and a dizzying array of military and NASA space projects soon followed.

Following President Kennedy's assassination, former first lady Jacqueline Kennedy suggested to President Lyndon Johnson that renaming the space center for JFK would be a fitting memorial. LBJ went her one better, and Cape Canaveral itself was renamed Cape Kennedy.

The new name held up until 1973, when the original name (which first appeared on a 1564 Spanish map) was restored, and the *facility* became the Kennedy Space Center. Which was all Jackie had ever asked for. —*TL*

Also July 24:

1911: Hiram Bingham "Discovers" Machu Picchu
1959: Moscow Exhibit of U.S. Kitchen Ignites Nixon-Khrushchev Debate on Technology

Also 1950:

January 22: Innovative Automaker Preston Tucker Acquitted of Financial Fraud Charges
October 1: BBC Airs First Live TV Broadcast from Airplane

July 25
Four Women Who Made a Difference

Science and technology are spheres of society where women are woefully underrepresented, but today offers a bountiful exception.

1865: "James Barry," the first woman physician of modern times, dies. She was compelled to disguise herself as a man in order to practice medicine. Barry finished medical school at age thirteen, already in disguise. Barry rose to the rank of inspector general in the army and also worked with the Royal Navy.

It was discovered at Barry's autopsy that he was really a she. (Or perhaps was intersex, with mixed male and female anatomy.)

1920: Rosalind Franklin, the unheralded codiscoverer of DNA, is born.

Three men shared the 1962 Nobel Prize for their research a decade earlier on the structure of DNA (see page 117). Rosalind Franklin, a chemist whose X-ray diffusion photographs of DNA molecules showed their essential structure and paved the way for the trio's work, received nothing.

Unhappily, Franklin died of cancer in 1958, only thirty-seven years old. This has been cited as the reason she was not included with the others: Nobel Prizes are not awarded posthumously.

1978: Louise Joy Brown, the world's first baby conceived in vitro, is born.

Louise's mother's fallopian tubes were blocked. Gynecologist Patrick Steptoe surgically removed an egg from one of her ovaries and fertilized it in a petri dish. (The term *test-tube baby* was a media invention.)

Steptoe implanted the embryo in Lesley Brown's uterus and hoped for the best. Louise Joy Brown was delivered by cesarean section and weighed five pounds, twelve ounces. In 2006, she gave birth to her own naturally conceived son.

1984: Cosmonaut Svetlana Savitskaya becomes the first woman to walk in space. She was serving as flight engineer aboard the Soyuz T-12 mission to the Salyut 7 space station. Her extravehicular activity came nineteen years after cosmonaut Alexei Leonov became the first person to leave an orbiting spacecraft, and she beat American astronaut Kathryn Sullivan out the door by three months.

Savitskaya was the second woman to go into space, preceded only by fellow cosmonaut Valentina Tereshkova. —*TL*

Also July 25:

1871: Perforated Toilet Paper Patented
1909: Bleriot Is First to Fly Plane Across English Channel
1959: First Hovercraft Flight Across English Channel
1965: Bob Dylan Plays Electric Guitar, Shocks Folkie Audience at Newport

July 26
1943: LA Gets First Big Smog

In the middle of World War II, Los Angeles residents believe the Japanese are attacking them with chemical weapons. A thick fog that makes people's eyes sting and their noses run takes hold of the city. Visibility is cut down to three city blocks.

As residents would later find out, the fog was not from an outside attacker but from their own vehicles and factories. Massive wartime immigration to a city built for cars had made LA the largest car market the industry had ever seen. But the influx of cars and factories—combined with the region's bowl-like topography, which trapped fumes—created trouble for Angelenos.

After the first big smog, public pressure temporarily shut down a Southern California Gas Company plant, but the smog episodes continued to get worse. The mayor announced the problem would be eliminated in four months. Ha! (Cough!)

The search for the culprit of the "gas attacks" wasn't found until the early 1950s. Caltech chemist Arie Haagen-Smit was the first to recognize ozone as smog's primary source. Ozone is created when partially unburned exhaust from automobiles and the hydrocarbons from oil refineries are hit by sunlight. Haagen-Smit also demonstrated that ozone was the cause of the bleach smell LA residents were reporting, as well as the source of their eye irritation and respiratory problems.

But LA built more and more freeways, and new industries came into town. The smog became commonplace, with dangerous levels as often as two hundred days a year. People got used to it. (Wheeze!)

It would take another twenty years for California to finally enact decisive standards to limit smog pollution from vehicles. The smog laws were ultimately mirrored across the country. —*JM*

Also July 26:
1989: First Indictment Under Computer Fraud Act

Also 1943:
October 19: Streptomycin Discovered (see page 294)

July 27
1866: Transatlantic Cable Connects Old World to New

After years of planning and development, and many snafus, the transatlantic cable is successfully laid and put into operation.

Though telegraphic communication was in its infancy (see page 173), the idea of a transatlantic cable had been proposed by many. It was wealthy New York merchant Cyrus Field who eventually arranged for the funding and saw it through.

Field came aboard in 1854, a year after the USS *Dolphin* completed a 1,600-mile sounding between Newfoundland and Ireland that revealed a smooth ocean-bottom plateau, ideal for laying cable. Field assembled investors, and the New York, Newfoundland, and London Telegraph Company was formed.

The cable was almost laid in 1855, but bad weather at sea and the refusal of a ship's captain to follow orders scuttled that attempt. Cable was laid across Cabot Strait in 1856, completing a linkup between New York and Newfoundland. The cable was a third of the way home.

The British and U.S. governments agreed to supply additional funding for the project, and in 1857 success was nearly at hand when the cable snapped in heavy seas. Hundreds of miles of cable were lost. That, coupled with a bank collapse in the United States, meant the project had to be shelved again.

A much-ballyhooed 1858 cable provided just a few weeks of weak, sporadic, fading service before failing entirely. More disruptions followed, including the U.S. Civil War, in which London sympathized with the Confederacy.

By 1866, however, the war was over, and the *Great Eastern*—the world's largest ship (see page 101)—was ready to lay cable. The shore end of the new cable was laid at Follhummerum Bay, Ireland, and the *Great Eastern* made a smooth passage to Heart's Content, Newfoundland.

Today's news in London was today's news in New York.—*TL*

Also July 27:

1888: First Electric Vehicle, a Tricycle, Drives
    Through Boston

Also 1866:

September 21: Birth of Science-Fiction Author
    H.G. Wells

July 28
1907: Tupperware's First Burp

Earl S. Tupper, inventor of the famous Tupperware "burping" plastic kitchenware, is born.

Young Tupper fancied himself an Edison (see page 211), Ford, or Leonardo, dreaming up hundreds of inventions: a better garter for stockings, a comb that clipped to a belt, permanent-press trousers, a fish-powered boat, a convertible top for car rumble seats, even a new method for emergency appendectomies.

Then he started Tupper Plastics, which made beads and cigarette containers. He tried using polyethylene (see page 88) after World War II, but he found DuPont's formulation too rigid for his big idea. After obtaining softer samples, he set to work on the Wonderbowl. The polyethylene bowl had an airtight, watertight cover: just lift the top a tad, burp out some air, and push down to seal. Tupper got the idea from the metal lids on paint cans.

He patented the Tupper seal in 1949. Tupper tried selling his wares in department stores, but they didn't fly off the shelves. Then a woman who was making lots of money selling Stanley Home Products at house parties realized she could make even more applying the home-party technique to Tupper products.

Brownie Wise started selling Tupperware in 1949, and by 1951 she was the company's general sales manager. Thousands of women across the country made a business out of Tupperware, and millions of customers bought it. Tupper was finally rich and famous, but he became jealous of Wise: she was the first woman to grace the cover of *Business Week*.

Tupper fired Wise, sold the company, divorced his wife, bought an island in Central America, and moved to Costa Rica to avoid U.S. taxes. He died in 1983, but Tupperware has stayed fresh, selling through Tupperware parties in over one hundred countries.

The secret of its success is simple: just keep a lid on it. —*RA*

Also July 28:
1858: India Starts Using Fingerprint ID
1948: German Chemical Plant Explosion Kills 200, Injures 3,800

Also 1907:
June 6: Persil, First Household Detergent, Hits the Market

July 29
1958: Ike Inks Space Law;
NASA Born in Wake of Russ Moon

President Dwight Eisenhower signs the National Aeronautics and Space Act, creating the National Aeronautics and Space Administration.

Americans' hope of being first in space had been shattered the previous October when the Soviet Union launched Sputnik, the first artificial satellite (see page 279). There was talk of a growing technology gap and fears in U.S. military circles that Soviet satellites could pinpoint targets for a nuclear-missile attack. The pressure intensified with a failed attempt to launch the navy's Vanguard TV3 satellite in December 1957. The army convinced the Pentagon to set Vanguard aside and bet the ranch on the army's still-untested Project Explorer.

The successful launch of Explorer 1 on January 31, 1958 (see page 31), and the subsequent launch of Vanguard 1 didn't stave off a comprehensive reorganization of the U.S. space program. Eisenhower directed science adviser James Killian to convene a committee and come up with a game plan.

The first step was to reinvigorate the National Advisory Committee for Aeronautics, a geeky and elitist civilian panel that had been around since 1915, by handing it all nonmilitary responsibilities connected to space exploration. As NACA's charter grew, the decision was made to expand it into a full-fledged government agency taking direct responsibility for the nation's space program.

President Eisenhower signed the legislation creating NASA on July 29, and it became an official agency October 1, with T. Keith Glennan as its first administrator. There were eight thousand employees (inherited from NACA), three research laboratories, and an annual budget of $89 million. That's about $700 million in 2011 money; compare that to its 2011 budget of $18 billion. —*TL*

Also July 29:
1994: Video-game Makers Propose Ratings Board to Congress

Also 1958:
September 12: Kilby Chips In, Integrates Circuit (see page 257)

July 30
1898: Car Ads Get Rolling

The Winton Motor Carriage Company places a magazine advertisement cajoling readers to "dispense with a horse." It's the earliest known automobile ad.

The ad ran in *Scientific American* because automobiles were specialized, high-tech geek toys then. Some still are.

Alexander Winton was a Scottish-immigrant bicycle maker who switched to building cars in 1896 and had the world's largest auto factory by 1900. Winton stopped producing cars in 1924 and started making stationary engines. The firm was bought by General Motors in 1930 and spun off in 2005 as Electro-Motive Diesel, still in business making engines and locomotives.

Did *Scientific American* and Winton Motor Carriage know what they were starting? Did they know we'd be assaulted by the likes of Cal Worthington and Jerry's Cherries? A world where there's no such thing as the soft sell and you feel like you got away fairly clean after paying only 2 percent more than list?

Most assuredly not. Alexander Winton simply wanted you to "save the expense, care and anxiety of keeping" a horse. Instead, you could drop a mere $1,000 on his car ($27,200 in 2012 money). The advertisement must have worked because Robert Allison of Port Carbon, Pennsylvania, bought a Winton after seeing the ad in *Scientific American*. Later that year, Winton sold a staggering twenty-one more vehicles, including one to James Ward Packard, who went on to found his own car company. So the next time you hurriedly reach for the mute button on your remote (see page 313) or grimace at the blaring full-page ad for the used-car dealer down the street, remember where it all started. —*TB*

Also July 30:

1869: First Oil Tanker, the *Charles,* Leaves Port
1935: Penguin Starts Publishing Literary Paperbacks
2003: Last Old-Style VW Beetle Manufactured

Also 1898:

December 21: Marie and Pierre Curie Discover Radium
   (see page 357)

July 31
1790: New Nation Issues First Patent

Samuel Hopkins of Vermont gets the very first U.S. patent.

Inventors used to get patents from state legislatures. But the new Constitution empowered Congress to "promote the progress of science and useful arts, by securing for limited times to authors and inventors the exclusive right to their respective writings and discoveries."

Congress ignored a bunch of individual petitions in 1789 but, at President Washington's urging, passed a law the next year. It directed applicants to file a petition with the secretary of state. He, with the secretary of war and the attorney general, would grant a patent to any "invention or discovery sufficiently useful and important."

The application required a written description, drawings, and, if practical, a model, with enough detail to allow a skilled workman to make and use the invention. Thus, the public would benefit when the patent expired after fourteen years. Fees ran from four to five dollars (roughly a hundred dollars in today's money). You had to pay a copying charge of ten cents per every hundred words, because patents for carbon paper, photography (see page 233), photocopying (see page 297), and optical scanning were yet to be applied for. (Today's patent fees start at ninety-five dollars but rapidly jump into the thousands.)

Hopkins's patent covered a method of making potash and pearl ash (potassium carbonate), which were important ingredients for making glass, china, soap, and fertilizer. It was issued in New York City (the national capital then), complete with President Washington's signature. The board issued only two more patents in 1790: one for manufacturing candles, and the other for flour-milling machinery. The potash process also received the first Canadian patent, in 1791.

Although Hopkins received the first U.S. patent, it wasn't patent no. 1. That's because the government issued 9,957 patents before starting a numbering system, in 1836. —*RA*

Also July 31:

1964: Ranger 7 Lunar Probe Takes Video Until Last
　　Second Before Impact

Also 1790:

March 12: Birth of J. F. Daniell, Inventor of Practical
　　Battery
November 17: Birth of Mathematician-Astronomer-
　　Physicist A. F. Möbius

August 1
1949: FCC Gets In On Cable TV

A secretary at the Federal Communications Commission sends a letter to cable pioneer Ed Parsons in Astoria, Oregon, asking him to explain his community-antenna television system. It's the first known FCC involvement in cable TV.

Parsons was a radio engineer and station owner. He and his wife saw television demonstrated at a 1947 broadcasters' convention. Mrs. Parsons wanted the new-fangled gizmo, and Ed complied when Seattle's KRSC-TV announced plans to go on the air.

Parsons had to figure out a way to receive TV signals in Astoria from Seattle, 120 miles away. He rigged a large antenna atop the Astoria Hotel and ran a coaxial cable across the street to his apartment. He got it working on November 25, 1948. Problem solved.

Problem created: the Parsonses' apartment was the only place in town that could pick up the signal from Seattle, and soon friends, neighbors, and total strangers were crowding into their living room to watch the modern marvel. Parsons was nearly driven out of his home: "People would drive for hundreds of miles to see television... you couldn't tell them no."

He ran another cable from the hotel roof to a TV set in the lobby, but the crowds there interfered with the hotel guests. So Parsons began running cable to other people's homes. Mrs. Parsons got her living room back. Problem solved; industry born.

Parsons caught the attention of the FCC, but Arkansas's Jim Davidson had set up a similar system about the same time. It carried the Tennessee vs. Mississippi college football game on November 13, 1948.

An FCC attorney eventually concluded that CATV was subject to FCC jurisdiction, but the commission didn't agree. It was 1965 before the FCC decided to regulate cable TV.—*RA*

Also August 1:

1774: Joseph Priestly Isolates "Dephlogisticated Air" AKA Oxygen

Also 1949:

January 10: RCA Releases First 45 RPM Records
March 2: B-50 Propeller Plane Flies Around World Nonstop
June 8: George Orwell's Techno-Dystopia *1984* Published

August 2
1873: San Francisco's First Cable Car Conquers Nob Hill

Andrew Hallidie tests the first cable car in San Francisco.

Hallidie is said to have conceived his idea in 1869 while watching a team of horses being whipped as they struggled to pull a car up wet cobblestones on Nob Hill. They slipped and were dragged to their deaths.

It so happened that Hallidie's father held the British patent for wire-rope cable, and when the son came to the gold-rush fields, he put the cable to use hauling ore-laden cars from mines. So it wasn't too much of a stretch for him to envision horse-less cable cars carrying passengers up San Francisco's steep slopes. He formed the Clay Street Hill Railroad and got a contract to build a line up Nob Hill. It was an unqualified success, and cable cars have been operating in San Francisco ever since.

A number of cable car lines and companies sprang up in the wake of Hallidie's success. Prior to the great earthquake and fire of 1906, fifty-three miles of cable car track stretched to virtually every corner of town. A vast underground pulley system moves a cable at a steady 9½ miles an hour, pulling the cable car along the tracks. The operator, or gripman, uses a lever to grip the cable moving beneath the street.

The system was nearly dismantled around 1950 by politicians who, some say, were in the pockets of vehicle, oil, and tire companies that wanted buses to replace cable. Friedel Klussmann founded the Citizens' Committee to Save the Cable Cars and battled City Hall at every turn. The issue finally made it to the ballot, and San Franciscans voted overwhelmingly to keep their cable cars.

The system's been rebuilt—the cars refurbished but still propelled by underground cables. Those cables, however, are now pulled by electric motors instead of coal-fired steam engines.—*TL*

Also August 2:
1790: First U.S. Census Begins

Also 1873:
April 29: Railroads Lock and Load (see page 121)
May 20: Riveted-Pocket Blue Jeans Patented

216

August 3
1977: The TRS-80 Is Bad, and That Ain't Trash Talk

Tandy Corporation announces it will manufacture the first mass-produced personal computer. The TRS-80 would become an early rock star in the PC era and give Radio Shack bragging rights as the "biggest name in little computers."

The TRS-80—lovingly called the Trash 80—was a desktop machine, woefully underpowered by today's standards: 4 KB of RAM (expandable to 16 KB), a twelve-inch monitor, a cassette-based data recorder, and BASIC interpreter. And, oh yes, blackjack and backgammon.

But the TRS-80 Model I was a staggering success at the time, since your only other choice was building your own computer from a kit or buying something for thousands of dollars. The Model I cost $600 ($2,270 in 2012 coin), and all you had to do was plug in...three separate plugs.

This was the dawn of the personal-computing age. HP, IBM, and Wang had personal computers out in the early 1970s. The Apple I and II were on the market. Commodore introduced the first self-contained computer earlier in 1977, but the top-selling Commodore 64 didn't arrive until '82.

Radio Shack thought it would sell six hundred to a thousand TRS-80s the first year. After all, it was the most expensive item Radio Shack stocked. A buyer had to preorder and put down a hundred-dollar deposit. But stores took ten thousand orders the first *month*. Instead of the weeks Tandy thought it would need, it took months to deliver the first machines. The company sold more than two hundred thousand units in four years.

It would not be long before Tandy made laptops that ran for hours on four AA batteries and had crisp black-and-white displays and enough expandable RAM and memory to satisfy caveman-era road warriors. —*JCA*

Also August 3:

1492: Columbus Sets Out to Discover...a Trade Route
1803: Birth of Joseph Paxton, Crystal Palace Architect

Also 1977:

May 31: Trans-Alaska Pipeline a Source of Oil...and
    Worry (see page 153)
June 4: VHS Comes to America
June 5: Apple II Personal Computer Goes on Sale

August 4
1693: Dom Pérignon "Drinks the Stars"

Champagne is said to have been invented on this day by Dom Pierre Pérignon, a French monk. The legend almost certainly isn't true.

Because Dom Pérignon lived at the Benedictine abbey in Hautvillers at the time of his "invention," the village in France's Champagne region is generally regarded as the birthplace of the bubbly. But like many historical claims, this account of the night they invented champagne appears more fanciful than factual. Sparkling wine certainly existed before Dom Pérignon arrived on the scene, although it would be unrecognizable today as champagne. But whether he invented the champagne method single-handedly is doubtful.

This much *is* true: he made an enormous contribution by discovering the technique that finally produced a successful white wine from red-wine grapes, something vintners had been trying to accomplish for years. That was probably the major step toward the development of modern champagne.

Even his famous quote "Come quickly, I am drinking the stars" appears to be apocryphal. The evocative declaration is plastered on a champagne ad dating from the 1880s but is hard to trace back any further, and certainly not to the seventeenth century. In any case, Dom Pérignon spent a lot of time trying to get the bubbles *out*. Bubbles were produced by refermentation, an undesirable process and a major problem for winemakers then. Generations of bon vivants, from Madame Pompadour and Napoléon to Dorothy Parker and Noël Coward, were no doubt grateful that he failed.

What is fair to say is that Dom Pérignon established the principles of modern champagne-making that are still in use. And because champagne, of all beverages, deserves a birthday, a myth is good for a smile. *Joyeux anniversaire! À votre santé!* — *TL*

Also August 4:
1922: Phone Service Shuts Down for One Minute for Bell's Funeral
1977: U.S. Department of Energy Established

Also 1693:
March 24: Birth of John Harrison, Inventor of Chronometer to Find Longitude at Sea

August 5
1962: First Quasar Discovered

A nearly botched astronomical observation leads to the identification of the first known quasi-stellar object, or quasar.

In 1960, astronomers Allan Sandage and Thomas Matthews had discovered a blue star-like object that sent out particularly intense radio waves. British radio astronomer Cyril Hazard booked time on Australia's 210-foot Parkes Radio Telescope to observe the source. Hazard was better at astronomy than geography: he took the wrong train from Sydney and missed the whole show.

Fortunately, science is a group endeavor, and observatory chief John Bolton and his staff took over. The radio source in question was low on the horizon, so they cut down some trees and even removed the giant radio telescope's safety bolts to aim the dish low enough to make the observation. The object of their attention, 3C 273, was emitting huge amounts of energy with an unusual, never-before-seen spectrum. Maarten Schmidt used the Hale optical telescope at California's Mount Palomar to puzzle it out the following year. He saw a visible jet rising from the optically faint object. Like a hydrogen jet.

Schmidt discovered a hydrogen spectrum that was shifted an astonishing 16 percent toward the red—which is why it hadn't been recognized. But a redshift of that magnitude meant it was moving away from Earth at almost one-sixth the speed of light and was three billion light-years away: farther and brighter than most known galaxies.

Astronomers soon were calling this new class of objects—of which 3C 273 is the granddaddy—quasi-stellar radio sources. A NASA scientist trimmed that to quasars. Today they're called quasi-stellar objects, or QSOs, because not all emit radio waves. After half a century of research, new discoveries about QSOs seem to raise new questions even as they answer old ones. —*RA*

Also August 5:

1963: U.S., Soviets, Britain Sign Nuclear Test Ban Treaty

Also 1962:

August 22: First Commercial Nuclear Ship, NS *Savannah*, Docks (see page 236)

August 6
1945: U.S. Drops Atomic Bomb on Hiroshima

The U.S. becomes the only country ever to use an atomic weapon in warfare, obliterating the Japanese city of Hiroshima and instantly killing 70,000 people. Thousands more will die later from radiation poisoning.

Several countries, including Germany, had worked on atomic weapons, but none matched the U.S. in the resources, energy, or scientific manpower devoted to making the bomb a reality. The discovery of nuclear fission in a Berlin laboratory in 1938 alarmed émigré scientists who had come to the United States to escape Nazism.

Fearing Germany might actually develop this ultimate weapon, they appealed to President Franklin D. Roosevelt to make nuclear research a high priority. Though skeptical at first, FDR was persuaded, and a civilian-military committee soon launched the Manhattan Project.

Development followed two paths, one strategy using uranium235, and the other plutonium. Both bombs were ultimately built: the uranium-based Little Boy was dropped on Hiroshima. Three days later, the plutonium-based Fat Man laid waste to the port city of Nagasaki with the ultimate loss of 140,000 lives. Japan surrendered, ending World War II.

The atomic scientists had favored demonstrating their weapon to the Japanese in an isolated area, but military and political planners argued the shock of total destruction would have more impact. The United States maintains that the decision to drop the bomb was made primarily to avoid the necessity of invading Japan, which would have resulted in enormous casualties on both sides. But many historians believe the real U.S. motive was to end the war before the Soviets could become involved, thus denying them a postwar stake in the Pacific.

Whatever the reasons, the bombs *were* dropped, and most Manhattan Project scientists later expressed remorse for what they had wrought. —*TL*

Also August 6:

1890: First Electrocution of Condemned Criminal
1932: Germany Opens First Autobahn, from Bonn to Cologne
1997: Surprise! Microsoft Invests $150 Million in Apple to Rescue Struggling Company

Also 1945:

April 14: Tweaky Toilet Costs Skipper His Sub (see page 106)
April 30: Germany's New-Generation U-Boat Is Too Little, Too Late

August 7
1944: Still a Few Bugs in the System

Harvard and IBM dedicate the Mark I computer. The IBM Automatic Sequence Controlled Calculator produced reliable results and could run 24-7.

The Mark I was a monster: fifty-five feet long and eight feet high. It weighed five tons and contained 760,000 components, including 3,000 rotating counter wheels and 1,400 rotary-dial switches, along with an assortment of shafts, clutches, and electromagnetic relays, all linked together by five hundred miles of wire. Its clickety-clack sounded like a "roomful of ladies knitting."

You fed it instructions on paper tape and loaded the data on punch cards. It could perform operations only in the precise linear order it received instructions. The tape could not run backward. The Mark I handled twenty-three-decimal-place numbers: adding, subtracting, multiplying, dividing. It also ran subroutines for logarithms and trigonometry. It was slow, taking three to five seconds for a multiplication. It gave results through teletypewriter and punch card.

Mathematician Grace Hopper joined Howard Aiken's team at Harvard and was instrumental in keeping the Mark I running. She repaired it one day by removing a moth that had fouled its electromechanical innards, becoming the first person to literally debug a computer. (The term *debug* was already in use for other mechanical devices, and Hopper noted the irony.)

Hopper and Aiken used the Mark I to help the U.S. Navy produce tables for aiming artillery shells and bombs during the final year of World War II. The electromagnetic machine remained in use until 1959, left in the dust by true electronic computers using vacuum tubes, then transistors (see page 359).

Aiken went on to build the Mark II, in 1947, the same year he founded the Harvard Computation Laboratory and predicted, "Only six electronic digital computers would be required to satisfy the computing needs of the entire United States." — *RA*

Also August 7:

1955: Sony Transistor Radios Go on Sale
1991: Ladies and Gentlemen, the World Wide Web
(see page 99)

Also 1944:

May 14: Birth of George Lucas, Pop-Culture Jedi
(see page 136)
June 6: Prefabricated Artificial Harbors Assist D-Day
Invasion of Normandy

August 8
1876: Run This Off on the Mimeo

Thomas Edison receives a patent for the mimeograph. It will dominate the world of small-pressrun publication for a century.

The machines were everywhere—in schools, offices, and the military. If you needed just a few copies of a document, you used carbon paper. If you needed thousands, you sent it to a print shop. But if you needed something in between, say thirty copies for a classroom handout or five hundred to a thousand for a church bulletin or incendiary revolutionary poster, you had the mimeograph.

The process is simple: cut a stencil, push ink (with its characteristic smell) through the holes onto paper, and repeat. The business model is also simple: sell the machine, sell the stencils, sell the ink—maybe even compete to sell the paper.

Edison's 1876 patent covered a flatbed duplicating press and an electric pen for cutting stencils. Chicago inventor Albert Blake Dick improved the stencils while experimenting with wax paper, and he merged his efforts with Edison's. The A.B. Dick Company released the Model 0 Flatbed Duplicator in 1887.

If you didn't want to use the electric pen, you could try cutting a stencil with one of those newfangled typewriters (see page 176). But hand drawing of stencils for diagrams and math formulas persisted well into the twentieth century.

Later models replaced Edison's original flatbed press and hand roller for the ink with a rotating cylinder and an automatic feed from the ink reservoir. Deluxe models included an electric motor, but cheaper ones you cranked by hand.

The A.B. Dick Company believes almost every U.S. military personnel order of World War II was run off on one of its machines. But the Xerox was around the corner (see page 297). —*RA*

Also August 8:

1576: Cornerstone Laid for Tycho Brahe's Danish Observatory

Also 1876:

March 10: "Mr. Watson, Come Here ..." (see page 71)
July 18: Britain Appoints Royal Commission on Pollution; Environmental Regulations Ensue

August 9
1854: Thoreau Warns, "The Railroad Rides on Us"

Henry David Thoreau publishes *Walden; or, Life in the Woods*. It offers deep insights into not just nature and humanity's place in it but how that relationship was being degraded by the Industrial Revolution. It remains a trenchant criticism of the excesses of technology.

Thoreau lived in a cabin on the shore of Walden Pond in Concord, Massachusetts, from 1845 to 1847. He was not a hermit, though he valued solitude and wrote movingly about it. A man is rich, Thoreau wrote, "in proportion to the number of things which he can afford to let alone."

Thoreau had little regard for the newfangled telegraph (see page 173): "Maine and Texas, it may be, have nothing important to communicate." And he calculated a trip by rail (see page 177) was a bad bargain, because it would take less time to walk somewhere than to earn the train fare. He also criticized the worsening economic conditions of the laborers who built modern industry: "We do not ride on the railroad; it rides upon us." Thoreau was contemptuous of fashion:

> [B]eware of all enterprises that require new clothes...I cannot believe that our factory system is the best mode by which men may get clothing. The... principal object is, not that mankind may be well and honestly clad, but, unquestionably, that corporations may be enriched.

Recognized today as a literary giant and a founder of environmentalism, he's less well known as an engineer. Thoreau modernized his family's pencil-making business, improving the product and streamlining the factory's production process. A contradiction? Perhaps we should heed Thoreau's own dictum:

> If a man does not keep pace with his companions, perhaps it is because he hears a different drummer. Let him step to the music which he hears, however measured or far away.

—RA

Also August 9:
1995: Netscape IPO Wows Wall Street

Also 1854:
September 8: Pump Shutdown Stops London Cholera Outbreak (see page 253)

August 10
1909: Leo Fender and the Heart of Rock 'n' Roll

Clarence "Leo" Fender is born.

The designer-engineer-inventor would found the Fender Electric Instrument Manufacturing Company, which created the first wave of commercially successful electric guitars, basses, and amplifiers. Fender didn't invent the electric guitar. The first real innovations came with two patents for magnetic pickups in 1937: one to Gibson's Guy Hart (see page 196), the other to Rickenbacker's George Beauchamp, on August 10—Fender's twenty-eighth birthday.

Fender's earliest commercial successes were in amplifiers, but his first hit was the Fender Precision Bass. The P-Bass, introduced in 1951, was meant for players in jazz and dance bands who needed more volume. The Broadcaster and Telecaster also hit big. The Telecaster is still in production today.

But nothing endured, influenced, or captured the imagination like Fender's Stratocaster. It's more than just an electric guitar. The Stratocaster is a hallmark of modern art, synonymous with its players and their music: Jimi Hendrix, Eric Clapton, Bonnie Raitt, David Gilmour, Buddy Guy, Jeff Beck, Buddy Holly, Stevie Ray Vaughan, Dick Dale.

The Stratocaster's slopes and swooshes perfectly connected the empty spaces between the dawning space age, the sleek modernism of Calder's floating sculptures, the flamboyance and heat of a California hot rod, the raw lust of the sexual revolution, and the angry rebellion of youth. It came in colors like orange sunburst, pearl white, and the ever-popular candy-apple red. It begged to be touched; it practically screamed *trouble*.

All of Fender's guitars were noted for their clean, bright sound. They were versatile and durable. The carved slabs of wood with bolt-on necks were made to be abused. They're prized by both players and collectors.

Fender sold his company to CBS in 1965. He died in 1991, never having learned how to play the guitar. —*MC*

Also August 10:

Also 1909:

August 11
1903: Instant Coffee, a Mixed Blessing

A Japanese chemist living in Chicago receives the first U.S. patent for instant coffee. Hundreds of millions of caffeine addicts will rue the day, but others have no grounds for complaint.

Water-soluble instant coffees first saw light of cup in Britain in 1771. But the product had a short shelf life and went rancid fast, so production had a short historical life and went away fast. An American attempt in 1853 was followed by a pre–Civil War cake of powdered coffee. Same deal: it wouldn't keep, so it didn't sell.

Satori Kato was the Japanese inventor of soluble tea. A U.S. coffee importer and a coffee roaster asked him to apply his dehydration method to coffee. With the help of an American chemist, Kato solved the rancidity problem by removing nonvolatile coffee fat. The Kato Coffee Company distributed free samples at the Pan American Exposition in Buffalo, New York, where an assassin's bullet, not instant coffee, killed President William McKinley in September 1901.

But the Buffalo product did not take wing. Another inventor, George C.L. Washington, started mass-producing Red E Coffee in 1909, supplying instant-coffee rations to U.S. troops during World War I. Nescafé, which started marketing an improved product in 1938, supplied the U.S. military in World War II. Say what you will about the taste, instant coffee is fast, easy, and doesn't take much equipment: big advantages to troops in the field.

The 1950s mania for quick, convenient, and modern gave instant coffee a heyday, but the Europeanization of American coffee tastes in the 1980s and 1990s created disdain for the product. Still, you'll find instant coffee on supermarket shelves everywhere, including the premium instant coffee so popular in, yup, Europe. — *RA*

Also August 11:

1874: Parmelee Patents First Practical Automatic Fire Sprinklers
1942: Actress Hedy Lamarr, Pianist George Antheil Patent Stealthy Electronic Torpedo
1978: First Atlantic Balloon Crossing Takes Off

Also 1903:

November 24: Starting Your Car Gets a Little Bit Easier (see page 330)

August 12
1888: Road Trip! Berta Takes the Benz

Berta Benz, wife of inventor Karl Benz, takes her husband's car on the first documented automobile road trip.

The trip included the first road repairs, the first automotive-marketing stunt, the first case of a wife borrowing her husband's car without asking, and the first motor-vehicle violation of highway laws.

Karl Benz had been working on the car since 1884. But Berta was fed up with Karl's patient, conservative ways: she decided to sensationalize his invention. She recruited their two sons—Eugen, fifteen, and Richard, thirteen—to help push the car up the steeper hills, where the 0.88-horsepower engine wouldn't make the grade. They departed at five o'clock on a peaceful August morning (some sources say August 5). The sixty-mile trip from Mannheim to Pforzheim was a challenge when there were no gas stations or repair shops. Benz and her boys had to follow wagon tracks, which was illegal for automobiles. Top speed might have reached 10 miles an hour—downhill.

They had to stop every fifteen to twenty miles to buy fuel: benzine (aka ligroin), usually available from pharmacies. The cooling system was open, so water boiled off, which meant they had to refill the water tanks about every twelve miles. They occasionally asked local shoemakers to replace the leather brake-shoe linings. Frau Benz even plugged a leaky valve with a garter and used hairpins to free clogged valves. Despite the challenges, the Benzes reached Pforzheim that night. They stayed a few days so the Benz boys could show off their father's car, and then returned by a different, more level route.

Karl Benz exhibited a fancier version of the car in September, winning a gold medal and lavish press attention. Berta's publicity stunt had worked.

—*DT (who's married to Berta's great-great-grandniece)*

Also August 12:

1883: Extinction — Last Quagga, a Zebralike
   Species, Dies at Amsterdam Zoo
1981: IBM Unveils 5150 Personal Computer

Also 1888:

August 14: I Sing the Meter Electric (see page 228)

August 13
1913: Great Alloyed Victory for Stainless Steel

English metallurgist Harry Brearley casts a steel alloy resistant to acidity and weathering. He's often credited as stainless steel's inventor, but there are more metallurgists than metals in this story.

Chemists knew by the 1820s that iron-chromium alloys resisted some acids. Two Englishmen patented an acid-resistant steel in 1872. Then, in 1875, a French researcher named Brustlein detailed the importance of low carbon content, and the race was on...slowly. Many attempts produced many failures over the next twenty years.

Hans Goldschmidt of Germany broke the logjam in 1895 by producing carbon-free chromium. French metallurgist Leon Guillet forged ahead with work on iron-nickel-chromium alloys, but he seemingly ignored their resistance to corrosion. In Germany, P. Monnartz and W. Borchers discovered in 1911 that having 10.5 percent chromium seriously increased steel's resistance to corrosion. Englishman Harry Brearley started working on a project in 1912 for a small-arms manufacturer that wanted to prevent its rifle barrels from eroding away quickly from the heat and friction of gunshot. Brearley needed to etch his steel-alloy samples to examine their granular structure under the microscope, but his high-chromium samples resisted acid-etching.

After trying various combinations of chromium and carbon, he made a new alloy, on August 13, 1913, containing 12.8 percent chromium and 0.24 percent carbon. It resisted nitric acid, lemon juice, and vinegar. He took his discovery of "rustless steel" to knife-making firm R.F. Mosley, where manager Ernest Stuart renamed it stainless steel.

Other teams were simultaneously working on corrosion-resistant steel: Germany's Krupp Iron Works; Elwood Haynes and two other Americans; Max Mauermann of Poland; and a Swedish claimant. Brearley, however, formulated the first alloy called stainless steel, and he sharply recognized its potential as a blade (see page 165). — *RA*

Also August 13:

1902: Birth of Felix Wankel, Rotary Engine Inventor
2004: Adam Curry Launches First Podcast, *Daily Source Code*

Also 1913:

February 18: Radiochemist Frederick Soddy Coins *Isotope*
November 18: U.S. Pilot Loops the Loop

August 14
1888: I Sing the Meter Electric

Oliver B. Shallenberger patents an electric meter.

When Thomas Edison started selling electricity for illumination, in 1882, he charged per lamp. He soon replaced that with a complicated chemical meter. It was an electrolytic jar, and workers had to remove the electrodes every month and weigh them to see how much zinc had been transferred from one plate to the other: messy, inefficient, and inaccurate.

Elihu Thomson devised a walking-beam meter in 1888. The Rube Goldberg apparatus used an in-line heating element to warm a small alcohol-filled bottle on a seesaw lever. The alcohol evaporated and flowed into a matching bottle on the other side. When there was more alcohol in the opposite bottle, it would sink and start heating up to reverse the process. Each time the bottles rocked, they ticked off a notch on the meter. Not exactly a robust design.

Shallenberger was working on a new arc lamp at Westinghouse, in 1888, when a spring fell out and landed on a ledge inside the lamp. Before an assistant could reach in to replace it, Shallenberger noticed the spring had rotated. He soon determined that the lamp's rotating electric fields had caused the spring to turn. Shallenberger decided to use the effect to turn wheels in a meter to measure electrical charge... and built an alternating-current ampere-hour meter in just three weeks.

The Shallenberger meter was a key part of George Westinghouse's AC electrical system (see page 29). It went into commercial use within months. Shallenberger's meter and Thomson's 1888 commutator watt-hour meter fell by the wayside in the late 1890s when the induction watt-hour meter came into general use, where it remained for more than a century.

There's no free lunch. You'll get an electric bill. —*RA*

Also August 14:

1877: Otto Patents Four-Stroke Internal Combustion Engine
1901: Whitehead Purportedly Flies One and a Half Miles on Birdlike Monoplane

Also 1888:

January 27: National Geographic Society Gets Going (see page 27)

August 15
1877: "Hello. Can You Hear Me Now?"

Thomas Edison suggests using the word hello as a telephone greeting. The salutation is now used the world over, in one form or another.

Alexander Graham Bell's famous first words, in 1876—"Mr. Watson, come here"—were delivered with no greeting at all (see page 71). Bell soon proposed *ahoy, ahoy,* the age-old seafarer's hail. And *ahoy* was the first greeting used, until Edison suggested *hello.*

The phone was conceived of as a business machine to connect two offices with a permanently open line. Some people toyed with the idea of an alarm bell at each end to alert one office that the other wanted to speak. On August 15, 1877, Edison wrote to a friend who was setting up a phone system in Pittsburgh: "I don't think we shall need a call bell as Hello! can be heard 10 to 20 feet away. What do you think?"

Edison did not coin the word. *Halloo* and variants had been used for ages to urge on hunting hounds and to shout to people at a distance. Edison was tinkering with a prototype phonograph in 1877 and used a shouted *halloo!* for testing. Early gramophones and telephones alike had pretty low signal-to-noise ratios.

Hello itself turns up in a number of places prior to 1877, including Mark Twain's *Roughing It,* published four years before Bell called Watson. Earlier references also exist, one dating back to at least 1826. In any case, *hello* caught on quickly and entered the dictionary in 1883. And when was the last time you had to look up *that* spelling?—*TL*

Also August 15:

1914: Panama Canal Opens
1977: False Alarm —"Big Ear" Observatory Picks Up
   Seemingly Intelligent Signal from Space

Also 1877:

March 4: The Microphone Sounds Much Better
   (see page 65)

229

August 16
1960: *Geronimo-o-o-o-o-o-o!*

U.S. Air Force captain Joe Kittinger parachutes from an open gondola tethered to a helium balloon from an altitude of 102,800 feet—more than nineteen miles. It's the highest-altitude jump ever made.

Kittinger's jump was no stunt but rather part of Project Excelsior, established by the U.S. Air Force to study the problems presented to pilots by high-altitude escapes. He'd already made two jumps—from 76,400 feet and 74,600 feet—before strapping it on for the big one.

The statistics from the August 16 jump even now make for eye-opening reading. Wearing a pressurized suit, Kittinger:

- Jumped at 102,800 feet
- Was in free fall for four and a half minutes and was clocked at 714 mph, faster than the speed of sound, before deploying his chute at 17,500 feet
- Experienced temperatures down to minus 94 degrees Fahrenheit
- Landed on the New Mexico desert after thirteen minutes, forty-five seconds
- Set several long-lasting records—highest balloon ascent, highest parachute jump, longest free fall, and fastest free-fall speed

But Kittinger, then twenty-nine years old, wasn't through with high places or risk-taking. Accompanied by a U.S. Navy civilian astronomer in 1962, Kittinger piloted a balloon into the upper atmosphere to 82,200 feet so a high-powered telescope could take a closer look at deep space. During the Vietnam War, Kittinger returned to combat flying and commanded the 555th "Triple Nickel" Tactical Fighter Squadron, shooting down a MiG-21 in a dogfight. He was subsequently shot down himself and spent eleven months as a prisoner of war at the infamous Hanoi Hilton.—*TL*

(As this book went to press, Felix Baumgartner was planning a record-breaking 120,000-foot jump. Kittinger was advising him.)

Also August 16:

1884: Birth of Sci-Fi Publisher Hugo Gernsback
1899: Bunsen Burns Out — Death of Chemist-Inventor Robert Bunsen

Also 1960:

May 10: USS *Triton* Completes First Submerged Circumnavigation
August 20: Back from Space, with Tails Wagging (see page 234)

1859: U.S. Airmail Carried by Balloon

U.S. mail is carried by air for the first time.

In the town square of Lafayette, Indiana, the postmaster handed a bag with 123 postmarked letters to balloon enthusiast and well-known aeronaut John Wise. Destination of the balloon *Jupiter* and its precious cargo: New York City.

Delivering letters by air had been attempted before: carrier pigeons have an ancient heritage. And a 1785 balloon flight from Dover, England, to Calais, France, had carried mail. George Washington even gave it a shot (see page 64). But this attempt was an official U.S. first. Wise, fifty-one, also hoped to set a record for the longest balloon flight. He took off at 2:00 p.m. But the weather wasn't on his side. The wind was blowing southwest, not east. Still, he went up to 14,000 feet. Five hours later, Wise gave up and landed

A crowd gathers in Lafayette, Indiana, to watch the mail balloon ascend.

in Crawfordsville, Indiana—just thirty miles away. The mail had gone partway by air but was then ignominiously put on a train to New York City to assure the swift completion of its appointed round. The *Lafayette Daily Courier* mocked the flight as "trans-county-nental."

Wise tried again a month later and made it as far as Henderson, New York— flying nearly eight hundred miles. A storm forced a crash landing, and he lost the mail in the crash. Wise went on to fly observation balloons for the Union army during the Civil War. He died in 1879, when a storm pushed his balloon into Lake Michigan.

A piece of mail from Wise's first flight has survived and now resides at the Smithsonian National Postal Museum in Washington. The first airmail flight in an airplane took place when three letters traveled a few miles between Petaluma and Santa Rosa, California, in February 1911.—*PG*

Also August 17:

1807: Fulton's Folly Steams Up the Hudson (see page 198)

2000: Half of U.S. Households Have Internet Access

Also 1859:

April 25: Big Dig Starts for Suez Canal

May 15: Birth of Pierre Curie, Radium's Codiscoverer (see page 357)

August 27: First Commercially Viable U.S. Oil Well

August 18
1947: Birth of the Cool (Company, That Is)

Eight years after it was founded, Hewlett-Packard incorporates. The tiny garage in Palo Alto, California, where the company originated is now regarded as the birthplace of Silicon Valley.

Bill Hewlett and Dave Packard met as engineering students at Stanford in the early 1930s and cemented their lifelong friendship during a postgraduation camping trip. With $500 in cash ($8,100 in 2012 money) borrowed from Stanford professor Fred Terman, plus a used Sears, Roebuck drill press, Hewlett-Packard swung into action.

The company's first product, released in 1938, was an audio oscillator for testing sound equipment. When the Walt Disney Company bought eight of them to develop the technically advanced movie *Fantasia,* HP was off and running. Packard and Hewlett made the partnership permanent January 1, 1939. The name order was determined by the gracious winner of a coin toss. Packard won but preferred the way Hewlett-Packard sounded, so they went with that.

Their electronics products were first-rate and eagerly embraced. Want became need with the coming of World War II, and HP quickly grew, moving out of Packard's garage in 1940. HP also innovated humanistic management techniques. The firm valued its workers and their happiness. Packard's 11 Simple Rules became famous: the open-management style was the prototype for how many tech companies, particularly in Silicon Valley, would operate decades later.

Packard was especially interested in fostering a nonauthoritarian working atmosphere. Once, when an engineer circumvented an order to stop work on an oscilloscope that later became a commercial success, Packard gave him a special medal inscribed *Extraordinary Contempt and Defiance Beyond the Usual Call of Engineering.* — *TL*

Also August 18:

1868: Helium Discovered During Total Solar Eclipse
1990: Psychologist B.F. Skinner Dies. Is His Coffin a
    Skinner Box?

Also 1947:

October 3: Birth of Palomar's "Giant Eye"
    (see page 278)

August 19
1839: Photography Goes Open Source

Louis Daguerre reveals the secrets of making daguerreotypes to a waiting world. Photography is a hit.

Using chemistry to make images with light wasn't new. Doing it fast was. J.N. Nièpce created a rough image using silver salts and a camera obscura, or "dark box," in 1816. The image faded quickly. A decade later, Nièpce shot the first permanent photographic image. It took eight hours.

Commercial artist Daguerre was using a camera obscura to sketch outlines for his giant dioramas. But why not create images directly with the camera? Daguerre wrote to Nièpce,

An early daguerreotype camera.

and they became partners. After Nièpce died, his son Isidore labored on. But it was Daguerre's advances with silver-plated copper sheets, iodine, and mercury that cut exposure time down to minutes and created permanent, positive images on metal.

Daguerre's process intrigued the French Academy of Sciences' François Arago. The academy sponsored the first public display of daguerreotypes on January 9, 1839. They created a sensation. Arago used the buzz to lobby the French parliament to grant pensions to the inventors so they could publish the process, and France would "then nobly give to the whole world this discovery." Daguerre got 6,000 francs (about $30,000 in today's money) annually and Nièpce 4,000 francs annually. Daguerre and Arago published the technical details on August 19.

Opticians and chemists in Paris immediately sold out of the supplies needed to make cameras and plates. Improvements followed within weeks. Daguerre's instruction manual was translated into a dozen languages within months. Nobody wanted oil portraits; they all wanted daguerreotypes. Studios opened all over Paris. Daguerreotypemania spread across Europe and then to America.

Innovation continued. A decade after Daguerre died, in 1851, the daguerreotype was supplanted by the albumen print, which produced photographs on paper. —*RA*

Also August 19:

1856: Gail Borden Patents Condensed Milk (see page 315)

1887: Dmitri Mendeleyev Observes Solar Eclipse from Balloon

Also 1839:

February 24: William Otis Patents Steam Shovel

September 9: John Herschel Makes First Permanent Photograph on Glass Plate

August 20
1960: Back from Space, with Tails Wagging

Belka and Strelka, stray mutts impressed into the Soviet space program, become the first living creatures to return alive from an orbital flight.

The Soviets had been using dogs for experimental high-altitude flights long before Belka ("squirrel") and Strelka ("little arrow"). Two other dogs died in July when their rocket exploded on launch. Laika, the first animal to orbit Earth, went into space on a one-way ticket in 1957; the Soviets knew she wasn't coming back.

The U.S. put the first animals into rocket-powered missiles to test the effects of rapid acceleration and weightlessness on a living organism: Albert, a rhesus monkey, was launched into suborbital space in 1948 aboard a V-2 Blossom rocket. He did not survive.

In the early 1950s, the Soviets succeeded in bringing their dogs home safely. One plucky pooch, Snowflake, made six separate flights in 1959 and 1960. The dogs went through rigorous training. Because the dog would be immobilized in a safety module inside the capsule, it was confined in gradually smaller boxes for days on end and trained to sit still for long periods. The animal grew accustomed to wearing a space suit and was placed in flight simulators and centrifuges that prepared it for space flight.

Belka and Strelka were accompanied on Sputnik 5 by forty mice, a couple of rats, and some plants. All were unharmed by their sixteen-orbit, one-day voyage. Strelka later became a mother, and one of her puppies was given to President Kennedy's daughter, Caroline. When Caroline's dog produced her own litter, JFK called them pupniks.

Eight months after Belka and Strelka's journey, cosmonaut Yury Gagarin became the first human to orbit Earth and return safely. — *TL*

Also August 20:

1831: Birth of Geologist Edward Suess, Coiner of the Word *Biosphere*
1953: Soviets Say, "We've Got the H-Bomb Too"

Also 1960:

January 23: Journey to the Deepest Place on Earth (see page 23)

August 21
1993: Mars Probe Disappears, Never to Be Found

NASA loses contact with the Mars Observer, a scientific probe sent to study the geology and climate of the red planet. Contact with the craft is never reestablished, and the reasons for its disappearance remain something of a mystery.

The Mars Observer, launched on September 25, 1992, was supposed to be the first of a three-probe series for NASA's Planetary Observer project—the others were going to go to Mercury and the moon—but the Mars failure resulted in scrubbing the other two missions.

Contact was lost with the Mars Observer only three days before it was to enter its Martian orbit. A Naval Research Laboratory investigative panel concluded the most likely cause of failure was inadvertent mixing of gases during pressurization of the fuel tanks. That probably ruptured the tubing, and the release of fluids either sent the spacecraft into a catastrophic spin or damaged vital electric circuits, or perhaps did both.

The spacecraft was supposed to determine the elemental makeup of the Martian surface, study the planet's magnetic and gravitational fields, establish the nature of its atmospheric circulation, and evaluate the topography, but all of that would have to be left to future missions, and the Mars Observer goes into NASA's annals as a major failure. The project cost $980 million (about $1.5 billion these days). If anything at all was gained from it, it was that the scientific instruments developed for Observer were used by subsequent orbiters. — *TL*

Also August 21:

1986: Cameroon's Volcanic Lake Nyos Explodes, Killing 1,700
1989: Voyager 2 Reaches Neptune's Moon Triton

Also 1993:

June 24: Severe Tire Damage Performs First Live Internet Concert
November 3: Leon Theremin Fades Out
(see page 309)

August 22
1962: First Nuke-Powered Cargo Ship Docks

NS *Savannah*, the world's first nuclear-powered cargo-passenger ship, completes its maiden voyage.

The NS *Savannah*.

To a world terrified by the prospect of nuclear war, the *Savannah* was meant to demonstrate the peaceful use of nuclear power. President Dwight Eisenhower conceived the idea as part of his Atoms for Peace program in 1955, when the U.S. and Soviets were routinely testing increasingly powerful nuclear weapons.

Worldwide, just four nuclear-powered merchant ships were eventually built.

The *Savannah*, named for the first steamship to cross the Atlantic Ocean, in 1819, was a showcase. It had a sleek, modernistic design that wasn't really compatible with stowing large amounts of cargo, a fact that would eventually shorten its career. Passenger accommodation was comparable to many conventional liners' of the day: thirty air-conditioned staterooms, dining room for a hundred people, swimming pool, library, and convertible lounge-cinema.

But the heart of the *Savannah* was its propulsion system. The Babcock and Wilcox nuclear reactor drove *Savannah*'s two steam-turbine engines cheaply and efficiently. But it wasn't economical enough to offset the tight forward cargo area and other deficiencies that made the ship expensive to operate. Its tapered bow not only put its cargo capacity well below competing vessels' but also made loading difficult, especially as ports automated (see page 118).

The *Savannah* also required a large crew, 124, all of whom needed additional training to work with the propulsion system. The Maritime Administration, *Savannah*'s owner, leased it out for cargo-passenger service, but the ship never turned a profit. It spent most of the 1970s tied up in Galveston, Texas. Following the ship's decommissioning, the nuclear fuel was removed.

The *Savannah* has been designated a National Historic Landmark, and the Maritime Administration hopes to see it converted into a floating museum. It's waiting in Baltimore. — *TL*

Also August 22:

565: Irish Monk Saint Columba Sights Loch Ness Monster

1939: Julius Kahn Patents Disposable Aerosol Spray Can

Also 1962:

February 20: John Glenn First U.S. Astronaut to Orbit Earth (see page 51)

August 23
1868: Birth of a Web Visionary

Paul Otlet is born. The information-science pioneer will dream of a global interlinked "web" of documents. The Belgian bibliographer's sprawling card catalog and decimal classification system proved woefully inadequate.

Some historians see in Otlet's work a prototype of the World Wide Web and the hyperlink. It was one of the first known attempts to provide a framework for connecting all recorded culture with flexible links that could rapidly lead researchers from one document to another, perhaps revealing new connections. Anticipating postmodern literary theory, Otlet posited that documents

The telegraph room at the original Mundaneum in Brussels, Belgium.

have meaning not as individual texts but only in relationship to one another.

Building on the work of Carl Linnaeus (father of binary genus-species classification and inventor of the three-by-five index card) and Melvil Dewey (of the Dewey decimal system), Otlet created the universal decimal classification. It lives on in some European libraries. He got some Belgian government support and by the late 1930s had amassed a catalog of fifteen million index cards and tons of documents. He piled these into a building he christened the Mundaneum, or city of knowledge.

Otlet was lionized as a visionary, but he had to scale back when he lost funding. The Nazis destroyed much of his archive during the 1940 invasion of Belgium. He died four years later, a disappointed man. But Otlet has won renewed recognition for his early contributions to information theory. Researchers have restored what remains of his work at a museum in Mons, Belgium. Scholars study his legacy, and his theoretical writings about information science have been recently reprinted.

In a 2008 *New York Times* article, *Wired*'s Kevin Kelly compared Otlet's work to a steampunk version of the Internet. The Mundaneum now seems a kind of Ozymandian empire of the intellect: a colossal ruin, and a failure. But a glorious one.—*EH*

Also August 23:

1899: First Ship-to-Shore Wireless Signal to a U.S. Station

1977: Pedal-Powered *Gossamer Condor* Flies One-Mile Course

Also 1868:

August 18: Helium Discovered During Solar Eclipse

August 24
2006: Pluto Deplanetized

Pluto, once the ninth planet from the sun, is downgraded to "dwarf planet." The solar system now has only eight planets, officially speaking.

You'd think that discovering a planet would grant you immortality, but astronomer Clyde Tombaugh may have been de-immortalized when Pluto was deplanetized. He discovered Pluto in 1930 by identifying a moving object against a background of stationary stars in sky photographs taken over several nights. The planet was officially named Pluto, after the Roman god of the underworld, on May 1, 1930.

Astronomers had trouble accepting Pluto as a full-fledged planet. Size (smaller than Earth's moon) is one issue, and its orbit around the sun doesn't match those of the eight "classical" planets. Astronomers also began discovering similar bodies near Pluto, diluting its cachet even more.

So, in 2006 the International Astronomical Union redefined *planet* thus:

A celestial body that (a) is in orbit around the sun, (b) has sufficient mass for its self-gravity to overcome rigid body forces so that it assumes a hydrostatic equilibrium (nearly round) shape, and (c) has cleared the neighborhood around its orbit.

Because Pluto crosses Neptune's orbit, it has not cleared the neighborhood, making it (along with Ceres, Eris, Haumea, and Makemake) a dwarf planet—which the IAU defined as (a) and (b) above, but "(c) has not cleared the neighborhood around its orbit, and (d) is not a satellite." (Our solar system has seven moons bigger than Pluto: Io, Europa, Ganymede, Callisto, Titan, Triton, and Earth's own moon.)

A lot of people reacted to the demotion like our solar system was losing a favorite kid brother. Pluto proponents still rail about its relegation to an inferior category. —*DC, TL*

Also August 24:

79 CE: Vesuvius Buries Pompeii, Herculaneum
1995: Say Hello to Windows 95

Also 2006:

January 26: Western Union Ends Telegram Service
March 21: Twitter Takes Flight (see page 82)

August 25
1973: More Than One Way to Slice a CAT

The CT scan goes into use in the United States.

Originally known as a CAT scan—for computed (or computerized) axial tomography, or computer-aided (or assisted) tomography—it uses a series of X-rays to create sequential images of slices of body tissue. Those can be integrated into a 3-D X-ray, so doctors know the precise position of diseased or otherwise abnormal tissue.

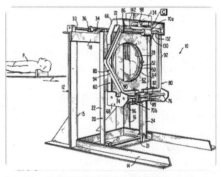

Patent drawing for the first whole-body CT scan.

Tufts University physics professor Allan Cormack theorized in the 1960s that you could take X-rays from varying angles; account for differences in the density of bone, muscle, and organs; and program a computer to assemble 3-D images.

English electrical engineer Godfrey Hounsfield was working on similar research at EMI, funded by massive profits from the Beatles' 1960s hits on EMI. Hounsfield began testing a CT brain scanner in 1971, sometimes carrying bulls' brains across London on public transit. His prototype took five minutes for the scan, and two and a half hours for the computer-processed image. The first EMI production model took four minutes to scan, and its Data General Nova minicomputer needed seven minutes per picture.

Meanwhile, in 1973, dentist-physicist Robert Ledley built a whole-body scanner at Georgetown University. It saved its first life while still in development: a pediatric neurosurgeon used it while Ledley was away for the weekend. Minnesota's Mayo Clinic and Boston's Mass. General also began using CT in 1973.

Hounsfield and Cormack shared the 1979 Nobel Prize in Physiology or Medicine. Ledley was inducted into the National Inventors Hall of Fame. CT scanners today are faster—four to ten images a second. Instead of taking discrete, individual slices as images, they use spiral, or helical, tomography, like a virtual HoneyBaked ham.

That's a lot of progress in 39 years...which is 273 in cat years.
—RA

Also August 25:

1958: Birth of Ramen Instant Noodles

Also 1973:

December 28: Endangered Species Get a Helping Hand (see page 364)

August 26
1346: First Cannon Fired in Battle, Maybe

Genoese mercenaries fighting under Philip VI of France are unpleasantly surprised when they come under cannon fire.

The Battle of Crécy, during the Hundred Years' War, may have been the first use of cannon on the battlefield. Like many claims that reach us through the mists of time, this one is subject to dispute. Arab historian Ahmad al-Hassan says the Mamluks employed the "first cannon in history" against the Mongols in 1260. It may depend on how you define *cannon*.

The French may have used their own version against England's Edward III at Cambrai, seven years before running into him again at Crécy. The actual damage caused by Edward's primitive cannon was negligible: the projectile was merely a stone that had been rounded into something resembling a cannonball. But the guns flashed and roared, and these soldiers had seen nothing like it before.

Despite the psychological shock, the cannon weren't decisive. The Genoese crossbowmen were too far away, and their arrows fell short. The English longbowmen shot farther and rained down destruction on the enemy. Then came the chaos and carnage of knights clashing in close-quarter combat. Crécy was a decisive English victory. The French suffered upwards of 30,000 casualties.

Although its battlefield debut was underwhelming, the cannon was here to stay. Within a few decades, most major combatants had powder-and-shot cannon. The French fired hundred-pound stone balls during a siege in 1375, Balkan gunners fired on Venetian ships in 1378, and the Ottoman Turks reportedly used cannon at Kosovo in 1389.

From Crécy onward, the cannon, along with handguns and rifles, would continue evolving, as man worked diligently to perfect a deadly art. — *TL*

Also August 26:

1883: Krakatau Eruption Begins; Volcano Affects
   Global Climate for Five Years
1939: Reds, Dodgers Split Doubleheader in First Major
   League Baseball Telecast

Also Fourteenth Century:

March 20, 1345: Triple Conjunction of Saturn, Jupiter,
   Mars Blamed for Black Death
January 9, 1349: Basel, Switzerland, Blames Black
   Death on Its Jews, Burns Them All

August 27
2003: The Lights Will Stay On in Fairbanks

Fairbanks is connected to the world's largest storage battery, built to provide Alaska's second biggest city with an uninterrupted power supply.

Fairbanks's remote location and subarctic climate made supplying reliable power to the city of 32,000 difficult. In deep winter, the temperature in Fairbanks is almost constantly subzero, dropping as low as minus 60 degrees Fahrenheit. The situation was complicated by the fact that Alaska isn't connected to the power grid that keeps the Lower 48 humming. As a result, Fairbanks used to experience a serious "cascading" blackout every two or three years, along with a number of smaller failures every month. It was necessary to devise another source of reliable local power.

The answer turned out to be a massive battery, the largest ever built, which now sits in a warehouse on the outskirts of Fairbanks. According to ABB Communications Services, the power-components specialist that built it, the battery can generate up to forty megawatts of power—enough to keep 12,000 people supplied with electricity—for seven minutes. That's long enough to fire up the city's backup diesel generators and restore the power supply. The battery energy-storage system, or BESS, which cost $35 million, contains 13,760 nickel-cadmium cells weighing a total of 1,400 tons and covering more than 10,000 square feet.

BESS is controlled by a Pentium PC–based platform programmed to provide all the essential services, including a complicated temperature-control system designed to withstand the rigors of the Alaskan winter.

In its first two years of operation, BESS reportedly prevented at least eighty-one power failures, an average of more than three per month. In a hostile environment like the area around Fairbanks, that can mean the difference between life and death.—*TL*

Also August 27:

1859: First Commercially Viable U.S. Oil Well
1874: Birth of Ammonia-Process Chemist Carl Bosch
1989: Brits Launch Direct-to-Home TV Satellite

Also 2003:

April 28: Apple Opens iTunes Music Store
July 30: Last Old-Style VW Beetle Manufactured
October 15: China Launches Astronaut Yang Liwei
    into Space

August 28
1963: Road to Redmond Walks on Water

The world's longest floating bridge opens. The Evergreen Point Floating Bridge connects Seattle with the east side of Lake Washington.

Pontoon bridges have been around since ancient times. Lash some boats together side by side, lay planks across them, and you've got a serviceable bridge. Armies love 'em because they're quick to build. For a large, permanent bridge, the concept is scalable, with difficulty. However, if you need to cross a deep body of water that has a soft bed, other designs might not be feasible. And state engineers had already built the shorter Lake Washington Floating Bridge, in 1940.

Starting in 1960, construction crews ashore built thirty-three hollow concrete boxes, about sixteen feet high and the length of a football field. These huge pontoons were floated and then towed into position, where they were linked by thick steel cables to anchors that hold them in place. At 7,578 feet, the floating bridge is essentially a 1.42-mile barge with a road on top of it.

That road links Seattle with Bellevue and Redmond, where a certain software company (see page 96) is headquartered. Seattle's tech-driven growth has put a huge load on the bridge. Designed to carry 65,000 vehicles daily, it now carries 115,000. Wear and tear, exacerbated by storm damage, has led to costly repairs.

Highway planners say if the bridge sank, the commute between Seattle and Redmond would jump from thirty-three minutes to fifty-five. But retrofitting it to current seismic and safety standards would be more expensive than simply starting over. So they're building a new floating bridge just north of the old. It'll have six lanes (plus a bike and pedestrian path) instead of four, cost almost $5 billion, and is scheduled to open in 2014.

They should call it Evergreen 2-Pont-0. — *RA*

Also August 28:

1845: *Scientific American* Publishes First Issue
1988: Ramstein Air Show Disaster Kills Seventy,
   Injures Hundreds

Also 1963:

November 9: Mining and Train Disasters Kill 600 in
   Japan
November 22: Zapruder Films JFK Assassination
   (see page 328)

1965: Long-Distance Calling... *Very* Long Distance

An astronaut in space holds a conversation with an aquanaut underwater, marking another milestone in human communication.

Astronaut Gordon Cooper, orbiting the earth with Pete Conrad in Gemini 5, hooked up by radiotelephone with an old pal, astronaut-turned-aquanaut Scott Carpenter, who was living and working 205 feet under the Pacific Ocean near La Jolla, California, aboard Sealab II.

Sealab II.

The two men had known each other since 1959, when they were among NASA's original seven Project Mercury astronauts. Carpenter, a former navy pilot, had already been in space, the solo astronaut on a mistake-plagued, three-orbit flight aboard *Aurora 7* that resulted in his effectively being grounded. He then joined the navy's Sealab II project as a training officer.

Cooper and Conrad, meanwhile, were nearing the end of an eight-day orbital mission to test human endurance in space. Eight days would be the time needed to travel to the moon and back (see page 203).

The radio hookup was partly a gimmick, taking advantage of Carpenter's astronaut status to publicize the Sealab II project. But it was also a method of testing the effectiveness of an underwater electronics lab installed aboard the submersible. Nor was it the only long-distance call made from Sealab II. The aquanauts also spoke with President Lyndon Johnson at the White House and with Jacques Cousteau's (see page 164) Conshelf III team conducting a similar underwater-habitat test in the Mediterranean.

Carpenter re-created his seabed-to-space call in 1995, chatting with astronauts aboard the space shuttle *Endeavour* while staying at Jules' Undersea Lodge off Key Largo, Florida. —*TL*

Also August 29:

1866: World's First Mountain Cog Railway Demonstrated on Mount Washington, New Hampshire

1949: First Soviet Atomic Test Stuns West

Also 1965:

April 19: How Do You Like It? Moore, Moore, Moore (see page 111)

July 14: Mariner 4 Takes First Close-Up Photos of Mars

August 30
1885: Daimler Gives World First True Motorcycle

Gottlieb Daimler patents what is generally considered to be the first true motorcycle.

The Daimler riding car.

Daimler, the automotive pioneer usually associated with major improvements to the development of successful internal combustion engines (and, subsequently, the first automobile), staked his claim of priority in the two-wheeler world a year before developing his famous auto.

However, the idea of a motor-driven two-wheeled vehicle did not originate with Daimler, nor was his the first such contraption to see the road. Sylvester Roper, who spent the U.S. Civil War working in a Union armory, built a primitive motorcycle as early as 1867. Roper's supporters—and he has more than a few—argue that he should be credited with building the world's first motorcycle.

What gives credibility to Daimler's claim of developing the first true motorcycle is the fact that it was gasoline-driven. Roper's post–Civil War hog, with a tiny two-cylinder engine, was powered by steam. Daimler's motorcycle was essentially a wooden bicycle frame (see page 48) with foot pedals removed, powered by a one-cylinder four-stroke internal-combustion engine of the type patented by Nicolaus August Otto eight years earlier. It may have also included a spray-type carburetor (then under development for use in the Daimler automobile that appeared in 1886).

Daimler's son Adolf took the two-wheeled riding car out for a two-mile test run between Cannstatt and Untertürkheim (near Stuttgart) in November 1885. Two miles! What a trip it must have been.—*TL*

Also August 30:

1954: President Eisenhower Signs Atomic Energy Act
1963: Telegraph Hotline Links Kremlin, Pentagon to Avoid Accidental War

Also 1885:

July 6: Rabies Vaccine Saves Boy — and Pasteur
September 5: Pay at the Pump (see page 250)

August 31
1920: News Radio Makes News

A Detroit station airs what's considered the first radio news broadcast: coverage of primary elections.

Radio's commercial prospects were still unappreciated, as wireless was considered primarily a "narrowcast" point-to-point medium (see page 348). The few radio shows were mostly music, created and operated by radio-set manufacturers to drum up business: "software" driving sales of "hardware." The *Detroit News* knew radio could also outrace newspapers in getting news to the public and could threaten their market dominance. It wanted in.

In a move that foreshadowed the computer age, the paper hired a teenager to build its setup and explain it to staff. It even instructed underage radio pioneer Michael Lyons to obtain government permission for the station in his own name, even though the station was conceived of, owned, and operated by the *Detroit News* and assembled in the newspaper building itself.

Lyons got permission to broadcast on August 20. He played music for ten days to work out the kinks. Station 8MK was poised to make history. The next day's *Detroit News* trumpeted the achievement:

> The sending of the election returns by The Detroit News' radiophone Tuesday night was fraught with romance and must go down in the history of man's conquest of the elements as a gigantic step in his progress. In the four hours that the apparatus, set up in an out-of-the-way corner of The News Building, was hissing and whirring its message into space, few realized that a dream and a prediction had come true. The news of the world was being given forth through this invisible trumpet to the waiting crowds in the unseen market place.

Station 8MK is now WWJ: all-news radio. You can listen to it live, on the Internet, 24-7. — *JCA*

Also August 31:
1909: First Chemotherapy Drug Treats Syphilis
1968: First Simultaneous Multi-Organ Transplant —
One Donor, Four Recipients

Also 1920:
September 29: Ready-Made Radio Receivers Go on
Sale (see page 274)
November 2: Gaston Chevrolet Dies in Race Crash

September 1
1902: First Sci-Fi Flick Explores the Moon

A *Trip to the Moon,* probably the first science-fiction movie, debuts in France.

Le Voyage dans la Lune created a pop-culture image: the man in the moon's disgruntled face sporting an oversize space capsule stuck in his right eye.

It's obviously not to scale, and it's somewhat gruesome when you consider the pain a celestial body would experience with a heavy chunk of metal forcibly injected into its eye socket. It could be legally considered satellite abuse. If it weren't science *fiction.*

Conceived by, produced by, and starring special-effects pioneer Georges Méliès and inspired by the words of Jules Verne, the black-and-white silent film tells the fanciful story of a group of dandy astronomers who journey to the moon via a rocket shot from a cannon.

After their projectile arrives, just south of the moon's frontal lobe, the explorers venture forth in waistcoats and leggings into a wondrous landscape filled with craters, volcanic geysers, and spiraling rock formations. After descending into a crater, the astronomers discover primitive, contortionist moon people. They immediately kill the first beings they encounter and then must flee an enraged mob of spear-bearing moon folk to return to Earth.

Méliès is the grand-old-man character in the Oscar-winning 2011 film *Hugo.* He was genuinely ahead of his time in his mastery of still-developing film techniques, a pioneer of cinematic art direction and visual effects. Referred to by contemporaries as the Cinemagician, Méliès developed the use of substitute edits, dissolves, time-lapse photography, and multiple exposures. He also produced some of the world's first color movies (see page 191) by hand-painting his frames.

Méliès produced an astounding 550 films in his lifetime. Most were early horror, fantasy, or sci-fi efforts—making Méliès the father of three genres.—*JSL*

Also September 1:

1939: Germany's *Wehrmacht* Puts the *Blitz* in *Blitzkrieg*

1974: Jet Flies New York to London in Under Two Hours

Also 1902:

May 17: Ancient Antikythera Calculating Mechanism Discovered (see page 139)

September 2
1969: First U.S. ATM Starts Doling Out Dollars

Six weeks after landing men on the moon (see page 203), Americans take another giant leap with the nation's first cash-spewing automated teller machine.

The Docuteller machine was installed in a wall of the Chemical Bank in Rockville Centre, New York. It marked the first time Americans could withdraw cash with reusable, magnetically coded cards. A bank advertisement announced "On Sept. 2, our bank will open at 9:00 and never close again!"

Don Wetzel, an executive at Docutel, a Dallas company that developed automated baggage-handling equipment, is generally credited as coming up with the idea for the modern ATM. His inspiration came while he stood in a bank line. Previous automated bank machines had allowed customers to make transactions only after purchasing a single-use voucher or card from a teller.

The new device was the first in the United States to dispense cash using a mag-stripe card that didn't require teller intervention. But the mag-stripe machines couldn't receive deposits or transfer money between accounts until the 1971 Total Teller. The ATM freed customers from the tyranny of banking hours, giving them access to dough 24-7. Because early ATMs were offline, a bank couldn't check a customer's balance to see if there was enough money to cover a withdrawal. So the banks imposed a $150 daily limit for ATM withdrawals.

Bank execs initially worried that customers wouldn't like the machines or that reducing face-to-face interaction would make it harder to sell them other services. But customers embraced ATMs, which eventually let banks cut costs, lay off tellers, and, of course, charge outrageous user fees.

But with the machines' ubiquity came security issues: hackers and scammers have kept banks on the defensive with ever more sophisticated schemes to steal cash through ATMs. Automated holdups, you might say. — *KZ*

Also September 2:

1985: Hey, Everyone, We Found the *Titanic*
1993: Space Race Ends as U.S., Russia Sign Pact to
      Cooperate in Space

Also 1969:

March 2: Concorde Takes to the Skies
April 7: Birth of the Internet (see page 99)

September 3
1976: Viking 2 Lands on Mars

Viking 2 lands on Mars and begins transmitting pictures and soil analyses.

Viking 2 took this self-portrait on Mars.

The Viking mission went to Mars to look for signs of life, study the soil and atmosphere, and take pictures. Viking 1 and Viking 2 descended to Mars within six weeks of each other. Viking 2's landing was more dramatic than NASA might have hoped: as the lander separated and began to descend, the orbiter's stabilization system went awry, blacking out for almost an hour.

The Viking landers were each approximately ten feet across and seven feet tall; they weighed roughly 1,270 pounds. Their descent engines were designed to disperse exhaust, in order to minimize disturbance of the landing sites. But a radar miscalculation caused Viking 2's engines to fire briefly just before landing, cracking the planet's surface.

Viking 2 performed a raft of experiments on the planet's soil, seismology, atmosphere, meteorology, and possible biology. The Viking orbiters mapped nearly all Mars's surface and observed Martian seasons. Neither Viking lander found significant amounts of water or ice, but both gave important information on Mars's iron-rich clay and Martian wind and climate. Viking 2's seismometer may even have recorded a Mars-quake, and both landers observed dust storms — local and global.

The spacecraft were rated for ninety days' work, but they kept going and going. Viking 2's orbiter sprang a leak and was shut down in 1978. The lander's batteries died in 1980. (Viking 1's lander worked until 1982.)

Recent analyses of crater impacts have shown that if Viking 2 had dug just three or four inches deeper, it would have discovered ice deposits, confirming the existence of large amounts of water on the planet decades sooner. — *JBJ*

Also September 3:

1803: Dalton Introduces Atomic Symbols
1925: *Shenandoah* Crash a Harbinger of Dirigibles' Grim Future
1935: Campbell Shatters 300 MPH Barrier at Bonneville

Also 1976:

March 24: President Ford Orders Swine Flu Shots for All
April 1: Jobs, Wozniak, Wayne Found Apple Computer
September 17: Prototype *Enterprise* Rollout Launches Shuttle Era

September 4
1957: The Short, Unhappy Life of the Edsel

Ford Motor Company introduces its newest make, the Edsel.

In an industry celebrated for its spectacular failures, the Edsel still takes the cake. Although as mechanically sound as other Ford products, the car was criticized from day one for being ugly, expensive, and overhyped.

The 1958 Edsel was intended to be an intermediate-level brand that bridged the gap between the cheaper Fords and pricier Mercurys and Lincolns. The most affordable Edsel (the Ranger) cost seventy bucks less than Ford's top-end Fairlane, while the most expensive (the Citation) cost more than a Mercury Montclair. Ford later cited the pricing as a big reason the car failed. And sales weren't helped by rolling the Edsel out just as a recession started. But there was more.

The Edsel (named for Henry Ford's son) was the subject of an intense marketing blitz while still on the drawing board. The company promised something revolutionary, baited the hook, and then failed to deliver. The Edsel was just another sedan on the basic Ford chassis. An ugly sedan. The superhype that had generated so much anticipation boomeranged almost immediately. Automotive writers roundly trashed the Edsel, even comparing the oval-shaped vertical grille to female genitalia—racy stuff for 1957.

During the Edsel's first year, four models were produced and barely more than 63,000 sold in the United States. Ford cut back to two models in 1959, but sales dropped further. On November 19, 1959, the company threw in the towel on the Edsel.

The Edsel today is a prized collector's item, fetching as much as $200,000 for a rare 1960 convertible. Another victim of this historic automotive fiasco was the name Edsel itself. Although never a particularly popular boy's name (rising to four hundredth on the 1927 list), Edsel has now almost entirely vanished.—*TL*

Also September 4:

1882: Edison Opens First Central Electric Power
Station in U.S., on Pearl Street in Manhattan
1888: George Eastman Patents First Roll-Film Camera

Also 1957:

October 1: Thalidomide Cures Morning Sickness,
But ... (see page 276)

September 5
1885: Pay at the Pump

Sylvanus Bowser delivers the first gasoline pump.

People in those days bought cooking and lamp fuel at the store. You brought your own can, and the storekeeper ladled the flammable fluid from a barrel. Wasteful. Messy. Dangerous.

To reduce spillage, Bowser built a pump in his Fort Wayne, Indiana, barn. He sold and delivered the first one to merchant Jake Gumper. The self-contained unit included a wooden barrel, marble valves, wooden plunger, hand lever, and upright faucet lever. It was a success. Bowser formed the S.F. Bowser Company and patented his pump in 1887.

Bowser's pump soon became known as a filling station, and Bowser started selling an improved model to the first auto-repair garages in 1893. Most places selling fuel to motorists still used the drum-and-measure method: gasoline was gravity-fed from a large steel drum into a five-gallon measuring can. The motorist carried the can to his automobile and filled the tank through a funnel lined with a chamois filter to remove grit.

In 1905, Bowser enclosed a square metal tank in a wooden cabinet equipped with a suction pump. A hand lever pumped the gas. This pump featured safety vents, stops to deliver a predetermined quantity, and—wonder of wonders—a hose to deliver the gasoline. He called it the Bowser Self-Measuring Gasoline Storage Pump. (Rival John Tokheim had fitted hose to pump in 1903.) *Bowser* soon became generic for any vertical gasoline pump. That usage has faded in the U.S. but lingers in Australia, New Zealand, and Canada. A bowser is also a tank truck that fuels airplanes on the tarmac. (In Britain, the term applies to tanks carrying other fluids as well.)

Also September 5:

1902: Death of Rudolf Virchow, Father of Cellular Pathology

1906: Death of Physicist Ludwig Boltzmann, Pioneer of Atomic Theory

1980: Switzerland's St. Gotthard Road Tunnel Opens

Also 1885:

March 17: Elephant Man's Physician Hazards a Diagnosis

August 30: Daimler Gives World First True Motorcycle (see page 244)

Bowser's later career was quirky: he invented and personally marketed a back-scratcher and a sit-down enema. —*RA*

September 6
1891: Risky Heart Surgery Saves Stabbing Victim

The victim of a stab wound becomes the first person to undergo heart surgery involving the suturing of the pericardium, or heart sac.

The victim, twenty-two, sustained a two-inch tear to the pericardium when he was stabbed during a fight. The wounded man was taken to City Hospital in St. Louis, where the decision was made to attempt surgery.

It was controversial, because opening the chest cavity to repair wounds to the heart was not yet accepted practice, owing to the excessive risks involved. Nevertheless, with the patient's temperature at 101 and his complaining of pain, faintness, nausea, and loss of feeling on his left side, Dr. H.C. Dalton made the decision to go in.

Dalton made an eight-inch incision and removed part of the fourth rib to get to the heart. The sliced pericardium was immediately evident, but an examination showed that the heart muscle itself was intact. Dalton then sutured the heart sac — no mean feat considering the heart was pounding at a rate of 140 beats per minute. (The heart-lung machine that lets surgeons stop the patient's heart was more than six decades in the future.) But following the motion of the beating heart, Dalton was able to stitch up the tear with catgut.

It was touch-and-go for a while: Dalton's account says it appeared that the patient came close to dying during the surgery, but hypodermic injections of whiskey and strychnia (called strychnine these days) revived him. The surgical team used sterilized warm water to irrigate the wound area, then stitched him up. Once he turned the corner, the patient made a full recovery. — *TL*

Also September 6:

1978: Genetically Engineered Insulin Announced

Also 1891:

July 14: John T. Smith Patents Corkboard

Other Surgical Milestones:

January 14, 1794: First Successful Cesarean in U.S. (see page 14)
December 13, 1809: First Removal of Ovarian Tumor
(see page 349)
September 30, 1846: Ether He Was the First or He Wasn't
(see page 275)
May 23, 1962: Give That Kid a Hand! (see page 145)
December 3, 1967: Patient Dies, but First Heart Transplant a Success
(see page 339)
August 25, 1973: More Than One Way to Slice a CAT (see page 239)

September 7
1948: Where the Rubber *Is* the Road

A mile-long stretch of West Exchange Street in Akron, Ohio, opens. It's the first U.S. road paved with a rubber-asphalt compound.

Rubber was everywhere in postwar Akron. As the home of B.F. Goodrich, Goodyear, Firestone, and General Tire, Akron dubbed itself Rubber Capital of the World, and the fortunes of the city rode with the tire industry (see page 346).

As early as the 1840s, scientists added natural rubber to pavement to create surfaces that resisted cracks and repelled water better. Goodyear president Paul Litchfield was so impressed by the rubberized roadways he'd seen in the Netherlands that he donated synthetic rubber for a test of rubber roads in Akron. A sign called it THE FIRST RUBBER STREET IN AMERICA, but the road surface contained only between 5 and 7 percent rubber. The rest was asphalt.

Rubber companies immediately put their show on the road with dry-powder or latex rubber additives sold under brand names such as Rub-R-Road and Pliopave. Roads from Ohio to Virginia got the rubber treatment. Engineers eventually questioned the benefits. Rubberized asphalt was more expensive, and studies didn't show any clear advantages. West Exchange Street was torn up and repaved in 1959.

But Charlie McDonald, a Phoenix city engineer, found a way in 1965 to blend shredded crumb rubber from *waste* tires into asphalt. With waste tires in abundant supply, rubber roads became popular again, especially in warm climates, because rubber roads are more resistant to cracking. These "quiet roads" also reduce road noise by up to 12 percent, sometimes negating the need for sound barriers.

Akron? It has a pedestrian walkway made of crumb rubber. — *KB*

Also September 7:
1998: If the Check Says "Google Inc.," "We're "Google Inc."

Also 1948:
March 17: Birth of William Gibson, Father of Cyberspace
October 26: Death Cloud Envelops Pennsylvania Mill Town (see page 301)

September 8
1854: Pump Closure Ends London Cholera Outbreak

Physician John Snow convinces London authorities to remove a pump handle. A deadly cholera epidemic comes to an end immediately, though perhaps serendipitously. Snow maps the outbreak to prove his point... and launches modern epidemiology.

Snow had published *On the Mode of Communication of Cholera* in 1849. His data showed the disease was spread by sewage pollution in drinking water. The work went largely ignored by the medical establishment, which blamed miasma, or bad air.

John Snow's map of the Soho cholera outbreak.

Cholera broke out in London's Soho district in 1854, killing 127 in the first three days of September. Three-quarters of Soho's population fled. Snow lived in the district and interviewed victims' families. Almost all lived near the Broad Street water pump. Streets nearer another pump had fewer fatalities. Snow examined water from Broad Street under a microscope (see page 148) and found "white, flocculent particles." Local authorities weren't convinced their water was polluted, but they were desperate enough to try shutting the pump down. With the handle removed, people couldn't draw water from the subterranean cause of home-sick blues. The pump don't work 'cause the council took the handle.

Numbers of new cases dropped almost immediately, but the epidemic had already been subsiding. It turned out, however, that three cholera patients who lived elsewhere had in fact drunk from the infected pump. Few local people who drank from a different well got sick, nor did brewery workers who drank free beer instead of water. The cause was positively Broad Street.

It was soon traced to a single source: a cholera-infected baby whose diaper wash was emptied in a cesspool that leaked directly into the well. Over six hundred people died.

Snow's maps illustrating the distribution of cholera cases relative to the pump are a milestone in the study of the geography of disease. Edward Tufte cites them in *Visual Explanations.* —RA

Also September 8:

1930: 3M Starts Marketing Scotch Tape
1966: *Star Trek* Debuts; Liftoff for Starship *Enterprise*

Also 1854:

March 15: Birth of Emil von Behring, Developer of Diphtheria Vaccine
August 9: Thoreau Warns, "The Railroad Rides on Us" (see page 223)

253

September 9
1982: 3-2-1... Liftoff! First Private
Spaceship Launched

A private company sends a rocket ship into space for the first time.

Space Services of America had to secure eleven separate U.S. government approvals and get a federal gun dealer's license to even think about acquiring the Minuteman rocket that powered Conestoga 1. Even then, NASA couldn't *sell* the Minuteman, which was used to launch nuclear missiles. But NASA offered to lease it and make the company pay full price if it wasn't returned in working condition. That was a subterfuge, because both parties knew the rocket would either blow up on the launchpad, crash somewhere, or fulfill its mission and sink to the bottom of the Gulf of Mexico.

Space Services of America had only seven employees. Mercury 7 astronaut Deke Slayton was company president and mission director. Financing came from fifty-seven dreamers who kicked in a total of $6 million (more than $14 million in 2012 dollars).

Some two hundred reporters covered the launch at Matagorda Island, Texas. The thirty-six-foot Conestoga 1 carried its payload—forty pounds of water—a mere 321 miles during a ten-and-a-half-minute suborbital flight that reached an altitude of 195 miles. It landed in the Gulf of Mexico.

In 1985, Space Services became the first company to get a federal license to provide commercial launch services. It launched two scientific payloads in 1989 out of New Mexico's White Sands Missile Range. The firm merged with Celestis in 2004 to become Space Services Incorporated. Space Services may be best known as the company that lets you name a star; subsidiary Celestis will take your loved one's remains on a memorial ride into space.

What goes up must come down. —*JCA*

Also September 9:
1926: RCA Sets Up NBC (see page 321)
1999: 9/9/99 No Big Deal for Computers

Also 1982:
September 19: Can't You Take a Joke? :-)
    (see page 264)

September 10
1984: DNA Leaves Its Print

English geneticist Alec Jeffreys is performing advanced but routine lab work when he has a eureka moment and discovers DNA "fingerprinting."

Jeffreys was working in his genetics lab at Leicester University, trying to trace genetic markers through families, looking for patterns of inherited disease-causing mutations in the repeated DNA (see page 117) segments carried by all humans. At precisely 9:05 a.m. Monday, September 10, as he removed an X-ray film from the developing tank and studied the image, he saw what looked at first like a complicated tangle of DNA strands. Then Jeffreys had what he called a "blinding flash."

Every individual (except identical twins, triplets, etc.) has a unique DNA profile. Therefore, DNA can be used to identify individuals as precisely as fingerprints do. What's more, half an individual's DNA comes from one parent and half from the other, so lineage can be traced too. All that in a flash of insight!

Jeffreys saw that the lack of uniformity, which was a *problem* in the research he had begun, was actually a *solution* to many other problems. Before the day was out, Jeffreys started a list of potential uses for his discovery, including criminal detective work, transplant biology, and establishing biological kinship in paternity and other cases.

Evidence from Jeffreys's lab helped convict a murderer-rapist and exculpate another suspect in 1986. The technology soon went global. Beyond the initial uses envisioned by Jeffreys, anthropologists today use DNA fingerprinting to study millions of years of human evolution and current global variation, and biologists use it to study nonhuman genetics.

Jeffreys was elected a Fellow of the Royal Society in 1986, and Queen Elizabeth II knighted him in 1994 for services to science and technology. — *RA*

Also September 10:

1846: Elias Howe Patents First Practical Sewing Machine

1941: Birth of Evolutionary Theorist Stephen Jay Gould

1977: France Performs Last Execution by Guillotine

Also 1984:

January 22: Apple Debuts Macintosh in "Big Brother" Super Bowl Commercial (see page 22)

September 11
1822: Church Admits It's Not All About Us

The College of Cardinals finally caves in to the hard facts of science, saying the "publication of works treating of the motion of the Earth and the stability of the sun, in accordance with the opinion of modern astronomers, is permitted."

It represented a major shift in dogma for Catholicism, a concession that Earth might, in fact, revolve around the sun. Unfortunately, it came 180 years too late to do Galileo Galilei any good. Still, it would be another thirteen years before Galileo's *Dialogue Concerning the Two Chief World Systems*, his defense of heliocentrism, was removed from the Vatican's list of banned books.

Heliocentric theory had existed since the ancient Greeks, who were the first to determine that Earth is a sphere (see page 172) in a sky full of spheres. The geocentric view propounded by Ptolemy and Aristotle, and embraced by Rome, placed Earth at the center of the universe.

Galileo was greatly influenced by Polish astronomer Nicolaus Copernicus, who posited that Earth not only revolves around the sun but makes a complete turn on its own axis every twenty-four hours. The Catholic Church considered this heresy, and Galileo was convicted by the Roman Inquisition in 1633 and remained under house arrest for the last eight years of his life.

Nearly two centuries later, however, the weight of scientific evidence was so overwhelming that the College of Cardinals finally reversed itself and allowed the teaching of heliocentrism. But it was 1992 before Pope John Paul II officially conceded that, yes, Earth isn't stationary in the heavens. John Paul apologized in 2000 for the way the Church treated Galileo. — *TL*

Also September 11:

1998: Starr Report on President Clinton Showcases
    Internet as Instant News Medium

Also 1822:

September 27: Deciphering Rosetta Stone Unlocks
    Egyptian History

September 12
1958: Kilby Chips In, Integrates Circuit

Jack Kilby shows his Texas Instruments colleagues a little something he's built. A very little something: a working integrated circuit on a piece of semiconductor material. The world will soon change.

Most TI employees took a two-week summer vacation that year, but new hire Kilby hadn't earned any time off yet. Using his solitude to good effect, Kilby dreamed up the integrated circuit and constructed a prototype by September. It was a sliver (a chip, you might say) of germanium with wires sticking out glued to a glass slide about the size of a thumbnail.

The stakes were high for the new guy. Top company execs came to see Kilby's September 12 demonstration. Kilby connected his device to an oscilloscope and threw the switch. A continuous sine curve pulsed on the screen, and a new era began.

As is often the case with great inventions, there was a prior claim. British radar scientist Geoffrey Dummer had presented the concept of a miniaturized integrated circuit at a 1952 electronics symposium in Washington, DC. But *his* prototype failed, and his bosses at the Ministry of Defence abandoned the idea.

And as also often happens, science advances with near-simultaneous discovery. Fairchild Semiconductor engineer Robert Noyce was working on an integrated circuit using silicon instead of germanium. Kilby filed his patent six weeks before Noyce but received his patent later. TI and Fairchild fought a lengthy legal battle before agreeing to cross-license their technologies. Noyce's silicon chip eventually triumphed over Kilby's germanium. Noyce went on to cofound Intel.

Kilby shared the 2000 Nobel Prize in Physics, but Noyce didn't. He'd died in 1990, and Nobel Prizes aren't awarded posthumously.

Earth to earth, ashes to ashes, germanium to silicon. — *RA*

Also September 12:

1933: Physicist Leo Szilard Conceives Nuclear Chain Reaction
1962: President Kennedy Pledges Manned Moon Landing and Return by Decade's End

Also 1958:

October 30: First Coronary Angiogram Performed (see page 305)

September 13
1833: Imported Ice Chills, Thrills India

Nearly a hundred tons of ice, cut in blocks from frozen New England lakes earlier in the year, arrives in Calcutta.

The transoceanic operation, undertaken by Tudor Ice Company, began in early May 1833, when approximately 180 tons of freshwater ice was loaded into the sailing ship *Tuscany* in Boston. Its hold was insulated with multiple layers of dry bark and hay to reduce melting. The historic four-month trip of the precious, perishable cargo was arranged by Frederic Tudor, Boston's "Ice King." The Tudor family had already built a lucrative business shipping northeastern ice to the Caribbean and Europe.

Workers had formerly used axes and saws to hack the frozen water from northern lakes during winter months. The labor-intensive undertaking had made ice a luxury item that only the wealthy could afford. But a new horse-drawn metal ice plow allowed mass production. It was invented by Nathaniel Jarvis Wyeth, of the family that would produce many artists and inventors (see page 299).

The arrival of pure U.S. ice in Calcutta signaled the end for Hooghly ice, a dirty, slushy substance made by freezing water in shallow pits on the Hooghly River in West Bengal. You could use it to cool containers, but it wasn't fit to drink—as in a gin and tonic. The imported ice, however, was pristine. Locals marveled at the giant, icy cubes as they were unloaded from the ship. As trade took off, ice was stored in icehouses in Bombay, Calcutta, and Madras. Over the next two decades, India became Tudor's most lucrative market.

But science marches on. The Bengal Ice Company, India's first artificial-ice manufacturer (see page 197), began production in 1878. The availability of cheaper domestic ice killed the Boston-to-India ice trade within four years.—*LW*

Also September 13:
1899: New Yorker Is First U.S. Pedestrian Killed by Car
1955: De Mestral Patents Velcro Fastener

Also 1833:
June 5: Ms. Software, Meet Mr. Hardware (see page 158)

September 14
1959: Moon Feels First Cold Touch of Humans

Luna 2 becomes the first artifact of humanity to strike the moon.

The Soviet Union launched the Sputnik-like probe from the Baikonur Cosmodrome on September 12. It took thirty-three and a half hours to reach its destination. Hitting the moon, a prestigious accomplishment for the Soviet space program, was not Luna 2's only objective. Prior to impact, the craft sent back data confirming that the moon had neither a magnetic field nor any radiation belts.

Luna 2.

Because Luna 2 itself lacked a propulsion system, it was guided along by the third stage of its SS-6 booster rocket until the two separated shortly before impact. The probe hit the lunar surface east of Mare Serenitatis, momentarily disrupting the serenity near the Sea of Serenity. The booster's third stage hit the moon about half an hour later.

Luna 2 was the second in the Soviet Union's ambitious and long-running Luna program, which collected information about the lunar environment at least in part to prepare for a Russian attempt to land a man on the moon, which never materialized. Lunas 2, 9, 13, and 15 all reached the moon before Neil Armstrong (see page 203), but the big prize eluded the Soviets. The program ended with Luna 24 in 1976.

Despite the Soviets' failure to get a man to the moon, the Luna program must be considered a success in the long view. Among its achievements: first lunar flyby (Luna 1), first landing (Luna 2), first photographs from the far side of the moon (Luna 3), first successful soft landing on the lunar surface (Luna 13), first analysis of lunar soil, and first deployment of a lunar rover (Luna 17). — *TL*

Also September 14:

1716: Boston Light Is First North American Lighthouse
1904: First Isle of Man Road Race Features Autos, Not Motorcycles

Also 1959:

February 6: Titan Launches; Cold War Heats Up (see page 37)
April 9: America Meets Its Seven Original Astronauts

September 15
1916: All Disquiet on the Western Front

The tank makes its debut as a battlefield weapon, attacking the Germans as part of a British assault near Delville Wood on the western front.

The concept of an armored assault vehicle dates back to 1770, with the first appearance of the caterpillar track. The British army used a few steam-powered tractors during the Crimean War, but they carried no offensive weapons. Frederick Simms developed an engine-driven "motor war car" in 1899. It was armor-plated and carried two Maxim machine guns, more of an armored car than a tank. Simms offered it to the British army, but he was turned down.

The British high command preferred to concentrate on infantry and cavalry. But the "landship" had a champion in the first lord of the Admiralty, Winston Churchill. In fact, the first tanks were manned by the Royal Navy, which was already responsible for the operation of armored cars. The vehicles were code-named tanks because their shells looked like water carriers. The name stuck.

A fourteen-ton D1 tank made the attack at Delville Wood. A larger attack, employing fifteen tanks, followed at Flers-Courcelette. The British had wanted to use all their tanks, but the other thirty-four broke down. The Germans were shocked, but they quickly rallied. Though small-arms fire and machine guns had little effect on the tanks, German artillery could knock them out easily.

The tankers themselves found the machines difficult to operate. Visibility from the viewing slits was poor, and the tanks crawled along at less than one mile an hour, often getting hung up in the trenches. Nevertheless, General Douglas Haig ordered construction of a thousand more. By 1918, the British had produced around twenty-eight hundred tanks. The French built four thousand, and the United States eighty-four.

Germany fielded just twenty tanks during World War I. That would change in the next war (see page 188).—*TL*

Also September 15:

1884: Ophthalmologists Learn of Cocaine as Local Anesthetic

1947: Association for Computing Machinery Gets Whirring

Also 1916:

March 25: Ishi, Last Survivor of Yahi American Indian Tribe, Dies in San Francisco

October 16: Sanger Opens First U.S. Birth Control Clinic (see page 291)

September 16
1736: One Degree of Separation — Fahrenheit Dies

Physicist and instrument maker Daniel Gabriel Fahrenheit dies. He will live on, to a degree.

Galileo (see page 256) had created a water-based thermoscope as early as 1593, and Santorio Santorio introduced a numerically graduated model in Florence in the mid-seventeenth century. The Florentine instruments used the expansion of an alcohol solution to register changes in temperature. But no two instruments were exactly alike. They were usually marked only with the high and low temps for Florence in the year they were made.

You could compare the temperature in the same place from day to day, or even year to year, but not place to place. Isaac Newton (see page 107) had considered standardizing thermometers by marking universal points that didn't vary. But Newton had other things on his mind that he didn't want to drop, and he never followed up.

Fahrenheit wanted to standardize the thermometer. In 1714, he produced two thermometers that gave identical readings, a major accomplishment. Next, he substituted mercury for alcohol as the measuring medium, and he introduced the cylindrical shape, replacing a spherical bulb.

He developed the scale that bears his name and that's still used in the metric-averse United States. Folklorically, the coldest day of that winter was 0 degrees, and 32 degrees was the freezing point of water, which resulted in 212 degrees for the boiling point of water at sea level. In fact, 0 degrees was the coldest he could make a concentrated solution of ice, water, and salt, and 96 degrees was supposed to be human body temperature. He was low, but striving for mathematical convenience: $96 = 3 \times 32$.

Fahrenheit also invented a hygrometer for measuring the moisture content of air. Remember that the next time you say "It's not the heat, it's the humidity." *—RA*

Also September 16:
1985: Steve Jobs Quits Apple
1997: Steve Jobs Returns to Apple

Also 1736:
January 19: Birth of James Watt, Improver of the Steam Engine

September 17
1908: First Airplane-Passenger Death

Lieutenant Thomas Selfridge becomes the first passenger to die in an airplane accident.

The crash that killed Thomas Selfridge.

Many pioneer pilots contributed to early aviation. But what of the unsung individuals who risked their lives as *passengers* on those flimsy contraptions? Thomas Selfridge was one...and the first to lose his life that way.

After their historic 1903 flight (see page 353), Wilbur and Orville Wright worked on building a business. They won a U.S. Army contract in 1908, subject to performance tests of their airplane at Fort Myer, Virginia. In addition to receiving the basic $25,000 (more than $600,000 in 2012 money), the brothers would get an extra 10 percent for every mile per hour faster than 40 mph.

Orville made a record-breaking seventy-four-minute flight on September 12 and then took Selfridge aboard on September 17. Selfridge was a member of the Aerial Experiment Association and had designed that group's first powered airplane.

Three or four minutes after the Wright Flyer took off, a wooden propeller blade split and hit a bracing wire. That pulled a rear rudder from vertical to horizontal. The plane pitched nose-down and hit the ground hard. Orville broke his leg and several ribs. Selfridge fractured his skull and died in the hospital a few hours later.

Despite the first airplane-passenger fatality, the army was impressed and let the brothers complete the trials. They won the contract, but because of Selfridge's death, army pilots were required to wear football-style leather helmets to minimize head injury.

The Wrights and other air pioneers kept flying. On the same date just three years later, pilot Cal Rodgers took off from New York for the first successful transcontinental flight across the United States. It took seventy-plus stops and eighty-two hours' flight time, spread over seven weeks. And you think *your* flight delay was something? —*JP*

Also September 17:

1683: Leeuwenhoek Writes About Animalcules (see page 148)

1976: Prototype *Enterprise* Rollout Launches Shuttle Era (see page 5)

Also 1908:

June 18: Prescient Letter Creates Concept of TV (see page 171)

1830: Horse Beats Iron Horse, for the Time Being

America's first native locomotive loses a smack-down race to a draft horse.

Baltimore thrived with wagon trade on the National Road linking it to the Ohio River, but merchants worried about competition from the Erie (see page 198) and Chesapeake and Ohio canals. Declaration of Independence signer Charles Carroll of Carrollton led the group founding the Baltimore and Ohio Railroad in 1828. They planned to beat the canals by jumping directly to a superior technology: steam engines (see page 177).

Iron horse versus bio-horse.

Engineer Peter Cooper's Tom Thumb was a mini-locomotive. The boiler was no bigger than those in large kitchen ranges, and Cooper repurposed musket barrels for boiler tubes. The B&O debuted the Tom Thumb on its thirteen miles of track. Hauling a car full of forty officials, dignitaries, and social notables, the little loco raced from Baltimore to Ellicott Mills in an hour, reaching the unheard-of speed of 18 miles an hour. Gracious!

On the return trip, the train was met, on the adjacent track, by a similar car hitched to the local stagecoach company's best steed. The iron horse and the bio-horse would race. The horse car pulled ahead first as the tiny engine struggled to build up steam. It finally got going: Tom Thumb came up even, and eventually passed the horse. The crowd cheered.

Then a drive band slipped in the locomotive's works. The engine slowed to a crawl. The horse car pulled ahead. By the time Cooper fixed things, it was too late: the horse was far past them and won the race.

But embarrassment does not alter the course of history. It's easier to increase the horsepower of a steam engine than to up the horsepower of a horse. An improved locomotive reached 30 miles an hour in 1831, and the railroad stopped using horses to pull its carriages. —*RA*

Also September 18:

1895: Palmer Performs First Chiropractic Adjustment
1998: Nonprofit ICANN Takes Over Internet
    Governance from U.S. Government

Also 1830:

August 31: Edwin Budding Patents Lawnmower

September 19
1982: Can't You Take a Joke?:-)

Computer scientist Scott Fahlman posts the following electronic message to a computer science department bulletin board at Carnegie Mellon University:

19-Sep-82 11:44 Scott E Fahlman:-)
From: Scott E Fahlman
I propose ... the following character sequence for joke markers:
:-)
Read it sideways. Actually, it is probably more economical to mark things that are NOT jokes, given current trends. For this, use:
:-(

With that post, Fahlman became the acknowledged originator of the emoticon. From those two simple emoticons (*emotion + icon*) have sprung scores of others that are the joy, or bane, of e-mail, text-message, and instant-message correspondence the world over.

Fahlman was not the first person to use typographical symbols to convey emotions. The practice goes back at least to the mid-nineteenth century, when Morse code symbols were occasionally used for the same purpose. In 1881, the American satirical magazine *Puck* published what we would now call emoticons, using hand-set type. No less a wordsmith than Ambrose Bierce suggested what he called a snigger point:

__/

to convey jocularity or irony. Baltimore's *Sunday Sun* suggested a tongue-in-cheek sideways character in 1967.

But none of those caught on. The Internet emoticon truly traces its lineage directly to Fahlman, who says he came up with the idea after reading "lengthy diatribes" on the message board from people who had failed to get the joke or sarcasm in a particular post.

Fahlman's original post was lost for a couple of decades and then retrieved from an old backup tape, thus cementing his claim of priority. —*TL*

Also September 19:
1991: Hikers Discover Ötzi, the Alpine Iceman

Also 1982:
December 2: Barney Clark Gets First Artificial Heart
December 26: *Time's* Top Man? The Personal
   Computer (see page 362)

1842: Dewar's Fortune Is Scotched

James Dewar is born, but not into a vacuum.

Scotsman Dewar worked with gases liquefied to temperatures approaching absolute zero. He figured there had to be a way to keep them really, really cold long enough to study them better.

Dewar built a machine in the early 1890s that could manufacture industrial quantities of liquid oxygen. But there was no way to store it, to keep heat out for extended periods. It occurred to Dewar that putting a bottle within a bottle—and having the interior bottle barely touch anything in the outside world—would slow heat transfer by conduction. Creating a vacuum between the two bottles would stop heat transfer by convection: no gas molecules available to move the heat. And applying reflective material to the interior bottle would stop heat transfer by radiation. Minimize all three forms of heat transfer, and that does the trick.

Indeed, that did the trick so well that a German company was formed to market it commercially. Thermos GmBH began selling its eponymous product in 1904, without any input from, or remuneration to, Dewar. The absent-minded professor had failed to patent his invention. He sued, but lost.

So successful was the idea of keeping cold things cold and hot things hot that Thermos itself, in 1963, lost the right to use the name exclusively, joining other genericized trademarks—including aspirin (see page 67), cellophane, linoleum, and even e-mail.

Dewar's research into high vacuums was used in atomic-physics experiments. He was knighted. There's even a lunar crater named for him. But the man who invented but didn't profit from the thermos, and who wasn't a member of the whisky-making dynasty, was *nominated* several times for a Nobel Prize and... well, you get the idea. Pass the Scotch, please. —*JCA*

Also September 20:
1952: Kitchen Blender Pegs DNA as Stuff of Life

Also 1842:
March 13: Henry Shrapnel Dies, but His Name Lives
    On (see page 74)
December 21: Birth of Peter Kropotkin, Anarchist and
    Darwin's Detractor

September 21
1756: McAdam Paves the Way

John McAdam is born. Along the road of life, he'll invent a new way to smooth the roads of our lives.

Road-building hadn't improved much since Roman times. Your choices were between dirt (alternately dusty or muddy) and rock or cobble paving (hard on carriage wheels and passengers' backs, and slippery when wet, which was frequently, thanks to bad drainage).

John Metcalfe introduced a three-layer system in the mid-seventeenth century to improve drainage: large stones, excavated earth, and gravel on top. Thomas Telford raised the center of his roads to let water drain off the convex surface toward the sides.

McAdam added a few ideas of his own: dispense with expensive, precisely cut foundation stones, elevate the road for drainage, and seal the top with successive levels of tightly packed small stones of varying sizes. You could obtain the stones by breaking rocks, then you would use sieves to sort the results by size. Road *surveying* became more expensive (to ensure good drainage), but you could actually *build* the road with low-paid unskilled labor instead of highly paid stonecutters.

His new system was such a success that from 1816 to 1818, he became consulting surveyor to thirty-four toll-road companies. Richard Edgeworth added an improvement that McAdam didn't like but almost everyone else did: mixing stone dust with water to create a smoother "water-bound macadam." McAdam's patents were widely infringed. Parliament gave him a small payment. He was offered a knighthood, but, with his health declining, he declined the honor.

By the time McAdam died, in 1836, the word *macadam* was in common usage for a cambered, paved road. When the stone-dust sealer was replaced by asphalt or tar in the 1850s, people referred to it as tar-macadam, which was eventually shortened to tarmac. — *RA*

Also September 21:

1866: Birth of Science-Fiction Author H.G. Wells
1937: Tolkien's *The Hobbit* Opens Up a Brave New
World

Also 1756:

November 30: Birth of Physicist E.F.F. Chladni, Father
of Acoustics

September 22
1792: Day One of Revolutionary Calendar

It's 1 Vendémiaire of An I in the French revolutionary calendar, the first day of the first month of the first year of the First Republic. But no one would know about it for another year, when the calendar was imposed retroactively.

Conversion to the new metric system was already under way (see page 130), and the Gregorian calendar (see page 1) was considered a vestige of the ancien régime. The new calendar became law in autumn 1793 but officially began September 22, 1792: birth date of the republic and, conveniently, the autumnal equinox.

The year comprised twelve egalitarian months of thirty days each. The week was abolished. Months were divided into three *décades* of ten days each, numbered from one to ten. The tenth day was a day of rest, leaving nonagricultural workers with only three days off a month. Some folks resented that. Five extra days (outside any month) at the end of each year were celebrated as holidays. Leap year (see page 61) added a sixth.

Revolutionaries even attempted a metric day: ten hours of one hundred minutes each, each minute being one hundred seconds. The metric second was 14 percent shorter than the one we know. It was even less popular than the *calendrier révolutionnaire:* clocks and watches are considerably more expensive than paper calendars and ledgers.

The republican calendar was a major bother to diplomats and international merchants. They needed cumbersome conversion charts that listed each day next to its counterpart in the other calendar. Emperor Napoléon, who'd already abolished the republic itself, abolished the republican calendar in 1805. The Gregorian calendar resumed throughout Napoléon's empire on January 1, 1806, aka 11 Nivôse An XIV.

Historians labored with conversion tables for another two centuries, until the World Wide Web (see page 99) came along with its handy conversion engines. — *RA*

Also September 22:

1791: Birth of Chemist-Physicist Michael Faraday
1953: First Four-Level Stack Highway Interchange Opens in Los Angeles

Also 1792:

February 20: George Washington Signs Postal Service Act

September 23
1889: Success Is in the Cards for Nintendo

Fusajiro Yamauchi founds Nintendo Koppai in Kyoto, Japan, to manufacture hanafuda, Japanese playing cards.

Western playing cards came to Japan in the sixteenth century with Portuguese traders, but over the next three centuries, a variety of card games were created in Japan. The most popular in the late 1800s were *hanafuda*, cards printed with beautiful, colorful images of flowers. The yakuza often used *hanafuda* in their illicit gambling halls.

The fact that the cards were often used for gambling was reflected in the name Yamauchi gave to his company. *Nin-ten-do* is written with characters that mean, roughly, "luck-heaven-hall," or the place where you put your fortune in the hands of the gods. Nintendo became the country's preeminent maker of playing cards, expanding into making *toranpu* (meaning "trump," for Western playing cards). The firm struck a deal to print cards with Disney characters, which widened the market for playing cards, turning a gambler's tool into a children's toy.

The company stayed in the hands of the Yamauchi family for over a century. Fusajiro Yamauchi's great-grandson Hiroshi took over in 1949 at the young age of twenty-two. One of his first acts was to have all remaining Yamauchi family members fired, to make it clear who was in charge. Yamauchi oversaw the expansion of the company into a wide variety of other products, all failures — until the company moved into electronic toys and games.

After the company forged a partnership with hardware maker Sharp, Nintendo engineers developed unique electronic toys, such as the Beam Gun, which used solar cells to let kids imagine they were firing guns and making targets explode. From there, the company expanded into video games (see page 293).

Nintendo is still the dominant playing-card maker in Japan, and it still produces *hanafuda* decks, although some now have Super Mario characters instead of flowers. — *CK*

Also September 23:

1846: Neptune Right Where They Said It Would Be
1869: Birth of Typhoid Mary Mallon

Also 1889:

June 3: First Long-Distance Transmission of Electricity, Fourteen Miles to Portland, Oregon
November 23: SF Gin Joint Hears World's First Jukebox (see page 329)

September 24
1979: First Online Service for Consumers Debuts

CompuServe begins offering a dial-up online information service to consumers.

The company opened in 1969, providing dial-up computer timesharing to businesses. It grew into a solid business that supplied online data to corporations. Offering the service to consumers was a bit risky in 1979, when personal computers still seemed like a wild and crazy idea to most people. Launched as MicroNET in 1979 and sold through Radio Shack, the service became surprisingly popular, thanks perhaps to Radio Shack's Tandy Model 100 computers, which were portable, rugged writing machines that dovetailed nicely with the 300-baud information service.

MicroNET was renamed CompuServe Information Service in 1980, around the time it began offering online newspapers, starting with the *Columbus (Ohio) Dispatch*. But it was chat that people used most. The CB Simulator was one of the first online real-time chat programs in the world, and CompuServe users loved it. CompuServe added a wealth of other features: stock quotes, weather reports, forums, even airplane-ticket booking. And, of course, there was e-mail, which CompuServe apparently trademarked as Email. CompuServe e-mail addresses were a strange collection of octal digits, like 72241.443@compuserve.com.

CompuServe's heyday was the early 1990s, when it was the largest and most popular online service in the United States. However, its text-centric interface and by-the-minute fees eventually made it an easy target for AOL, which offered a prettier face and unlimited service for a flat monthly fee. The Internet overtook both services by the late 1990s.

As Internet service providers began offering easy ways to connect, and as resources on the web grew, the closed garden of carefully selected services offered by CompuServe grew less attractive. The consumer service was eventually sold to AOL, which closed the proprietary features in 2009 but still offers a CompuServe-branded monthly ISP. —*DT*

Also September 24:

1947: President Truman Allegedly Authorizes MJ-12 UFO Panel

1960: First Nuclear Carrier, USS *Enterprise*, Launched

1993: Beautiful Myst Ushers in Era of CD-ROM Gaming

Also 1979:

October 10: Pac-Man Brings Gaming into Pleistocene Era (see page 285)

September 25
1878: Yes, Smoking Is a Health Hazard

Eighty-six years before the U.S. surgeon general issues a report confirming the dangers of smoking tobacco, a letter in the *Times* of London condemns its use.

Charles R. Drysdale, senior physician to London's Metropolitan Free Hospital, had already published *Tobacco and the Diseases It Produces* when he wrote the letter that described smoking as "the most evident of all the retrograde influences of our time." Drysdale had been on an antismoking crusade since 1864, when he published a study documenting tobacco's effects on young men. It reported cases of jaundice and "most distressing palpitations of the heart."

Drysdale's book pinpointed nicotine as the dangerous agent and reported its ill effects on the lungs, circulatory system, and skin. He also warned against exposure to secondhand smoke:

> Women who wait in public bar-rooms and smoking-saloons, though not themselves smoking, cannot avoid the poisoning caused by inhaling smoke continually. Surely gallantry, if not common honesty, should suggest the practical inference from this fact.

The prolific Drysdale also wrote on medicine as a profession for women (see page 273) and on issues related to population control.

Though physicians and scientists understood numerous health hazards were associated with smoking, the number of smokers increased dramatically in the first half of the twentieth century. Thank you, Madison Avenue. Thank you, Hollywood.

The turning point probably came in 1957, when surgeon general Leroy Burney reported a causal link between smoking and lung cancer. Burney's successor, Luther Terry, published *Smoking and Health: Report of the Advisory Committee to the Surgeon General* in 1964. It was released on a Saturday to minimize effect on tobacco stocks, but it began a massive change in people's attitudes toward smoking.

It took only eighty-six years. — *TL*

Also September 25:

1929: Doolittle Proves Instrument Flying Works Takeoff to Landing

2002: Mysterious Meteorite Dazzles Siberia

Also 1878:

December 18: Swan Demonstrates Electric Bulb (see page 354)

September 26
1956: First Interstate Highway Paving Begins

The first paving of the new Interstate and Defense Highway System is laid west of Topeka, Kansas, on what will become I-70.

Urged to ease congestion on America's roads and inspired by Germany's autobahns, President Dwight Eisenhower signed the Federal Aid Highway Act in 1956. The new law poured $33 billion (about $250 billion in today's money) into overhauling the country's roadways. America would never be the same. Before the act, U.S. highways were narrow, meandering, stop-and-start affairs, passing right through cities and towns. Connecting roads sprouted off haphazardly, and bazaar-like marketplaces lined the shoulders. They linked the nation in organic, mom-and-pop fashion.

After the act, interstate travel was defined by the massive, multilane, high-speed funnels we know today. It was General Motors' 1939 Futurama exhibit (see page 122) come true: coast to coast without a stoplight. "When we get these thruways across the whole country, as we will and must," wrote John Steinbeck, "it will be possible to drive from New York to California without seeing a single thing."

Some older roads were improved and incorporated into the system, but small towns that were bypassed withered and died. New towns flourished around exits. Fast-food and motel franchises replaced small businesses. Trucks supplanted trains for shipping goods cross-country. In ever greater numbers, Americans fled from inner cities (often scarred by new interstates built over old neighborhoods) to suburbs. Old forms of traffic congestion gave way to new.

Not everyone holds with Steinbeck's nostalgia. They see the 43,000-mile system as the arteries and veins of the nation's economy. "More than any single action by the government since the end of the war, this one would change the face of America," said Eisenhower in 1963. Before the information superhighway came the superhighway. —*BK*

Also September 26:

1960: JFK, Nixon Open the Era of Presidential TV Debates
1983: Soviet Officer Ignores False Alarm, Avoids World War III

Also 1956:

April 26: The Containership's Maiden Voyage (see page 118)
May 21: U.S. Tests Hydrogen Bomb on Bikini Atoll

September 27
1941: First Liberty Ship Launched, More to Follow

SS *Patrick Henry*, the first Liberty ship, is launched at the Bethlehem-Fairfield shipyard, near Baltimore.

The SS *Patrick Henry*.

Originally referred to as emergency vessels, these cargo ships were among the first to be mass-produced. Numbers were critical as the Allies hustled to recover from the staggering losses wrought by German submarines during the Battle of the Atlantic. The vessels became known as Liberty ships after President Franklin D. Roosevelt, christening the *Patrick Henry*, quoted the ship's namesake: "Give me liberty, or give me death."

Liberty ships represented the assembly line fully realized. The keel was laid in traditional fashion, but the ship was then constructed from prefabricated sections welded together in the graving dock. Although it took 244 days to build the *Patrick Henry*, the average building time dropped to a mere 42 days per ship by the middle of the war. One Liberty, the SS *Robert E. Peary*, was built in an astounding four days at the Kaiser shipyard in Richmond, California. It was a publicity and morale stunt, however, and not repeated.

The *Patrick Henry* slid down the ways before the United States entered the war, but the U.S. Navy was already escorting merchant convoys through the U-boat–infested waters of the North Atlantic. Around 2,700 Liberty ships were built during World War II, and many found their way into merchant fleets after the war. Two fully operational Liberty ships remain afloat: SS *Jeremiah O'Brien*, in San Francisco, and SS *John W. Brown*, in Baltimore.

The *Patrick Henry*, meanwhile, survived the war and was scrapped in 1960. —*TL*

Also September 27:

1825: World's First Commercial Passenger Railway Opens, Stockton to Darlington, England
1892: Joshua Pusey Patents the Matchbook

Also 1941:

March 29: U.S. Radio Stations Shuffle Frequencies (see page 90)

September 28
1865: England Gets Its First Woman Physician, the Hard Way

Elizabeth Garrett becomes the first woman in England to receive a medical license. It didn't come easy.

Daughter of a London pawnbroker, Garrett was inspired to enter medicine after meeting Elizabeth Blackwell, the first practicing woman physician in the United States. But first, Garrett had to overcome the opposition of her parents as well as Victorian gender and class restrictions. She tried applying to medical schools. All turned her down. Garrett enrolled as a nursing student at Middlesex Hospital and sat in on some medical classes. She was booted after the male students complained.

She hung in there and continued studying independently. No rule specifically barred women from taking the medical-license examination, so Garrett took the exam on September 28, 1865, and was one of three successful candidates (out of seven). It enabled her to obtain a certificate to begin practicing medicine. The licensing authority immediately changed its rules to prevent other women from trying this.

Garrett opened a dispensary for women and later became a visiting physician to the East London Hospital. Still lacking a formal medical degree, Garrett learned French and slipped across the Channel to the University of Paris, where more enlightened attitudes prevailed. She earned her degree, which the British Medical Register refused to recognize.

Undaunted, Garrett (now Elizabeth Garrett Anderson after marriage) opened the New Hospital for Women in London, staffed entirely by women. Elizabeth Blackwell joined as a professor of gynecology. Garrett's persistence, and subsequent success, shook the British medical establishment to its foundations. The old-boy network finally cracked in 1876, when all-male med schools began admitting women. (See page 208 for another early woman doctor.) —*TL*

Also September 28:

1858: First Ever Comet Photo — George Bond Photographs Comet Donati

1925: Birth of Engineer Seymour Cray, Builder of Supercomputers

1998: Internet Explorer Leaves Netscape Browser in Its Wake

Also 1865:

February 8: Mendel Reads Paper Founding Genetics (see page 39)

April 9: Civil War Ends, but Not the Horror of Mechanized Warfare

September 29
1920: Radio Goes Commercial

A Pittsburgh department store advertises ready-made radio receivers that can pick up a local broadcast station.

Just like early homebrew computer enthusiasts later in the century, radio aficionados had to build their own sets from scratch or with kits. Electrical engineer Frank Conrad built a receiver and transmitter. Under license 8XK, he started broadcasting from his garage in 1916. In fact, Conrad was a pioneer in using the word *broadcasting* in radio. It's a farmers' word for spreading seeds far and wide. Radio had been mainly a two-way, point-to-point medium. Using one transmitter to reach a broad audience equipped only with receivers was a new idea.

Tweaking his equipment for hours on end, Conrad tired of constantly announcing his call letters and location, so he started playing gramophone records to rest his voice. Conrad was radio's first DJ, and he was building an audience.

Horne's department store had something new: the first shipment of ready-to-use radio receivers. Nothing to build; just plug and play. The store advertised in the *Pittsburgh Sun* that you could listen to music over the air:

> Mr. Conrad is a wireless enthusiast and puts on these wireless concerts periodically for the entertainment of many people in this district who have wireless sets. Amateur wireless sets are on sale here $10 and up. [Ten bucks equals $115 in 2012 cash.]

Harry Davis, Conrad's boss at Westinghouse Electric, saw business possibilities. Davis applied for a commercial license to supplant 8XK and received the arbitrary call letters KDKA. The station went on the air November 2 and broadcast the results of the Harding-Cox presidential election over its mighty 100-watt transmitter.

In 1922, the United States had thirty radio stations, and one hundred thousand consumer radios were sold. Just a year later, 556 stations were on the air, and half a million receivers were sold. Radio was on its way. —*RA*

Also September 29:

1898: Birth of Trofim Lysenko; His Stalinist Science Will Impede Soviet Genetics, Agronomy
1901: Birth of Nuclear Physicist Enrico Fermi, Discoverer of Neutrino

Also 1920:

July 6: U.S. Navy Pilots Start Using Radio Compass
August 31: News Radio Makes News (see page 245)

September 30
1846: Ether He Was the First or He Wasn't

Dentist William Morton uses ether to anesthetize a patient in Boston. It was not the first use, but it led to the widespread adoption of ether for surgical anesthesia.

Dr. Crawford Long of Jefferson, Georgia, removed a tumor from a patient's neck under ether anesthesia in 1842, but he didn't publish his results until 1848.

Pre-med student Morton was practicing dentistry in Boston, apparently without the benefit of a formal dental education. In 1845, he'd arranged a demonstration of nitrous oxide, or laughing gas, as an anesthetic. It failed, perhaps because he didn't use enough gas. Morton and his tutor Charles Jackson tried a different gas, ether. Morton secretly experimented on small animals and himself at home. Then, on September 30, 1846, he used ether to painlessly extract a tooth from Eben Frost.

Word spread. Then it got in the newspapers. Just sixteen days after ether's first dental use, Morton anesthetized a surgical patient for John Warren in a well-attended demonstration at Massachusetts General Hospital. Afterward, the patient, who'd just had a congenital vascular formation removed from his neck, announced, "I did not experience pain at any time, though I knew that the operation was proceeding."

Surgeon Warren declared, "Gentlemen, this is no humbug." Surgeon Henry Bigelow published the procedure in the *Boston Medical and Surgical Journal* on November 18.

Morton and Jackson applied for a patent on October 27, and it was granted a swift sixteen days later. They called their anesthetic Letheon and tried to keep the formula secret. That drew angry protests from the medical profession, but doctors soon identified ether's distinct smell.

For two decades, Morton petitioned Congress to compensate him for the widespread use of "his" discovery. Nada. He died in poverty at age forty-nine in 1868. —*RA*

Also September 30:

1861: Birth of Morgan Robertson, Novelist Who Foretold *Titanic* Disaster

1882: Birth of Physicist Hans Geiger, Inventor of Geiger Counter

Also 1846:

June 28: Adolphe Sax Patents Saxophone (see page 181)

September 23: Neptune Right Where They Said It Would Be

October 1
1957: Thalidomide Cures Morning Sickness, But...

Thalidomide, a drug developed to treat morning sickness, is first marketed in West Germany. Forty-six countries approve its use before thalidomide's terrible side effects become apparent.

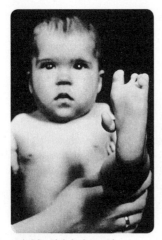

A thalidomide baby, born with incomplete arms and an extra appendage on the foot.

Thalidomide was a powerful synthetic tranquilizer, originally developed by Ciba. Unable to make the drug commercially profitable, the Swiss pharmaceutical company gave up on it. German firm Chemie Gruenenthal took over and eventually began marketing thalidomide as a "completely safe" method for resolving morning sickness. Except thalidomide wasn't completely safe. It was quite the opposite.

Inadequate testing procedures were to blame for what followed. Had the pharmaceutical labs done a better job of testing thalidomide, they would have discovered that the drug could penetrate the placenta, especially during the first trimester of pregnancy, when the fetus is developing. This invasion of the womb resulted in profound birth defects, including deformed and missing limbs, deafness, blindness, cleft palate, and a slew of internal problems.

By the early 1960s, more than ten thousand thalidomide babies had been born. Although many died in infancy (an estimated 40 percent died before their first birthday), quite a few survived into adulthood. Most of them suffer from a variety of chronic health problems. Thalidomide's impact in the United States was minimal, because the Food and Drug Administration had held up the drug's approval for reasons having nothing to do with birth defects. Thalidomide was finally pulled off the world market in 1962.

In a reversal of fortune, thalidomide resurfaced thirty years later; it improves the survival rate of bone-marrow-transplant patients. It's also used to treat multiple myeloma and Hansen's disease. — *TL*

Also October 1:

1908: First Production Model T Rolls Out of Ford Plant
1950: BBC Airs First Live TV Broadcast from Airplane
1982: Sony Sells First CD Players

Also 1957:

October 4: Soviets Orbit Sputnik, First Artificial
Satellite (see page 279)

October 2
1608: Up Close and Personal with Hans Lippershey

Hans Lippershey, a German-born Dutch spectacle maker, demonstrates the first refracting telescope, forerunner of the modern optical telescope.

The modern refracting telescope uses two lenses, a convex objective lens (nearer the object being observed) and a convex eyepiece (or ocular) lens. Together, they bend light and focus parallel light rays at a single point. That has the effect of magnifying distant objects for the viewer. Zacharias Janssen and Jacob Metius have also been credited with inventing the first telescope, but Lippershey's demonstration in front of the States-General (parliament) of the Netherlands is the earliest documented evidence, so priority generally goes to him.

Lippershey seized upon the idea after one of his assistants found that lining up a long-focus lens and a short-focus lens in front of the eye made distant objects appear closer. Lippershey mounted the lenses in a tube at the optimum distance of separation, liked what he saw, and applied for a patent. The Dutch government, appreciating the refracting telescope's military value, became a customer.

Galileo (see page 256) also got wind of the new device, built his own version, and turned it to the sky. The word *telescope* (or *telescopio*) was reportedly coined in 1611 by a guest at a banquet honoring Galileo.

Galileo's design featured a convex objective lens and a concave ocular lens, which represented an improvement over Lippershey's original design. But it was the German astronomer Johannes Kepler who first used convex lenses for both the objective and ocular. That configuration is still used in modern devices like binoculars and telephoto lenses. — *TL*

Also October 2:

1866: J. Osterhoudt Patents Sardine Can with Self-Contained Key Opener
1996: Freedom of Information Amendments Mandate Federal Documents Online

Also Seventeenth Century:

November 14, 1666: Watching a Transfusion, and Taking Notes (see page 320)
June 22, 1675: Greenwich Observatory Established (see page 288)
October 29, 1675: Leibniz ʃums It All Up, Seriesly (see page 304)
May 26, 1676: Leeuwenhoek Observes Bacteria Under Microscope (see page 148)

October 3
1947: Birth of Palomar's "Giant Eye"

After thirteen years of grinding and polishing, the Palomar Observatory mirror is completed at Caltech.

At the time, it was the largest telescope mirror ever made in the United States, measuring two hundred inches in diameter. The disk was mounted in Palomar's Hale Telescope and first used in January 1949 to take pictures of the Milky Way. Edwin Hubble was the first astronomer to make images using the new scope.

The mirror began as a twenty-ton piece of molten Pyrex, a new glass blend, at the Corning Glass Works in upstate New York. Pyrex expands and contracts far less than regular glass, making it less prone to distortion, a problem that plagued the 100-inch mirror already in operation at Palomar.

After being heated to 2,700 degrees Fahrenheit, the Pyrex was poured into a ceramic mold. It was carefully cooled at an average rate of one or two degrees per day for eleven months, and then allowed to reach room temperature. After being shipped to Caltech in Pasadena, the glass was painstakingly ground to perfection in a process lasting more than a decade.

Telescopes began in the early 1600s (see page 277), and giant telescope lenses a century later, when astronomers recognized that the bigger the lens (or reflecting mirror), the better the image. The Palomar Observatory opened in the 1930s; astronomer George Hale (for whom the telescope is named) proposed the site after he determined that the Mount Wilson Observatory was no longer ideal because of the encroaching lights of a growing Los Angeles. The new site he chose was atop Mount Palomar, a hundred miles southeast of Pasadena.

The Hale Telescope is one of seven operational scopes at Palomar. It was the largest optical telescope in the world until the completion of Hawaii's ten-meter Keck I telescope, in 1993. —*TL*

Also October 3:

1283: First Recorded Execution by Drawing and
　　Quartering
1803: Birth of Refrigeration Pioneer John Gorrie
　　(see page 197)

Also 1947:

November 2: *Spruce Goose* ...or an Expensive Turkey?
　　(see page 308)

October 4
1957: Soviets Put Man-Made Moon in Orbit!

The space age dawns a little sooner than expected with the Soviet Union's successful launch of Sputnik 1. It's a pivotal moment, the kind that still has people asking, "Do you remember where you were when...?"

Elementary Satellite 1, nicknamed "Sputnik," was a twenty-three-inch aluminum sphere circling Earth every ninety-six minutes in an elliptical orbit about 550 miles up. But it may not have been quite the world-beater it seemed at the time. In interviews fifty years later, Boris Chertok, one of the founders of the Soviet space program, admitted Sputnik was something of a lash-up, a hastily-put-together gamble that used a spare rocket to launch a satellite assembled from parts at hand.

Nevertheless, the American public was caught off guard by Sputnik's *beep, beep, beep* and frightened by the implications of a successful Soviet rocket launch. If the Soviets could put a beach-ball-size artificial satellite into orbit, they could certainly land a nuclear-tipped missile on a U.S. target. With Sputnik 1's successful deployment, the political, military, and technological relationship between the Soviet Union and United States changed dramatically. The U.S. immediately put the Explorer project on a fast track ahead of Vanguard, a satellite smaller than Sputnik.

Not only did Sputnik herald the beginning of the space age, it also constituted the opening salvo in the U.S.-Soviet space race. Within four months, the United States placed Explorer 1 into orbit (see page 31). And NASA was established (see page 212) as a direct result of Sputnik's success.

Since Sputnik, an entire generation has come of age that can't remember when space flight was only the province of dreamers and science-fiction authors. — *TL*

Also October 4:

1582: Last Day of Julian Calendar in Italy
(see page 1)
1958: BOAC's DeHavillands Start First Transatlantic Jet
Service

Also 1957:

December 2: Nuclear Power Goes Online
(see page 338)

October 5
1931: First Nonstop Transpacific Flight Ends in Cloud of Dust

More than forty-one hours after departing Japan, Clyde Pangborn and Hugh Herndon Jr. perform a controlled crash landing in central Washington state. After the dust settles, they emerge from the airplane, completing the first-ever nonstop flight across the Pacific Ocean.

Pangborn and Herndon were after the $25,000 prize (worth about $375,000 in today's money) offered by a Japanese newspaper for the first nonstop flight between Japan and the United States. The small, single-engine Bellanca Skyrocket aircraft, named *Miss Veedol,* for a brand of motor oil, took off from Misawa, north of Tokyo, on October 4. It was heavily modified to carry 930 gallons of fuel, but Pangborn still worried they wouldn't make it across without perfect weather.

He decided to save fuel by jettisoning the landing gear after takeoff, but his release mechanism malfunctioned. Former barnstormer Pangborn had to walk out on the wing struts fourteen thousand feet over the Pacific to free the gear. Without it, the more aerodynamic airplane could fly 15 miles an hour faster and go six hundred miles farther. Of course, it would be a rough landing.

Miss Veedol crossed the Washington coast after forty hours. Not content with flying across the Pacific, Pangborn wanted to set a long-distance record by continuing to Idaho. Fog ended those plans, so Pangborn flew to Wenatchee, Washington. He'd grown up nearby, and his mother still lived there. Pangborn managed a near-perfect belly landing in an open field. The propeller struck the ground, but the airplane was largely undamaged.

Pangborn and Herndon had covered more than 5,500 miles. It was longer than Lindbergh's 1927 nonstop Atlantic flight, but the duo won little fame or fortune beyond the cash. Wenatchee's airfield is now called Pangborn Memorial Airport. — *JP*

Also October 5:
1895: First Bicycle Time Trial Held in London
1986: Israel's Secret Nuke Arsenal Exposed

Also 1931:
March 18: The Schick Hits the Fans (see page 79)

October 6
1887: An Architect for the Machine Age

Charles-Édouard Jeanneret-Gris, better known as Le Corbusier, is born. He will help pioneer the International Style of architecture and will be one of the most influential proponents of the machine aesthetic.

His formative work reflected his reverence for mechanical beauty. Le Corbusier admired the design of well-built automobiles and the great transatlantic steamships of interbellum Europe. He summed it up: "A house is a machine for living in." (The man was born and raised in a Swiss watchmaking town.) Signature buildings like the Villa Savoye outside Paris embody the style he called purism. The lines are clean and sharp, and the interior functions are precisely laid out in modules. It's built on reinforced concrete stilts, another Le Corbusier trademark.

But Le Corbusier didn't limit himself to single structures. He's also known for his theories on urban planning and renewal. He tried to interest Parisian officials in bulldozing the Marais and replacing the district with a forest of egalitarian sky-scrapers surrounded by tracts of open space. He believed a complete break with the past was necessary for society to advance. His Radiant City idea was rejected, but some *unité* apartment blocks were built around Europe, notably one in Marseilles.

Le Corbusier's acolytes worshipped him with blind, near-religious fervor. His critics were less sanguine. The harshest argued that his urban designs were cold and sterile. And lesser architects certainly turned out some ugly giant ice-cube trays as the style proliferated across the globe.

Nevertheless, by the time Le Corbusier died, in 1965, his enormous influence on twentieth-century architectural sensibilities was indisputable. He ranks in a select pantheon alongside such worthies as Mies van der Rohe, Walter Gropius, and Frank Lloyd Wright. — *TL*

Also October 6:
1927: *The Jazz Singer* Gives Movie Audiences the Talkies (see page 73)
1956: Sabin Polio Vaccine Ready to Test

Also 1887:
March 2: Birth of Master Locksmith Harry Soref (see page 63)

October 7
1954: IBM Gets Transistorized

IBM builds the first calculating machine to use solid-state transistors instead of vacuum tubes.

IBM already had a business selling calculating machines (see page 129), and it was humming along quite nicely. The IBM 604 Electronic Calculating Punch was a desk-size cabinet that ate and spat out punch cards in its single-minded mission of calculating math problems. Under the hood, more than 1,400 miniature vacuum tubes (see page 322) made up the guts of the machine.

IBM's engineers saw a way to improve things. The newly invented transistor (see page 359) was just then beginning to replace the vacuum tube in radios. The transistor could also replace vacuum tubes in a computer, letting one circuit control the flow of electricity in another circuit, so that switches could be turned off and on. From that basic function, clever engineers could build a whole host of logical and arithmetic processing functions.

The modular design of the 604 let field engineers swap out defective circuits for working ones. Presumably the same modularity let the design team replace all the vacuum tubes with transistors. The resulting calculator was neither smaller nor faster, but it consumed just 5 percent of the power of the old model, despite having 2,000 transistors inside.

Egged on by this success, IBM went on to develop the first commercial calculator based entirely on transistor technology: the IBM 608, released in 1958. With more than 3,000 germanium transistors inside, the 608 was too pricey to be a commercial success. Eventually, transistors became cheaper than vacuum tubes, and computing moved into a newer, faster, lower-power age. — *DT*

Also October 7:
1806: Carbon Paper Patented
1959: Space Probe Luna 3 Takes First Photos of Far
      Side of Moon

Also 1954:
March 25: RCA TVs Get the Color for Money
      (see page 86)
June 2: Convair XFY-1 Pogo Aircraft Takes Off and
      Lands Vertically

October 8
1823: New York Gets That Erie Feeling

DeWitt Clinton inaugurates the Erie Canal, opening the section going from Albany to beyond Rochester. It will soon link the Atlantic to Lake Erie and build New York City into an economic powerhouse.

Railroads were experimental, wagon roads mostly unpaved. To reach inland, merchants had to ship upriver from New Orleans. The idea of a canal linking the Hudson River and Great Lakes had been floated as early as 1724, but politics delayed the project. New York's Governor Clinton finally convinced the legislature in 1817 to construct a 363-mile canal linking lakeside Buffalo with Albany on the Hudson. Built largely by laborers digging soil by hand, it provided employment for thousands.

Forty thousand people attended the opening ceremony. The brightly decorated craft in the river included some of those newfangled steamboats (see page 198). Local traffic produced profits for the canal almost immediately, spurring completion of the full system. The canal was forty feet wide and four feet deep, passing over eighteen aqueducts and through eighty-three locks to overcome a 568-foot gain in height.

Once completed in 1825, the canal offered freight rates at 10 percent of the cost of shipping by road. Wheat shipments multiplied from 3,640 bushels in 1829 to a million in 1841. Tolls covered the entire cost of construction by 1834. By 1840, New York City's port shipped more than Boston, Baltimore, and New Orleans combined. Competition from railroads began in the 1840s, but the canal was widened, deepened, and given branches. Only after World War II did it finally succumb to railroads and highways.

Clinton's prediction that all Manhattan would become "one vast city" was more than fulfilled. Manhattan, Brooklyn, and three other counties merged to create the current New York City in 1898. The Erie Canal had made New York New York.—*RA*

Also October 8:

1872: Birth of Food-Safety Chemist Mary Engle Pennington

1906: Karl Ludwig Nessler Demonstrates Chemical "Permanent Wave" Hair Styling

1958: Patient Gets First Fully Implantable Human Heart Pacemaker

Also 1823:

October 12: Charles Macintosh Starts Selling Rubberized Raincoats (see page 365)

October 9
1992: My Insurance Agent Will Never Believe This

A meteorite smashes into Michelle Knapp's 1980 Chevy Malibu — fortunately parked and unoccupied — in Peekskill, New York.

Final score: Meteorite 1; Malibu 0.

The stone meteorite was a fragment of a fiery meteor. The spectacular fireball was described as flaring brighter than a full moon. It was strikingly visible in the night sky from West Virginia to New York in a forty-second blaze of glory. The fireball broke up and sent this particular meteorite on a southwest-to-northeast trajectory that would take it straight into Ms. Knapp's car.

The significance of this particular meteorite strike, aside from its direct hit on the car's trunk, is that it was only the fourth instance for which accurate data on a meteorite's trajectory was recovered, thanks to many videos and photographs. In this case, dark flight began at an altitude of approximately nineteen miles, and velocity dropped to about two miles per second without further vaporization.

This incident entered the scientific annals as the Peekskill meteor. The resulting meteorite was composed of dense rock and had the size and mass of an extremely heavy bowling ball. It was later determined to have weighed twenty-seven pounds on impact.

As for the Malibu — well, it was totaled. But the wreck, rather than being hauled off to a junkyard, began a new life: Created by a shooting star, the smashed Malibu became a different kind of star. It toured the United States, Germany, France, Switzerland, and Japan for curiosity seekers to behold.

No word on what Ms. Knapp's insurance company had to say, but it probably wasn't encouraging. *—TL*

Also October 9:

1855: Joshua Stoddard Patents Steam-Powered
Calliope
2000: Ozone Hole Exposes Punta Arenas, Chile
(see page 286)

Also 1992:

November 5: Oldest Beer Ever (see page 311)

October 10
1979: Pac-Man Brings Gaming into Pleistocene Era

A new game is shown at an arcade trade show: before Halo, before World of Warcraft, before Myst, there was Pac-Man.

It wasn't the first video game; arcade games (see pages 295, 335), including video ones, had existed for years. But Pac-Man turned video-gaming into a phenomenon by burning itself into the collective consciousness.

The brainchild of Toru Iwatani, a designer for Japanese software firm Namco, Pac-Man is a model of complex simplicity: the player controls a hungry-mouthed blob that navigates a two-dimensional maze, eating dots and ghosts while trying to avoid being eaten. The concept could have been dreamed up by a ten-year-old. But try racking up big points; there's the rub.

After its release May 22, 1980, the game received a lukewarm reception in Japan (originally sold under the name Puck-Man). But it became an instant hit when it arrived in the United States. The name was supposedly changed to Pac-Man out of fears that some bright wit might alter the spelling into an obscenity.

Regardless of the name, Pac-Man quickly left every existing arcade game in its wake. Versions were made to accommodate virtually every platform out there, and spinoffs of the game itself, such as Ms. Pac-Man, were marketed to feed off the popularity of the original. Pac-Man is still being sold and remains one of the most popular video games of all time.

As for racking up points, it was twenty years (and millions of quarters) before anyone scored a perfect game: 3,333,360 points. Billy Mitchell took six hours on July 3, 1999, to navigate 256 boards (or screens), eating every single dot, blinking energizer blob, flashing blue ghost, and point-loaded fruit, without losing a single life. His reaction: "It was tremendously monotonous." —*TL, LK, CK*

Also October 10:

1846: Lassell Discovers Neptunian Moon Triton Just Seventeen Days After Galle Discovers Neptune

1861: Birth of Fridtjof Nansen, Arctic Explorer, Zoologist, Diplomat

Also 1979:

January 25: Robot Kills Human (see page 25)

October 11
1995: "We're Trashing the Ozone Layer"

Two Americans and a Dutch scientist win the Nobel Prize in Chemistry for their research showing that the release of nitrogen oxide through man-made chlorofluorocarbons damages Earth's natural ozone layer.

The groundwork for the Nobel was laid by Dutch chemist Paul Crutzen of the Max Planck Institute in Germany, who released a landmark 1970 study of the effects of nitrogen oxides on the accelerated decomposition of the ozone layer.

Four years later, professors Mario Molina of MIT and F. Sherwood Rowland of the University of California at Irvine followed up with their own study, published in *Nature*. It described the threat to the ozone layer from chlorofluorocarbon gases (CFCs), or Freons, being released into the atmosphere through their use in plastics and aerosol sprays and as refrigerator coolants. The hard science led to legislation limiting CFC release into the atmosphere.

The erosion of the ozone layer is not only a contributing factor to global warming but a threat to life itself: without the ozone layer to absorb most of the sun's ultraviolet rays, life as we know it is not possible. Besides the danger to humans of sunburn, skin cancer, and eye damage, ozone loss also threatens livestock — less sensitive than humans to UV, but outdoors longer. And UV damages the DNA of marine plankton, which could lead to die-offs and local or regional food-chain collapses that destroy the ocean fisheries for human use.

The Antarctic ozone hole extended so far north in 2000 that health officials in Punta Arenas, Chile, started warning residents not to go out in the midday sun. The problems are also severe in neighboring Argentina and in New Zealand and Australia — which have the highest skin-cancer rates in the world. — *TL*

Also October 11:

1983: Last U.S. Hand-Cranked Telephones Taken Out of Service

Also 1995:

February 15: FBI Busts Computer Hacker Kevin Mitnick

March 2: Yahoo! Incorporated

March 22: Longest Human Space Adventure Ends (see page 83)

March 25: First Internet Wiki Makes Fast Work of Collaboration

May 26: Gates, Microsoft Jump on "Internet Tidal Wave"

October 12
1928: Iron Lung, Savior to a Generation

A young polio sufferer at Children's Hospital in Boston becomes the first person to use the iron lung artificial respirator. Her recovery from respiratory failure is nearly instantaneous.

Sustained artificial respiration had been attempted before. As far back as the late 1700s, physicians experimented with bellows systems. That was discarded, and during the 1800s, several other methods were tried. None achieved much success.

The iron lung was invented by Philip Drinker, an industrial hygienist from the Harvard School of Public Health. Its cylindrical chamber encases a person's entire body, save for the head, and uses regulated air pressure to help a patient breathe when the muscle control necessary for normal breathing has been lost. The dramatic recovery of the first patient immediately invested Drinker's machine with credibility, although he didn't ride the crest of success for very long. An improved version of Drinker's iron lung, built by inventor Jack Emerson, soon arrived on the scene. It was cheaper, lighter, and more efficient, and—following some legal wrangling—it replaced Drinker's lung.

Poliomyelitis, which can paralyze the victim's diaphragm and make normal breathing impossible, was widespread in the 1930s and 1940s, and most of the people placed in iron lungs were polio sufferers. It was the Emerson lung that filled hospital wards during the polio outbreaks that preceded the arrival of vaccines in the mid-1950s. Those with mild cases were able to keep breathing long enough to recover and leave the iron lung after a few weeks, months, or years. Some people spent their lifetimes in the apparatus.

Today, with polio nearly eradicated and more sophisticated breathing devices available, the iron lung has pretty much been sidelined as a respiratory therapy. It is still indicated, however, for some rare conditions, such as the congenital respiratory disease Ondine's curse. —*TL*

Also October 12:

1823: Charles Macintosh Starts Selling Rubberized Raincoats (see page 365)
1920: Work Begins on First Underwater Tunnel for Autos, Holland Tunnel from New York to New Jersey

Also 1928:

June 9: Four-Man Crew Completes First Transpacific Airplane Flight — in Eight Days
November 6: All the News That's Lit to Print (see page 312)

October 13
1884: Greenwich Resolves Subprime Meridian Crisis

The International Meridian Conference selects Greenwich as the prime meridian, the global standard for zero degrees longitude.

Despite the widespread adoption of the metric system (see page 130) and time zones (see page 324), navigation at sea—and charting stars in the heavens—often remained a matter of local, national, or even religious preference. Maps were based on longitude east or west of Jerusalem, St. Petersburg, Rome, Pisa, Copenhagen (think Tycho Brahe), Oslo, Paris, Greenwich (in eastern London), El Hierro (in the Canary Islands), Philadelphia (former U.S. capital), and Washington, DC. These divergent reference meridians ranged over 112 degrees of longitude.

You *could* do the math, but that meant *you* did the math. These were the days before computers and even the bulkiest of mechanical calculators (see page 129). Got abacus?

In the interests of global amity and commerce, President Chester Alan Arthur convened delegates from twenty-five countries in Washington in 1884. But the prime meridian was really a done deal. The United States had already adopted the Greenwich meridian for navigation, and 72 percent of the world's commerce used nautical charts based on Greenwich.

Britain had first solved the problem of longitude, Britain had the world's largest navy, and the sun indeed did not set on the British Empire. Britannia ruled the waves, and was not about to waive its rules.

Thus, the conference established the meridian passing through the Royal Observatory at Greenwich as the world's prime meridian, with all longitude calculated east and west up to 180 degrees. The conference also established Greenwich mean time as a standard for astronomy and setting time zones. The vote to select Greenwich passed twenty-two to one. The Dominican Republic voted against. France and Brazil, diplomatically, abstained. —*RA*

Also October 13:

1937: National Safety Council Offers Guidelines for Walk/Don't Walk Signals for Pedestrians
1972: Rugby Team Plane Crash Leads to Grueling Ordeal in the Andes

Also 1884:

May 1: First Steel-Frame High-Rise, Chicago's Home Insurance Building
June 16: A Technology with Plenty of Ups and Downs (see page 169)

October 14
1858: This History Might Ring a Bell

Manual labor hoists the great hour bell into place in the Houses of Parliament in London. People are already calling the 14.33-ton bell Big Ben.

Fire destroyed most of the ancient Palace of Westminster in 1834. The giant tower of the rebuilt neo-Gothic Houses of Parliament was to have a giant clock (with a twenty-three-foot-diameter face on each side of the tower) and a giant bell. The clock — with fourteen-foot minute hands — was ready in 1854, but the 314-foot-high tower wasn't.

The first bell cracked when they tested it. The bell was broken up, and the pieces taken to East London's Whitechapel Bell Foundry (birthplace of Philadelphia's Liberty Bell). The metal was melted down and poured into a new mold on April 10, 1858. Once the bell arrived at Westminster, it took eight men eighteen hours to raise it into place. They turned a giant windlass hauling an 1,800-foot chain over huge drums. Guide wheels ran along restraining timbers inside the tower to steady the bell's cradle.

After Big Ben was hung in the belfry, the clockworks could finally be installed below it. The bell first rang the hours on May 31, 1859, and officially entered service in July. In September, it cracked. A lawyer on the committee had insisted on a bell hammer twice the weight recommended by the foundry. Big Ben got a lighter hammer in 1862 and was rotated an eighth of a turn so the hammer would hit a different spot. The crack remains, giving the bell its distinctive tone.

Big Ben is probably named for "Big Ben" Caunt, a famous 238-pound professional boxer. His nickname was a catchphrase for the biggest of any particular category. Striking the bell has been controlled by electric motor since 1912, but the clock itself is still hand-wound thrice weekly. — *RA*

Also October 14:

1947: Chuck Yeager Machs the Sound Barrier
(see page 49)
1985: Reference Guide for C++ Programming Language
Published

Also 1858:

July 1: Darwin and Wallace Shift the Paradigm (see page 184)
December 4: Birth of Earmuff Inventor Chester Greenwood

289

October 15
1900: Boston Embraces the Sound of Music

An acoustical marvel, Boston's Symphony Hall opens with a concert by the Boston Symphony Orchestra.

Unlike most American concert halls, which tend to favor a wide, fan-shaped configuration, Symphony Hall was built along European lines—deep, narrow, and high. Architectural firm McKim, Mead & White modeled the Boston hall after Leipzig's Gewandhaus (destroyed during World War II).

But the architects also did something unprecedented: They hired Wallace Clement Sabine, a young assistant physics professor from nearby Harvard University, to act as acoustical consultant. For the first time ever, scientifically proven acoustical principles were applied to concert-hall design. On the basis of Sabine's work, the hall was built using brick, steel, and plaster, with wooden flooring as the only soft material. The side balconies are narrow to avoid trapping sound, and—to help focus it—the stage walls are banked inward. The architects also carved niches into the walls and topped the hall with a coffered ceiling, which, in acoustical terms, guarantees nearly every seat an optimum aural experience.

From maestro Wilhelm Gericke's opening downbeat at the inaugural gala, Symphony Hall was a resounding success with musician and concertgoer alike. When great acoustics are discussed, three halls in the world are almost always mentioned: Vienna's Musikverein, Amsterdam's Concertgebouw, and Boston's Symphony Hall.

The Renaissance-style building is home to both the Boston Symphony Orchestra and the Boston Pops and sits only a block from the New England Conservatory of Music. Symphony Hall was declared a National Historic Landmark in 1999.

The proscenium arch carries a single inscribed plaque. The intention was to make the hall a pantheon of composers, but Beethoven alone was deemed worthy. The other plaques remain empty. —*TL*

Also October 15:
1956: FORTRAN Computing Language Unveiled
2003: China Launches Astronaut Yang Liwei into Space

Also 1900:
November 3: First Big U.S. Auto Show Opens in New York
December 14: Max Planck Gives First Paper on Quantum Mechanics

October 16
1916: Sanger Stakes Everything on Birth Control

The first birth control clinic in the United States opens for business in New York City.

Margaret Sanger, founder of the American Birth Control League (now Planned Parenthood), established her clinic in the Brownsville section of Brooklyn. It took police nine days to figure out what was going on. Then they raided the joint and arrested her.

Well known from her column "What Every Girl Should Know" in the *New York Call*, Sanger was charged with maintaining a public nuisance and was jailed for a month. Released, Sanger reopened her clinic and got jugged again. Sanger's clinic stood in defiance of the Comstock Act, which banned birth control outright and made it a crime to send contraceptives through the mail. Sanger's opposition brought the legal system down on her head, forcing her to flee the country at one point to avoid prosecution. While in Europe, she learned plenty more about contraception and sexual politics.

Sanger saw birth control as not only a woman's issue but a class issue. Although contraception was technically illegal for everyone, wealthy Americans practiced it freely, obtaining condoms and spermicidal jelly from abroad. With Sanger once again forcing the issue in court, a judge lifted the federal ban on birth control devices in 1938. This pretty much ended the Comstock era. Almost immediately, the diaphragm became a popular method of contraception.

Nevertheless, state laws and Puritan morality kept the U.S. from completely legalizing birth control methods (see page 131) until 1965. A year before Sanger's death, the Supreme Court struck down a Connecticut law banning married couples from using birth control, citing the right to privacy. It took another seven years to extend that ruling to unmarried sex partners. —*TL*

Also October 16:
1987: First Successful Newborn Organ Transplant
2002: Second Great Library Opens in Alexandria

Also 1916:
April 30: Birth of Claude Shannon, Father of
    Information Theory
September 15: All Disquiet on the Western Front
    (see page 260)

October 17
1855: Bessemer Becomes the Man of Steel

Englishman Henry Bessemer receives patents for a new steelmaking process that revolutionizes the industry.

The Bessemer converter was a squat, ugly, clay-lined crucible that simplified the problem of removing impurities (excess manganese and carbon, mostly) from pig iron through the process of oxidation. Once the impurities were removed, either as gas or as solid slag, the molten iron was bolstered with other elements to create steel alloys, then poured into molds and given shape. Depending on the size of the converter, as much as thirty tons of molten iron could be processed in one go. Air was blown into the converter through a number of small channels and forced through the liquid to remove the impurities.

The Bessemer process, which could take as little as thirty minutes to complete, resulted in better quality steel that could be mass-produced. This made steel a viable (read: cheaper) building material, and it soon became the standard for heavy construction projects, like skyscrapers and bridges.

The first Bessemer steel mill in the United States opened outside Detroit in 1855. As a Great Lakes port city, and given its proximity to the fertile iron-ore-producing fields in the upper Midwest, Detroit became an early steel-producing town. Bessemer, meanwhile, moved his mill operations to Sheffield in England's industrial Midlands, which became the British equivalent of Germany's Essen, seat of the Krupp steel dynasty.

Bessemer wasn't alone in working on this process. In fact, an American, William Kelly, had independently developed a similar oxidation technique a few years earlier. He held a patent but was forced, through bankruptcy, to eventually sell it to Bessemer.

The Bessemer process was used into the 1960s, when it was finally replaced by newer technologies, including the Linz-Donawitz process. — *TL*

Also October 17:

1604: Johannes Kepler First Sees Kepler's Supernova
1973: Arab Nations Impose Oil Embargo on West

Also 1855:

October 9: Joshua Stoddard Patents Steam-Powered
    Calliope

October 18
1985: Nintendo Entertainment System Launches

Nintendo releases a limited batch of Nintendo Entertainment Systems in New York City, quietly launching the most influential video-game platform of all time.

The American video-game market was in shambles. Sales of game machines by Atari, Mattel, and Coleco had risen to dizzying heights, then collapsed even more quickly. In America, video games were dead, dead, *dead*. Personal computers were the future, and anything that just played games but couldn't do your taxes was hopelessly backward. To get away from the term *video game*, Nintendo (see page 268) took its marketing emphasis off the controller and focused on two accessories it had released for Famicon, the Japanese version.

The Zapper light gun played the target-shooting game Duck Hunt. And R.O.B. the Robot Operating Buddy whirred and spun around, taking commands from the television, helping you play complex games like Gyromite. This was light-years ahead of Atari, went the message: It has a *robot!*

Nintendo launched the system with seventeen games, including Baseball, Golf, Tennis, Pinball, and Donkey Kong Jr. Math. (The trump card, Super Mario Bros., had just been released in Japan but wasn't ready for America.) At this point, you're expecting to hear that the Nintendo Entertainment System was a huge surprise hit, flying off the shelves. But that's not what happened. In fact, Nintendo only sold about 50,000 consoles that holiday season — half of what it had manufactured.

But it was enough to convince retailers that Nintendo had a viable product. In early 1986, Nintendo expanded into Los Angeles, then Chicago, then San Francisco. At the end of that year, NES went national, with Mario leading the charge. Video games were back. — *CK*

Also October 18:
1870: Benjamin Chew Tilghman Patents Sandblasting Process
1945: Red Spy Steals U.S. Atom Bomb Secrets

Also 1985:
January 17: Britain's Red Phone Boxes Are Put on Hold (see page 17)

October 19
1943: A Wonderful Discovery, and a Helluva Row

A biochemistry grad student discovers streptomycin, an antibiotic that will be used to treat tuberculosis and other infectious diseases.

Sole credit for the discovery initially went to Selman Waksman, who ran the laboratory at Rutgers University where the research was performed. But it was Albert Schatz, a twenty-three-year-old graduate student under Waksman, who actually isolated the antibiotic, after months of feverish work.

The key that unlocked streptomycin was Schatz's isolation of two active strains of actinomycete bacteria. Both could stop the growth of other, stubbornly virulent strains of bacteria that had proven resistant to penicillin, itself a new wonder drug.

Years later, Schatz told the *Guardian* about his moment of discovery:

On October 19, 1943, at about 2 p.m., I realized I had a new antibiotic. I named it streptomycin. I sealed the test tube by heating the open end and twisting the soft, hot glass. I first gave it to my mother, but it is now at the Smithsonian Institution. I felt elated, and very tired, but I had no idea whether the new antibiotic would be effective in treating people.

It proved the *most effective* way to fight tuberculosis, a deadly and often fatal infectious disease still widespread at the time.

Waksman, who had once described Schatz as his most gifted student ever, took full credit—and the 1952 Nobel Prize in Physiology or Medicine—after getting the young man to sign over his royalty rights to Rutgers. Schatz said he'd agreed because he believed that streptomycin should be made available quickly and cheaply.

But feeling slighted and discarded, Schatz sued his former mentor and the university in 1950, winning an out-of-court settlement. It was 1990 before Schatz finally received the official credit he had spent four decades pursuing. — *TL*

Also October 19:

1941: Electric Turbines Get First Wind

Also 1943:

June 10: The Ballpoint Pen — Ink Dry for Me, Argentina (see page 163)

October 20
1975: Atari Sits Down on Hi-Way

Atari patents a sit-down cockpit arcade cabinet, ushering in a new era of realism for video games. The design makes Atari's new game Hi-Way a big hit.

Pong (see page 335) fever had the U.S. and the world in its silicon grip throughout 1973, as adults and kids alike rushed to play the latest sensation: coin-operated electronic video games. But Atari realized it would have to get away from iterations like Superpong, Quadrapong, and Pong Doubles and create entirely new games to keep making money. Gran Trak 10, the company's first car-racing game, featured realistic controls (gas and brake pedals, steering wheel and gearshift), but you had to stand up to play.

Hi-Way featured more sophisticated graphics and a cabinet in which you could sit down, as if you were driving a real car. Atari engineer Regan Cheng applied for the design patent on the cabinet October 20, 1975. Hi-Way incorporated both seat and screen into a single molded form, heightening the feeling of sitting inside a vehicle.

Hi-Way still had one problem: Although its cabinet was uniquely designed to enhance realism, the game itself was still played from a third-person point of view. You were driving the car, but you watched from overhead, as if you were having an out-of-body experience. In 1976, Atari solved that problem with Night Driver: the player looked through the windshield. The graphics still left a lot to be desired — just a few small dots suggested a road weaving its way toward the viewer — but it set the standard for all future driving games.

Later games like Sega's motorcycle-racer Hang-On added even more realism to the experience, building the game's monitor into a life-size motorcycle that the player sits and leans on to control the onscreen action — *vrrRROOM!* — CK

Also October 20:
1984: Monterey Bay Aquarium Opens

Also 1975:
January 30: Rubik Applies for Patent on Magic Cube
   (see page 30)
July 17: Apollo and Soyuz Spacecraft Dock in Orbit

October 21
1879: Edison Gets the Bright Light Right

Thomas Edison burns an incandescent electric lightbulb for thirteen and a half hours.

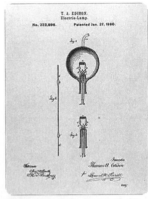

Thomas Edison received a patent for his lightbulb in January 1880.

Edison wanted to replace the harsh arc lighting then available with the soft radiance created by electricity heating up a filament until it glowed. Edison was riding high on the fame and profits from his gadgets for telegraph printing, multiplex telegraphy, telephone improvements, and the brand-new phonograph. In 1878, he figured this project would take three or four months.

The lab was working on a platinum filament, and gas bubbles in the platinum had led Edison to develop an efficient vacuum pump to remove air from inside the bulbs. And that suggested carbon as a filament material. Carbon conducts electricity, has a high resistance, and can be shaped into thin filaments. And it's cheap. But it burns very easily, unless there's no oxygen around. Vacuum bulbs were ideal for carbon.

Edison pushed hard on his research assistants, whom he called muckers. After testing hundreds of materials, they baked a piece of coiled cotton thread until it was all carbon. Inside a near-vacuum bulb, it stayed alight for more than half a day. The three-to-four-month project had taken fourteen months. The lab soon got a carbon-filament bulb to last forty hours. It had cost $40,000 (nearly a million in today's money) and taken 1,200 experiments.

On New Year's Eve, 3,000 people visited the New Jersey lab to witness forty electric lightbulbs glowing merrily. Edison switched them on and off at will, dazzling and delighting his guests. In England, Joseph Swan had beaten Edison by a year (see page 354). But Edison's bulbs lasted longer than Swan's. That, and a talent for self-promotion, won him fame and fortune. — *RA*

Also October 21:

1833: Birth of Chemist Alfred Nobel, Inventor of Dynamite and Nobel Prize

Also 1879:

March 14: Birth of Albert Einstein (see page 327)
April 8: The Milkman Cometh ... with Glass Bottles (see page 100)

October 22
1938: Xerox This

Inventor Chester Carlson produces the first Xerox copy.

Engineer Carlson worked in a corporate patent department, where hand-copying documents or sending them out to be photographed caused delays. Carlson conceived a dry-copying process using photoconductivity: Light striking the surface of certain materials increases electron flow. Project an image onto a photoconductive surface, and current will flow only where light strikes.

Four years of tinkering in his kitchen and in his mother-in-law's beauty salon yielded results in October 1938. Research assistant Otto Kornei put a sulfur coating on a zinc plate and rubbed it with a handkerchief to produce an electrostatic charge. A glass slide marked *10-22-38 ASTORIA* was placed on the plate in a darkened room and illuminated with a bright lamp for a few seconds. Powder from waxy moss spores was sprinkled on the sulfur and then blown off. There it was: a near-perfect mirror image of the writing.

Carlson patented his invention but needed development money. After rejections from IBM, Kodak, General Electric, RCA, and others, he ceded a 60 percent stake to Battelle Memorial Institute, which assigned physicist Roland Schaffert to perfect electrophotography. Battelle licensed the technology to Haloid of Rochester, New York. Battelle and Haloid publicly demonstrated the process on October 22, 1948, precisely ten years after Carlson's dry run. A professor devised the word *xerography* from the Greek for "dry writing," and Haloid soon became Xerox.

The photocopiers introduced in 1949 were a logistical mess: it took fourteen steps and forty-five seconds to make one copy, maximum twelve copies per exposure. The company had its first big hit in 1959 with the Xerox 914.

The Xerox machine had a profound cultural influence. It also made Carlson rich, earning today's equivalent of a billion dollars, two-thirds of which he gave to charity. — *RA*

Also October 22:

4004 BCE: World Created, According to Bishop James Ussher

1905: Birth of Karl Guthe Jansky, Father of Radio Astronomy

1962: Cuban Missile Crisis, and the Brink of War

Also 1938:

December 31: Drunkometer Breath Tests Begin (see page 367)

October 23
2001: Now Hear This...the iPod Arrives

Apple announces the iPod, eventually propelling the company to dominance in the digital-music field and changing the music industry forever.

Despite some conspicuous flaws—a wonky scroll wheel, no Windows compatibility, short battery life, and a whopping four-hundred-dollar price tag—the innocuous-looking device was a game-changer. The iPod was not the first MP3 player, but its simple interface and internal hard drive (replaced by flash memory in later models) set a new standard. Another advantage was integration with easy-to-use iTunes software. Later support for Apple's massive iTunes library provided iPod customers with a vast trove of music to populate their players.

From conception to completion, the original iPod took Apple engineers and designers just under a year to develop. It featured a 5 GB hard drive and was capable of playing music in several audio-file formats.

Apple CEO Steve Jobs said the iPod "puts a thousand songs in your pocket." It did that more efficiently and elegantly than any previous MP3 player. Released November 10, it wasn't an overnight success. Early sales were sluggish, and it wasn't until 2004 that the millionth iPod was sold. Things took off after the release of a Windows version and the rapid introduction of new models: the Mini, Shuffle, and Nano.

The iPod allowed Apple to blow up the music industry's CD-based business model by making downloading singles both cheap and easy. And music execs grumbled that people could rip previously purchased CDs into their iTunes libraries without having to pay extra. With iTunes now easily the world's biggest music retailer, the music industry is still reinventing itself.

Sales of the iPod peaked in early 2008, as people shifted to the iPhone. Still, Apple had sold 300 million by iPod's tenth birthday. —*TL*

Also October 23:

1911: Aero-Plane Makes Its Debut Above the
   Battlefield
1995: Judge Okays First Wiretap of Computer
   Network

Also 2001:

December 15: Leaning Tower of Pisa Reopens with
   New Angle (see page 351)

October 24
1911: Birth of an Inventive Wyeth

Nathaniel Convers Wyeth is born. He'll become a mechanical engineer and inventor of the plastic beverage bottle.

Nathaniel was the son of artist and illustrator N.C. Wyeth; the brother of artists Andrew Wyeth, Carolyn Wyeth, and Henriette Wyeth Hurd; father of drummer-pianist Howie Wyeth; and uncle of artists Jamie Wyeth and Michael Hurd. Nathaniel's original first name was Newell (same as his father), but it was changed at age ten, after the youngster showed an interest in science and engineering. Ice-plow inventor Nathaniel Jarvis Wyeth (see page 258) was an ancestral uncle. Some family, eh?

This Wyeth was a DuPont engineer when he began work in 1967 to produce a lightweight plastic bottle that wouldn't chemically contaminate its contents or split under the high pressure of carbonated beverages. He invented a process called stretch-blow molding that used biaxially oriented polyethylene terephthalate (aka DuPont Mylar).

The new PET bottle was light, strong, and flexible, and it had uniform wall thickness that could withstand pressure of more than a hundred pounds per square inch. It also met Food and Drug Administration standards for purity. DuPont patented it in 1973.

PET beverage bottles are everywhere—including lots of places we don't want them, like our oceans. Over 400 billion PET bottles (400 gigabottles, if you like) are manufactured each year around the globe. PET can be recycled, and it is. In Europe, almost half of it is recycled, and the U.S. rate has risen to about 30 percent. Recycled PET can be turned into carpets, backing for carpets and upholstery, other fibers, sheets and films, and—increasingly—more bottles.

So make sure that plastic bottle goes into the recycling bin. It's far too valuable to throw away. After all, it's a Wyeth. —*RA*

Also October 24:

1861: U.S. Transcontinental Telegraph Makes Pony
Express Obsolete After Just Eighteen Months
1960: Soviet Rocket Explodes, Killing Top Engineers,
Technicians

Also 1911:

January 18: Airplane Lands on Ship for First Time
(see page 18)
January 23: French Academy of Sciences Rejects
Marie Curie (see page 357)

299

October 25
1955: Microwave Arrives; Time to Nuke Dinner

The first domestic microwave oven is introduced. It's an appliance born of the radar systems used in World War II—and the labs of U.S. defense companies.

Raytheon engineer Percy LeBaron Spencer was fiddling with a 1940s radar set when it accidentally melted a candy bar in his pocket. Figuring microwaves could heat other food, he popped corn and cooked an egg. Let's get this show on the road.

Raytheon built Radarange, the first microwave oven in the world, in 1947. It was as large as a refrigerator, but heavier. Its magnetron tubes had to be water-cooled. Result: this test unit required a plumbing hookup, weighed seven hundred and fifty pounds, and was nearly six feet tall.

The commercial microwave oven didn't arrive until 1954. Raytheon's 1161 Radarange was expensive: $2,000 to $3,000 ($17,000 to $25,500 in today's cash). Tappan used Raytheon technology to introduce a large 220-volt wall-unit home microwave in 1955. It sold for $1,295. It had two cooking speeds (500 and 800 watts), stainless steel exterior, glass shelf, top-browning element, and a recipe-card drawer. Consumers stayed away from the strange device. Sales were slow.

Litton Industries developed the short, wide microwave shape we now know. It also created an oven that could survive even when there was no object inside to heat up. Prices fell. Amana—a Raytheon subsidiary—introduced the first popular home model in 1967. The countertop Radarange cost $495 ($3,200 today).

Finally, consumer interest grew. About forty thousand units sold in the United States in 1970. Five years later, that number hit a million. The addition of electronic controls made microwaves easier to use, and they became a fixture in most kitchens. Roughly 25 percent of U.S. households owned a microwave oven by 1986, and almost 90 percent do today. —*PG*

Also October 25:
1671: Cassini Spots a Two-Toned Saturnian Moon

Also 1955:
March 8: The Mother of All Operating Systems
    (see page 69)
July 9: Scientists Call for Nuclear Disarmament
August 7: Sony Sells First Transistor Radios

October 26
1948: Death Cloud Envelops Pennsylvania Mill Town

An inversion layer settles over Donora, Pennsylvania (near Pittsburgh), trapping industrial pollution in the atmosphere. When it clears, six days later, twenty people are dead, fifty are dying, and hundreds will live out their days with permanently damaged lungs.

Inversion occurs when the air near the ground is cooler than the air above it, a reversal of normal atmospheric conditions. When that happens, man-made pollutants are trapped, resulting in smog. The geography of Los Angeles, for example, leads to frequent inversion layers over the basin. Combined with heavy automobile pollution, that gives LA the nation's worst big-city air quality (see page 209).

But LA has never seen anything like the Donora Death Fog. Pollution from the nearby U.S. Steel smelting plants and Donora Zinc Works was the main culprit. Trapped in a temperature inversion, the pollutants blanketed the town during the night of October 26.

The companies connived with the U.S. Public Health Service to cover up the facts of the incident and succeeded in keeping it quiet for half a century. Whistleblowers were silenced; records disappeared. It wasn't until 1994 that a full accounting of what had happened in Donora was finally published. To Philip Sadtler, an industry consultant sent to evaluate the disaster who tried without success to expose the corporate cover-up, U.S. Steel was guilty of murder: "The directors of U.S. Steel should have gone to jail for killing people," Sadtler said shortly before his death, in 1996.

In the end, the death fog sickened 40 percent of Donora's population of fourteen thousand, and the town joined a growing list of other places hit hard by industrial pollution. — *TL*

Also October 26:

1825: Erie Canal Completed, Albany to Buffalo (see page 283)
1984: Baby Fae Gets Baboon Heart in First Cross-Species Transplant to Human
1992: Software Glitch Cripples London Ambulance Service

Also 1948:

June 21: Columbia's Microgroove LP Makes Albums Sound Good (see page 174)

October 27
1946: First Sponsored Television Show Debuts
1994: First Banner Ad Runs on World Wide Web

1946: The first TV show to have a sponsor debuts. The show doesn't last long. Commercials will.

The first legal TV commercial, a ten-second spot for Bulova clocks, ran on New York's WNBT-TV (now WNBC-TV) during a baseball game on July 1, 1941. The audience for the single ad (it was not a sponsorship of the whole broadcast) numbered four thousand television sets. When commercial television picked up again after World War II, audiences were still small, and programmers were desperate for moving images.

Carveth Wells hosted the show *Geographically Speaking*: 16mm home movies of her extensive travels with her explorer husband. It was like watching your neighbors' home movies on a small, grainy black-and-white screen. NBC didn't get a sponsor for the show until November 11, and *Geographically Speaking* ran just six weeks. Wells ran out of her travel movies. Sponsors didn't run out of money.—*RA*

1994: Wired.com, then known as HotWired, invents the web banner ad.

The Mosaic browser (see page 114) was just morphing into Netscape in 1994. And if you think ads slow down page loads now, readers had to download the first banner ads over thin dial-up connections. Despite those handicaps, the gaudy banner ad took over the web, 468 pixels wide by 60 deep. HotWired launched with banner ads from fourteen companies, including MCI, Volvo, Club Med, 1-800-Collect, and Zima. The very first one was likely AT&T, prophetically asking "Have you ever clicked your mouse right here? You will."

Also October 27:

1780: Harvard Astronomers Miscalculate, Miss Solar Eclipse Totality
1931: Outbreak of Dutch Elm Disease Reported

Also 1946:

November 12: The Abacus Proves Its Might (see page 318)

Also 1994:

April 12: Immigration Lawyers Invent Commercial Spam (see page 104)

Banner ads powered the web's first explosion. For the first time, a marketer could actually know how many people saw an ad and, even further, know how many people interacted with it. Online advertising is now a $26 billion business. Banner ads have mutated into pop-unders, pop-overs, and full-site takeovers.—*RS*

October 28
1989: Critter Crunch, Mother of All Robot Battles

MileHiCon, a sci-fi and fantasy gathering in Denver, hosts a truly epochal moment in the history of geekdom: the birth of robot battles.

MileHiCon hosted Critter Crawl in 1988, a sort of beauty pageant for windup toys and remote-control gizmos, but no official winner was declared, and there were no prizes.

"This year, all of that will change radically and violently," wrote event organizer Bill Llewellin in a mailing before the 1989 convention. "The winner will be the last critter standing (rolling, crawling) on the field of combat." Critter Crunch competitors would face off on a folding table. "Some potential entrants are discussing critters capable of significant mayhem," warned the mailer. "So don't get too attached to your entry."

Llewellin devised the parameters of the game with input from fellow members of the Denver Mad Scientists Club. (Their standard uniform was a white lab coat and a hard hat.) Combatants had to be lighter than twenty pounds and smaller than twelve by twelve by twelve inches, though they "may deploy appendages beyond these dimensions."

The Crunch attracted five or ten competitors, says Llewellin. His own forklift creature named Fluffy Bunny upended numerous opponents before being outmaneuvered by a tiny Radio Shack radio-controlled car. (The lack of weight classes made for some chaotic David-and-Goliath matches.) The ultimate winner was Mad Scientist member Pat Thompson's Thing One, a nineteen-pound behemoth armed with a can of Silly String that it sprayed at foes—and the audience.

The Critter Crunch gave birth to robot-fighting leagues around the world, as well as to TV shows and video games. It also reminded us all that 'bots aren't simply toys or industrial helpmates. They are dangerous and terrifying creatures... that will someday conquer and enslave mankind. — *CB*

Also October 28:

1955: Birth of Tech Mogul Bill Gates (see page 96)
1998: President Clinton Signs Digital Millennium
    Copyright Act

Also 1989:

March 23: Scientists Skeptical About Cold Fusion
    Announcement
March 24: *Exxon Valdez* Spill Causes Environmental
    Catastrophe
July 26: First Indictment Under Computer Fraud Act

October 29
1675: Leibniz ∫ums It All Up, Seriesly

Gottfried Leibniz writes the integral sign ∫ in an unpublished manuscript, introducing the calculus notation that's still in use today.

Leibniz was a German mathematician, philosopher, lawyer, and alchemist who fancied himself a poet. He also conducted diplomatic missions. In London he showed an unfinished calculating machine to the Royal Society, which elected him a fellow. Leibniz discussed with his English colleagues his interest in summing series and the geometry of infinitesimals, and corresponded with them from France. They apprised him of the latest books and also told him about Isaac Newton's unpublished work on the subject.

Newton wrote to Leibniz through an intermediary, and they exchanged letters that took weeks or even months to reach the recipient. The muddled back-and-forth led to bad blood, with Newton claiming that Leibniz had stolen his work in founding calculus. Newton's letters, however, described results, not methods. Leibniz's legal and philosophical formalism let him create his own symbolic system: the integral sign *and* the notation of differentials we still use today. Newton published slightly before Leibniz, but the German's notation was superior.

It's another example of simultaneous discovery (see page 184). The scientists were of the same era, associated with the same circles, read the work of the same precursors, and shared some of their own ideas. It should amaze no one that they reached the same results in slightly different language at nearly the same time.

Does Newton deserve credit? Maybe, but it's Leibniz's language you learn in calculus class. And ol' Isaac gets his props for many other discoveries (see page 107), so don't overestimate the gravity of the situation. Happy Integral Day, Gottfried! —*RA*

Also October 29:

1942: 1,500-Mile Alaska Highway Completed in Eight Months as Wartime Necessity

1945: Ballpoint Pens Go on Sale in New York City (see page 163)

Also 1675:

June 22: Greenwich Observatory Established (see page 288)

October 30
1958: Medical Oops Leads to First Coronary Angiogram

A cardiologist accidentally injects a large amount of dye into the small vessels of a patient's heart during a routine test. To the doctor's great surprise—and relief—the dye doesn't send the heart into a fatal spasm, and this happy accident marks the birth of modern cardiac imaging.

Doctors had used dye to view heart valves and chambers, but fear of killing their patients kept them from attempting to visualize the smaller vessels. Conventional wisdom held that injecting contrast dye into the coronary arteries would instantly cause deadly ventricular fibrillation.

F. Mason Sones was attempting to look at the heart valves of a twenty-six-year-old man with rheumatoid arthritis. To his horror, right before an assistant injected forty to fifty milliliters of dye into the aorta, the tip of the tube flipped and more than half went squirting into the right coronary artery. Sones cried, "We've killed him!" and rushed to the patient's side to get ready to open his chest and massage the heart by hand if necessary.

But the patient's heart simply skipped a few beats and recovered. Sones realized almost immediately he'd made a very important discovery: Patients could easily survive even large injections of dye into their heart vessels. And doctors could see blockages that might cause heart attacks.

Over the next few years, Sones and colleagues at the Cleveland Clinic developed the procedure for cardiac catheterization: inserting a flexible tube into the coronary arteries, injecting a small amount of dye, and viewing the arteries with an X-ray camera. By 1967, Sones had performed this procedure on more than 8,200 patients. He's known today as the father of modern cardiac imaging, which has saved the lives of countless heart patients. —*HL*

Also October 30:

1938: "War of the Worlds" Radio Drama Induces Panic
1961: Soviets Detonate 100-Megaton Hydrogen Bomb

Also 1958:

January 31: First U.S. Satellite Discovers Van Allen Belt (see page 31)

October 31
1951: We'll Cross That Street When We Come to It

The first official zebra crossing starts protecting pedestrians at Slough, just west of London.

Postwar Britain had only 10 percent of the road traffic it has now, but fatalities were mounting. The typical pedestrian crossing was marked with nothing more than metal studs in the road: easy for pedestrians to see, but difficult for the motorist. By the time a driver felt the bumps under his tires, it was usually too late to stop.

The government's Transport Research Laboratory ran visibility experiments on new types of crossings using one-twenty-fourth-scale model roads (a half an inch to the foot). Starting in 1949, the lab tested a variety of designs at a thousand locations. Broad black-and-white stripes had the most visual impact.

The new striped crossings were made the legal standard in Britain and widely introduced in late 1951. Pedestrian deaths dropped 11 percent the first year. Future prime minister Jim Callaghan visited the lab in 1948 and is sometimes credited with first noting the crossing's resemblance to a zebra. Despite Callaghan's saying in 1951 that he didn't remember that, nobody else ever claimed credit, and the name zebra crossing caught on.

Cities around the world have adopted this crosswalk of a different stripe. The old-fashioned, two-stripe crossing (with only its edges marked by full-length stripes perpendicular to the direction of traffic) cannot be seen from farther than a hundred feet or so away. Going 30 miles an hour, a driver has about two seconds before he arrives at the crossing. Zebra crosswalks can be seen from greater distances. And pedestrians crossing the street are highly visible as they move against the striped background.

The Beatles brought international fame to the zebra crossing in 1969 with the album cover for *Abbey Road*. Beatles producer Sir George Martin even has a heraldic badge of a zebra carrying an abbot's crook. —*RA*

Also October 31:

1913: Lincoln Highway Links New York and San Francisco, First Transcontinental Auto Road
1917: History's Last Successful Cavalry Charge Captures Beersheba, Palestine

Also 1951:

February 1: TV Shows Atomic Blast, Live (see page 32)

November 1
1909: Literary Giant Delves into Science Fiction with a Cautionary Tale

E.M. Forster publishes "The Machine Stops," the chilling story of a futuristic information-oriented society that literally grinds to a bloody halt. Some of it no longer seems so distant in the future.

The people in Forster's story live alone in small podlike rooms in a honeycomb of vast, multilayered underground cities spread across the globe. The physical comforts of food, clothing, shelter, and medical care are all taken care of by the global Machine. People communicate electronically with voice and pictures, one-to-one and one-to-many. All at the touch of a button.

They're not without anxieties, especially when dealing with the prospect of nonvirtual meetings: "People never touched one another. The custom had become obsolete, owing to the Machine." Each pod comes with an instruction manual, the Book of the Machine. It is the only book: All other info is electronic. The Book not only details which button to push when. It's regarded with reverence: "Vashti was afraid. 'O Machine!' she murmured, and caressed her Book, and was comforted."

Vashti's son criticizes life underground and virtual communication. He makes illegal trips to the surface, where even scholarly travel is forbidden. Experts are expected to do research by redigesting what they can already learn from the vast archives of the Machine: "Beware of first-hand ideas!"

But the music is losing fidelity. The air is no longer fresh. The water begins to smell. The fly in the ointment is that the device that's supposed to remove flies from the ointment has more than a few bugs itself: The Mending Apparatus needs mending. So the Machine and the civilization it supports spiral down rapidly to apocalyptic collapse. That's what happens in a homeostatic system if repair demand chronically exceeds repair capacity.

Consider, for instance, our planet... — *RA*

Also November 1:

1895: American Motor League, First U.S. Auto Club, Founded in Chicago
1952: U.S. Tests First Hydrogen Bomb at Enewetak Atoll in Pacific

Also 1909:

April 6: Robert Peary and Matthew Henson Reach North Pole but Claim Is Disputed
August 10: Leo Fender and the Heart of Rock 'n' Roll (see page 224)
August 31: First Chemotherapy Drug Treats Syphilis

November 2
1947: *Spruce Goose* ... or an Expensive Turkey?

The *Spruce Goose*, with Hollywood producer-aviator-tycoon Howard Hughes at the controls, makes its only flight, skimming the waters of Long Beach Harbor in California for roughly one minute.

That short hop, made mostly for the press and newsreel cameras, climaxed a story that began at the height of World War II. Appalled at the heavy toll German U-boats were taking on Allied shipping, Liberty ship (see page 272) builder Henry J. Kaiser proposed a fleet of gigantic flying transports to move men and materiel across the Atlantic. After Kaiser enlisted Hughes's support, the two men got an $18 million government contract (about $250 million in today's money) to build three flying boats.

Aircraft designer Hughes and his engineers came up with the Hughes H-4 Hercules, an eight-engine behemoth with a wingspan of 320 feet, wider than a football field is long. It was supposed to carry 750 troops. Because of wartime restrictions on critical materials, Hughes built the prototype, HK-1, not out of steel or aluminum but out of wood. The seaplane gained worldwide attention as the *Spruce Goose* (a name Hughes despised), but it was actually constructed largely of birch.

The project bogged down in cost overruns and red tape. Kaiser withdrew in 1944, but Hughes continued. When the government cut off funding and investigated Hughes for misappropriation of funds, he plowed $7 million of his own into the H-4.

Hughes finally got his plane off the ground (or, more accurately, off the water) in 1947. Following its short flight, the *Spruce Goose* was stored in a custom-built hangar and maintained in a state of flight-readiness. After Hughes's death, in 1976, the plane passed from owner to owner and is now the centerpiece of Oregon's Evergreen Aviation and Space Museum. — *TL*

Also November 2:

1815: Boole Born; Boolean Logic Logically Follows
1895: Cars Can't Get to First Gasoline Race
(see page 334)

Also 1947:

December 23: Transistor Opens Door to Electronic
Future (see page 359)

November 3
1993: Theremin Dies — Cue the Spooky Music

Leon Theremin dies in Moscow. He leaves behind a legacy in broadcasting, espionage, circuit design, and, most famously, music. He invented the theremin, the first fully electronic musical instrument, notable for its whooping, sliding high-pitched squeal.

As a young physicist, he attached a speaker to a charged antenna and discovered that when he waved his hand in and out of the electrical field, the speaker emitted a tone. The pitch rose as he moved his hand closer and dropped as he moved it farther away. Harnessing his childhood training as a cellist, Theremin soon mastered a few simple melodies. He could play with a vibrato effect by waggling his hand, and he added a second antenna to his invention to control the master volume. He named it the etherphone, but everyone called it the theremin.

The theremin became an international sensation. RCA began commercially producing theremins in the late 1920s. The Soviets, however, forced him to work on listening devices for government spies. One gizmo, the Thing, was a microphone and transmitter; it was fitted inside a wooden carving of the Great Seal of the United States and given to the U.S. ambassador in Moscow. He hung it in his office, allowing the Soviets to eavesdrop from nearby.

The passive design of the Thing served as the starting point for the technology in today's radio-frequency identification chips. Theremin also originated the technique of using interlaced scan lines to produce a cleaner television picture.

Robert Moog, inventor of the Moog synthesizer, claims Theremin as his greatest influence. The theremin's freaky tones can be heard on spooky film soundtracks like *The Day the Earth Stood Still*, Hitchcock's *Spellbound,* and Tim Burton's *Ed Wood.* —MC

Also November 3:

1900: First Big U.S. Auto Show Opens in New York
1957: Soviets Launch "Muttnik," Sputnik 2 Satellite with Dog Laika Aboard

Also 1993:

July 9: DNA Tests Confirm Identity of Czar's Family's Bones
December 8: Global Positioning System Opened to Civilian Use (see page 344)

November 4
1952: Univac Gets Election Right but CBS Balks

Computers make their first foray into predicting a presidential election on television. Univac makes an amazingly accurate projection that the network doesn't think credible.

The Univac computer.

The Univac, or Universal Automatic Computer, was the next-gen version of the pioneering Eniac built by J. Presper Eckert and John Mauchly at the University of Pennsylvania in the 1940s. The eight-ton, walk-in computer was the size of a one-car garage. Univacs cost about $1 million apiece (more than $8 million in today's money). The computer had thousands of vacuum tubes, which processed a then-astounding ten thousand operations per second. (Today's top supercomputers measure operations in quadrillions per second.)

Remington Rand (now Unisys) approached CBS News in the summer of 1952 with the idea of using Univac to project the election results. CBS was skeptical but thought it might speed things up and be entertaining to use an "electronic brain." Mauchly and mathematician Max Woodbury wrote a program to compare the 1952 returns with previous elections. Pre-election polls had predicted a Democratic victory: anything from a landslide to Democrat Adlai Stevenson barely edging out Republican Dwight Eisenhower.

So it was a surprise at 8:30 p.m. eastern time when Univac predicted Eisenhower would pile up 438 electoral votes to Stevenson's 93. The odds were a hundred to one in favor of Eisenhower winning the election. CBS scoffed at airing the prediction. But the Eisenhower landslide gathered momentum. The final vote was 442 to 89. Univac was less than 1 percent off.

Late at night, the network made an embarrassing confession to millions of viewers: Univac had made an accurate prediction hours before, but CBS hadn't aired it. The public was now sold on this computer stuff. By 1956, all three networks (yes, there were just three) used computer analyses of presidential results. It was here to stay. —*RA*

Also November 4:

1879: Ka-ching! The World's First Cash Register
1939: Packard Introduces Air-Conditioned
   Automobile

Also 1952:

December 1: Ex-GI Becomes Blond Beauty
   (see page 337)

November 5
1992: Discovery of Oldest Beer Ever

Scientists report in the journal *Nature* that evidence of beer has been found in a five-thousand-year-old jug at Godin Tepe, in the central Zagros Mountains of Iran. It's the earliest trace of beer ever discovered.

Researchers just the previous year had confirmed trace evidence of wine from around the same time at the same site, which became a fortress on the Silk Road. But later texts from the area suggest beer was the more popular beverage in lower Mesopotamia and was drunk by common folk as well as the upper class. The discovery of residue from beer-brewing in the interior grooves of a jug from the site supported the idea that beer was indeed the preferred fermented beverage among the Sumerians.

The yellowish substance found in the grooves of the jug, which was in the Royal Ontario Collection, proved to be calcium oxalate, also known as beerstone, a common by-product of brewing with barley. To confirm that it was beerstone, the scientists compared the chemical composition to residue scraped from the inside of a brew kettle at Philadelphia's Dock Street Brewery, as well as to scrapings from an ancient beer vessel from the museum's Egyptian New Kingdom collection.

The grooves may have been intended to collect the beerstone, which can be very bitter and even poisonous, and keep it from ruining the beer. And early Sumerian symbols for beer jugs have similar crisscross markings on them. Further possible evidence of beer: the floor of the supply room where the jug was found had barley on it, likely grown locally and used in the brewing.

It's unclear what type of beer was in the jug, but it is known that residents of Mesopotamia enjoyed many different varieties, including light, dark, amber, sweet, and specially filtered beers. Drink up! —*BM*

Also November 5:

1893: Birth of Industrial Designer Raymond Loewy
1895: George Selden Receives First U.S. Automobile
Patent
1955: Time-Travel Day in *Back to the Future*

Also 1992:

January 12: HAL of a Computer (see page 12)

November 6
1928: All the News That's Lit to Print

The *New York Times* begins flashing headlines to pedestrians outside its offices at 1 Times Square, using an electronic news strip that wraps around the fourth floor of the building.

Crowds watch wartime headlines, 1944.

The Motograph News Bulletin, aka "zipper," was a technological marvel of its day. It extended 380 feet around the Times Tower and displayed five-foot-tall moving letters visible from several blocks away. The first neon sign (see page 347) in Times Square appeared in 1924. But the zipper, with its streaming headlines, was something new and arresting. A *Times* column from 2005 said inventor Frank C. Reilly calculated there were 261,925,664 flashes an hour from the zipper's 14,800 bulbs.

The first use anywhere of the zipper happened to be election day, and the zipper's first streaming headline announced a new president: HERBERT HOOVER DEFEATS AL SMITH. Historic moments throughout the twentieth century became frozen as zipper headlines in the national consciousness:

PRESIDENT ROOSEVELT IS DEAD

TRUMAN ANNOUNCES JAPANESE SURRENDER

PRESIDENT KENNEDY SHOT DEAD IN DALLAS

MAN ON MOON

NIXON RESIGNS

Between monumental news events, the zipper churned out more prosaic headlines—even weather forecasts and sports scores. When the *Times* moved to West Forty-Third Street, in 1963, *Newsday* took over running the zipper. It's since been acquired by Dow Jones, which gave it a complete face-lift.

Also November 6:

1944: First Weapons-Grade Plutonium Produced

1971: Largest U.S. H-Bomb Tested Underground in Aleutian Islands

Also 1928:

July 2: America's First TV Station Goes on the Air (see page 185)

November 7
1905: Remote Control Wows Public

Spanish engineer Leonardo Torres-Quevedo uses a stationary radio remote controller to pilot a boat to more than a mile away in the Bilbao estuary. The crowd is amazed.

Nikola Tesla patented a wireless device for "controlling mechanism of moving vessels or vehicles" in 1898 and demonstrated a radio-controlled boat at New York's Madison Square Garden the same year. Torres-Quevedo began *his* work as a way of testing dirigibles without risking human life. He built a prototype of his "Telekine" and obtained patents in 1902 and 1903. He could soon control a tricycle almost a hundred feet away, using a telegraph key to make it go back and forth and to steer it left and right.

He tested small boats in a Madrid pond in 1905. But the big public demonstration came in Bilbao. Torres-Quevedo stationed himself on the balcony of the yacht club. The boat, the *Vizcaya*, carried eight passengers. The Telekine aboard would receive radiotelegraph commands to control the *Vizcaya*'s electric engine. Using just a wireless telegraph station, Torres-Quevedo guided the boat from the yacht club to mid-estuary, executed turns and reverses, and brought it back in. Triumph.

Tesla notwithstanding, the Institute of Electrical and Electronics Engineers recognizes Torres-Quevedo as originator of "modern wireless remote-control operation principles." Torres-Quevedo's advances in dirigible engineering helped British and French armies counter aerial domination by Germany's zeppelins during World War I. The inventor also designed the Aero Car cable ride over the Niagara Whirlpool in Canada.

The remote control you know best is for the television. The first wireless TV remote was the 1956 Zenith Space Commander, It relied on ultrasonic tones and was a big hit, even though it boosted the TV set's price 30 percent. Ultrasonic remotes were superseded in the 1980s by infrared control. — *RA*

Also November 7:

1932: Space Adventurer Buck Rogers Debuts on CBS Radio

1940: "Galloping Gertie" Tacoma Narrows Bridge Collapses

Also 1905:

November 21: It Was a Very Good Year, If You Were Einstein (see page 327)

November 8
1895: Roentgen Stumbles upon X-Rays

German physicist Wilhelm Roentgen is working in his laboratory in Würzburg when he accidentally discovers the X-ray.

Roentgen was conducting experiments with a Crookes tube, which is basically a glass gas bulb that gives off fluorescent light when a high-voltage current is passed through it. He noticed that the beam turned a screen nine feet away a greenish fluorescent color, despite the tube's being shielded by heavy black cardboard.

Roentgen concluded, correctly, that he was dealing with a new kind of ray, one that cast the shadow of a solid object even when passed through an opaque covering from its point of origin. Not exactly knowing what kind of ray he was working with led him to call it an X-ray. The name stuck.

To test his discovery, Roentgen made an X-ray image of his wife Bertha's hand, which clearly showed her bones, as well as a pretty hefty wedding ring.

In the next couple of months, Roentgen published a paper about his discovery: "On a New Kind of Rays." He made a presentation before the Würzburg Medical Society and X-rayed the hand of a prominent anatomist, who proposed naming the new ray after Roentgen. You don't hear them called Roentgen rays much these days, but the term *roentgenology* is still current, and the roentgen is a radiological unit of measure.

Roentgen received the first Nobel Prize in Physics in 1901 for his discovery. X-rays are no longer a mystery but a major tool of medical diagnosis. — *TL*

November 9
1801: Birth of Pioneer Food Technologist Gail Borden

Inventor Gail Borden Jr. is born. His method of condensing milk to preserve it without refrigeration will create an early and enduring industrialized, mass-market foodstuff.

After an itinerant childhood in New York, Kentucky, and Indiana, Borden moved to Mississippi and then to Texas shortly before it declared independence from Mexico. He worked as a surveyor and mapmaker, then as an influential newspaper publisher and politician. He served as the Republic of Texas's collector of customs in the port of Galveston and made a fortune in local real estate there.

Borden turned to inventing. His "terraqueous machine" was an amphibious wagon that could—theoretically, at least—travel overland or on rivers. Then he invented meat biscuits. *Scientific American* and other publications commented favorably on these combinations of flour with dehydrated beef or chicken. Borden tried to develop markets for his "portable desiccated soup-bread," especially as military rations, but failed and lost his fortune.

Undaunted, Borden applied his considerable inventiveness to the recent invention of canning (see page 323). He patented a technique in 1856 for condensing milk by boiling it in a vacuum. Adding sugar to the milk before sealing it in cans created a nearly indefinite shelf life. Operating out of New York and Connecticut, Borden tried repeatedly to market the product, but he lost money for his investors. It was the Civil War, with its rapidly moving armies, that created the demand that finally led to Borden's success. He enjoyed a prosperous later life. Borden also tried condensing apple juice, coffee, and tea, but those never took off like condensed milk.

After many mergers and acquisitions of the business, the Borden brand, but not the company, lives on. And classic recipe books are replete with kitchen goodies like grandma used to make with Borden's condensed milk. —*RA*

Also November 9:

1842: U.S. Issues First Patents for Designs
1925: R.A. Millikan Announces Discovery of Cosmic Rays

Also 1801:

December 10: Robert Hare Introduces Oxyhydrogen Blow-Pipe, Precursor of Blowtorch

November 10
1983: Computer Virus Is Born

Fred Cohen, a University of Southern California graduate student, offers a prescient peek at the digital future when he demonstrates a computer virus during a security seminar at Lehigh University in Pennsylvania. A quarter of a century later, computer viruses have become a pandemic for which there's no vaccination.

Cohen inserted his proof-of-concept code into a Unix command, and within five minutes of launching it on a mainframe computer, he had gained control of the system. In four other demonstrations, the code managed to seize control within half an hour on average, bypassing all security mechanisms. Cohen's academic adviser, Len Adleman, likened the self-replicating program to a virus, thus coining the computer term.

Cohen's malware wasn't the first. Others had theorized about self-replicating programs that could spread from computer to computer, and a couple of tinkerers had already successfully launched digital infections. But Cohen's proof-of-concept program put computer scientists on notice about the potential scourge, and gave it a name.

A fifteen-year-old actually beat Cohen to the punch in 1982. Rich Skrenta's Elk Cloner program infected Apple II computers through a floppy disk but didn't spread widely outside his circle of friends. The first virus spotted in the wild was the Brain, written in 1986 by two brothers who claimed they intended to infect only IBM PCs running bootleg copies of their heart-monitoring program. The virus included a copyright notice with the brothers' names and phone numbers so people with infected computers could contact them for a "vaccination."

The Love Bug virus that struck in 2000 spread more widely than any malware before it, hitting 55 million computers and infecting 2.5 to 3 million. It caused up to $10 billion in damage, but the Filipino student who unleashed it escaped prosecution because his country had yet to enact a computer-crime law. — *KZ*

Also November 10:

1903: Mary Anderson Patents Windshield Wiper
1983: Bill Gates Announces Windows ...Two Years
   Early

Also 1983:

January 26: Spreadsheet as Easy as 1-2-3
   (see page 26)
June 23: Domain-Name Test Sets Stage for Internet
   Growth

November 11
1930: Einstein Gets Ice-Cold

Albert Einstein and fellow nuclear scientist Leo Szilard receive an American patent for a new refrigerator that requires no electricity.

The most famous physicist of the twentieth century (see page 327) wasn't a prolific inventor à la Thomas Edison (see page 296): the fridge would prove to be one of Einstein's few forays into the world of commonplace engineering. The refrigerator uses chemical reactions of ammonia, butane, and water to turn a heat input into a cold output.

Ammonia gas is released into a chamber with liquid butane in it. This reduces the boiling point of the butane, causing it to evaporate and draw in energy from the environment, cooling the area outside the evaporator. The mix of gases passes through to a condenser filled with water. The ammonia dissolves into the water, and the butane condenses into liquid, which sits atop the water-ammonia mixture. The butane runs back into the evaporator, a heat source is used to drive the ammonia back into gas, and the ammonia heads to the evaporator to begin the cycle again. (Got that? The test is Friday.)

The fridge never became a commercial product, but Swedish company Electrolux licensed some of the patents, and academics have recently built coolers based on the Einstein-Szilard cycle. The design is intriguing because it doesn't use Freon or electricity, which could make it a simpler alternative in poor countries. The only problem is that compared to the modern refrigerator, it isn't very efficient at cooling per unit of energy input. At last report, however, Oxford researchers were working to quadruple the cooling output.

Einstein's only other U.S. patent was for a "light-intensity self-adjusting camera," which did come along eventually. Szilard's other patents mostly related to nuclear reactors and were assigned to the government. *—AM*

Also November 11:
1572: Tycho Brahe Observes "Tycho's Nova"

Also 1930:
January 6: First Diesel Auto Trip, Indianapolis to New York
February 18: Pluto Discovered (see page 238)
May 15: Stewardesses Make the Skies a Little Bit Friendlier
September 8: 3M Starts Marketing Scotch Tape

November 12
1946: The Abacus Proves Its Might

The U.S. Army holds a contest pitting a Japanese abacus against an electric calculator. The abacus wins.

The *soroban,* or Japanese abacus, is a handy calculating tool that hasn't changed since the nineteenth century. Despite the ubiquity of digital calculators, the *soroban* is still used in Japanese schools and banks today, and many users are faster on it than on calculators. One of the secrets behind the *soroban*'s popularity: it proved itself in that epic battle against a calculator.

A *soroban* has a rectangular frame with an odd number of vertical rods. Each column has five beads. The frame is traversed with one horizontal bar, which splits the beads into a set of four and a single bead below the horizontal fold. The single bead is called a heavenly bead and is valued at 5, and the other four, called earth beads, are each valued at 1.

A standard-size *soroban* has thirteen rods, though some have as few as nine. Having more rods allows for calculation of more digits or representations of several different numbers at a time. Most Japanese *sorobans* are made of wood and have metal or bamboo rods for the beads to slide on. What sets the *soroban* apart from its Chinese progenitor, the *suanpan,* is a dot marking every third rod.

The *soroban*'s biggest moment was its face-off against an electric calculator. At the Ernie Pyle Theater in Tokyo in 1946, Private Thomas Nathan Wood of the U.S. Army sat with an early electric calculator against Kiyoshi Matsuzaki from Japan's postal ministry. Scoring in the contest was based on speed and accuracy of results in addition, subtraction, multiplication, and division—and problems that combined all four. The abacus scored four points against just one point for the electric calculator.

Some abacus users were still winning contests like this as late as 1980.—*PG*

Also November 12:

1912: Searchers Discover Bodies of Scott's South Polar
    Team
1935: Portuguese Neurosurgeon Performs First
    Frontal Lobotomy

Also 1946:

November 13: Artificial Snow Falls for the First Time
    (see page 319)

November 13
1946: Artificial Snow Falls for the First Time

Artificial snow is produced for the first time in the clouds over Mount Greylock, Massachusetts.

While not exactly a blizzard—in fact, no snow ever hit the ground—it was the harbinger of a new industry and an overnight sensation. Using pellets of dry ice (frozen carbon dioxide), Vincent Schaefer, a scientist working for General Electric, seeded the clouds from an altitude of fourteen thousand feet. He was carrying out the first field experiment resulting from lab work in which he had successfully created precipitation by placing dry ice in a chilled chamber.

Flying over Mount Greylock (the highest point in the Berkshires of western Massachusetts), Schaefer dropped his pellets and produced the effect in the clouds, which resulted in snow that fell an estimated three thousand feet before evaporating in the dry air.

Artificial snow, like so many other scientific innovations, was born out of wartime necessity. In this case, development began during World War II experiments to create artificial fog, meant to conceal ships at sea. Schaefer, a research associate under Nobel Prizewinner Irving Langmuir, began examining the physics of cloud formation. This work led him to his postwar experiments with cloud seeding and the eventual creation of artificial snow.

Despite protests that artificial snow shouldn't be used because it messed with Mother Nature's design, ski resorts soon began looking for ways to create the fake stuff for use during low- (or no-) snow years. Nowadays, artificial snow is made by a variety of machines and seeding methods. In addition to being used by the ski industry, artificial snow is popular on movie sets and in locales where snow doesn't normally fall. — *TL*

Also November 13:

1460: Death Stills Prince Henry the Navigator
1982: Teen Sets *Asteroids* Record in Three-Day
    Marathon

Also 1946:

October 27: First Sponsored Television Show Debuts
    (see page 302)

November 14
1666: Watching a Transfusion, and Taking Notes

Samuel Pepys, writing in his famous diary, records the first description of a blood transfusion.

Pepys's observations of the dog-to-dog transfusion came barely four decades after English physician William Harvey (see page 156) demonstrated the circulation of the blood. Before Harvey, knowledge was so limited that in 1492 as Pope Innocent VIII lay dying, his physician suggested introducing fresh blood to the pontiff — orally. It didn't work.

The idea of replenishing blood through transfusion caught on shortly after Harvey. Physician Richard Lower experimented on animals, devising instruments and studying ways to get around the problems of clotting. Lower performed the first successful blood transfusion between dogs in 1665. Or partially successful: the donor dog bled to death.

Pepys observed an experiment a year later "upon a little mastiff and a spaniel with very good success, the former bleeding to death, and the latter receiving the blood of the other, and emitting so much of his own, as to make him capable of receiving that of the other." He added another's remark that it might "be of mighty use to man's health, for the amending of bad blood by borrowing from a better body."

Within a year, Lower and Jean-Baptiste Denys, personal physician to France's Louis XIV, performed the first transfusions involving human subjects. In Denys's case, a fifteen-year-old boy received the blood of a sheep and somehow survived, probably because of how little blood was used.

Subsequent sheep-blood transfusions were not as successful, and the practice was banned. British obstetrician James Blundell performed the first successful transfusion using only human blood in 1818. It was not until the first decade of the twentieth century that Viennese physician Karl Landsteiner discovered agglutinins and iso-agglutinins in the blood, and the four major blood groups were identified. — *TL*

Also November 14:

1889: Journalist Nellie Bly Travels Around the World in Seventy-Two Days
1910: Plane Takes Off from Ship for First Time
(see page 18)
1922: BBC Starts Broadcasting Its First Daily Radio Service

Also 1666:

December 22: French Academy of Sciences Founded

November 15
1926: NBC Starts Broadcasting

The National Broadcasting Company takes to the air.

Early radio networks were ad hoc and experimental arrangements, linking stations in various cities for special events like presidential speeches. AT&T was a leader, because it owned Western Electric (which made radio components), stations in several cities, and the telephone lines to link them. The Radio Corporation of America had ambitions to tie its local stations into a national network but was hamstrung by the poor audio quality of telegraph lines it leased from Western Union.

When AT&T decided to concentrate on phones, it sold its New York and Washington radio stations and allowed RCA to lease AT&T phone lines—a huge audio upgrade. RCA's new division, the National Broadcasting Company, officially launched programming November 15, 1926. Everything was live, of course; most of the entertainers are long forgotten, but over the decades they included Al Jolson; Jack Benny; Bob Hope; Fred Allen; Burns and Allen; and Edgar Bergen, a ventriloquist (think about it).

The NBC radio network was so big, it was actually two networks: NBC Red, the flagship, featured established shows and advertisers. NBC Blue carried shows without regular sponsors, such as news and cultural programs. (On the West Coast, NBC Orange carried Red Network programming in the early years, and NBC Gold carried Blue Network programming.)

NBC needed a way to alert everyone simultaneously and instantly when it was time for a local station break. This led to the birth of the three-tone chime NBC still uses. It was the nation's first audio trademark.

NBC got so successful the FCC split it in two. The Blue Network became the American Broadcasting Company (ABC) in 1945.—*JCA*

Also November 15:

1864: Sherman's March to the Sea Changes Warfare with Deliberate Strike at Infrastructure
1904: Gillette Patents Safety Razor with Disposable Blades

Also 1926:

March 16: Goddard Launches Rocketry (see page 77)

November 16
1904: Vacuum Tube Heralds
Birth of Modern Electronics

British engineer John Ambrose Fleming invents and patents the thermionic valve, the first vacuum tube. With this advance, the age of modern wireless electronics is born.

Although the Supreme Court eventually invalidated Fleming's U.S. patent—ruling that the technology he used for his invention was already known—he remains the acknowledged inventor of the vacuum tube, a diode (having two electrodes) that had far-reaching applications. The tube was standard equipment in radio receivers (see page 274), radar sets, early television sets, and other forms of electronic communication for at least fifty years, until it was replaced by solid-state electronics (see page 359) in the mid-twentieth century.

The principle of thermionic emission, essentially the transmission of a charged current using a heated conductor, was certainly well understood before Fleming incorporated it into his tube. It was first reported in 1873, and a number of other engineers and physicists—including Thomas Edison (see page 296)—had experimented with it.

Fleming's vacuum tube, however, represented a major breakthrough in the technology. For his work, he received a knighthood in 1929 and was awarded the Medal of Honor by the Institute of Radio Engineers (now the Institute of Electrical and Electronics Engineers) in 1933.

Fleming lived long enough to see the fruits of his labor literally save Britain during World War II. Radar sets using Fleming's diodes proved decisive in the Battle of Britain, allowing a relatively small number of British fighter planes to effectively turn back the German Luftwaffe's onslaught against the home island.

Fleming died in 1945 at age ninety-five, just three weeks before the fall of the Third Reich. —*TL*

Also November 16:

1932: New York's Palace Theater Switches to Movies; Vaudeville Is Dead

2000: Internet Overseers Okay Seven New Top-Level Domains

Also 1904:

May 4: U.S. Agrees to Complete the Panama Canal

July 21: All Aboard for Siberia (see page 204)

September 14: First Run of Isle of Man Tourist Trophy Road Race

November 17
1749: Father of Modern Canning Born

Nicolas Appert is born. He will invent the modern food-canning process while trying to help Napoléon conquer Europe.

By 1795, France was in an expansionist mood and quarreling with its neighbors. As the army and navy found themselves increasingly embroiled in foreign entanglements, the realization that an army travels on its stomach began forcefully hitting home. Looking for a way to efficiently provision its troops in the field, the revolutionary government offered a prize of 12,000 francs to whoever could devise a way of doing just that.

Nicolas Appert, an experienced chef living on the outskirts of Paris, took up the challenge. More than a decade later, he had the solution. Through experimentation, Appert eventually concluded that the best method of preservation was to heat the food to the boiling point of water, then seal it in airtight glass jars.

Appert's principles were tested successfully by the French navy, which found that everything from meat to vegetables to milk could be preserved at sea using his method. Napoléon was running things by now and immediately recognized the benefit to his far-flung armies. He was so grateful to have the problem of victualing solved that in 1810 he had the imperial government award Appert the 12,000 francs (worth perhaps $34,000 these days).

Appert took the money and opened the world's first cannery (*entrepreneur* is a French word). The cannery was destroyed in 1814 as Napoléon's world came crashing down. A few years later, Englishman Peter Durand refined the process even more by switching from glass to the tin containers we associate with modern canning.

Fortunately for Appert, Napoléon had *not* retained his services as chef on France's ill-fated invasion of Russia. Thus well preserved, Appert survived until 1841, dying at the ripe old age of ninety-one. — *TL*

Also November 17:

1790: Birth of Mathematician-Astronomer-Physicist
    A.F. Möbius
1869: Suez Canal Opens
1970: Douglas Engelbart Patents Computer Mouse
    (see page 345)

Also 1749:

May 17: Birth of Edward Jenner, Father of Vaccination
    and Immunology

323

November 18
1883: Railroad Time Goes Coast to Coast

U.S. and Canadian railways adopt five standardized time zones to replace the multiplicity of local times in communities across the continent. Everyone would soon be operating on railroad time.

Localities used to set their clocks by local noon—whenever the sun was highest. A town a few hundred miles west of yours might set its clocks ten or fifteen minutes later. No big deal: you couldn't get there fast enough for it to matter…until the railroads came.

Thousands of municipalities all worked to their own local times. Railroad timetables used about a hundred different standards. A single railroad that traveled east to west would use multiple noons. A big-city station might have five or six different clocks, one for each railroad serving the station, each running on its own time.

England, Scotland, and Wales standardized to Greenwich mean time (see page 288) on December 6, 1848. After decades of confusion, North American railroads finally agreed to the General Time Convention, on October 11, 1883. They adopted five time zones: intercolonial time (in eastern Canada, now called Atlantic time) and the eastern, central, mountain, and Pacific time zones. The U.S. zones were based on solar noon at 75, 90, 105, and 120 degrees west of Greenwich.

The new system took effect at noon on November 18. Conductors across North America resynchronized their watches from their individual railroads' times to the new standard times. Some folks objected, thinking they were being robbed of minutes of life. But businesses soon followed the lead of the railroads, and people everywhere began setting their clocks and watches to railroad time.

So convenient was the system of time zones that it thrived entirely on the say-so of the railroads for thirty-five years. Congress didn't enact standard time until 1918. —*RA*

Also November 18:

1894: *New York World* Prints First Sunday Comics
Section
1913: U.S. Pilot Loops the Loop
1963: First Pushbutton Phones in U.S. Go into Service

Also 1883:

January 19: Edison Lights Up New Jersey Town
(see page 19)
August 26: Krakatau Eruption Begins; Volcano Affects
Global Climate for Five Years

November 19
1996: Canadian Bridge Crosses Eight Miles of Icy Ocean

The structure of the Confederation Bridge is finished. It connects Prince Edward Island to New Brunswick on the mainland. Eight miles long, it is the world's longest bridge over icy waters.

When Prince Edward Island became a province of Canada, in 1873, the terms of the union, enshrined in the constitution, included year-round ferry service to guarantee trade and communication. But a winter voyage across the Northumberland Strait routinely involved getting stuck in the ice, often leading to casualties. Bridges, tunnels, and causeways were regularly proposed, with varying degrees of credibility. Technological advances finally made a permanent link possible. A 1992 plebiscite on Prince Edward Island and a 1993 constitutional amendment permitted a privately financed bridge to replace the ferry.

Construction began in late 1993. The design consisted of sixty-five piers in water often more than a hundred feet deep. The piers support a series of prestressed-concrete box frames, designed to resist a hundred years of high winds, strong currents, and the continual crush and release of ice floes. The bridge is essentially 176 different pieces organized into three parts: the west approach bridge, east approach bridge, and navigation span. Sections for the east and west spans were built in New Brunswick, while the central navigation span was constructed on Prince Edward Island.

A massive catamaran, designed specifically to build bridges, was brought over from Denmark to move the pieces into place. The same elements the bridge has to withstand—changing ice conditions, high winds, and unpredictable currents—complicated the building process but were successfully overcome.

The final concrete box was put into place on November 19, 1996, marking the completion of the bridge's structure. It opened to traffic the following May 31, and Prince Edward Island saw an immediate increase in visitors.—*ZR*

Also November 19:
1981: Philippine President Marcos Bans All Video Games as "Socially Destructive"

Also 1996:
December 4: GM Delivers EV1 Electric Car
(see page 340)

November 20
1923: Traffic-Signal Patent Has GE Seeing Green

Garrett Morgan patents his version of the traffic signal.

Morgan's signal resembled the semaphore signals seen at rail junctions. The T-shaped pole had three positions; stop and go are obvious, but Morgan included a cycle for all traffic to stop. That reduced accidents and gave pedestrians half a chance. Another innovation was the half-mast position, indicating the signal was not operating: proceed with caution.

It wasn't the first traffic light or the first patent for a traffic signal. But Morgan's design eventually attracted the attention of General Electric, which bought the rights from him for $40,000 (worth about $500,000 these days). Armed with the patent, GE went on to monopolize the manufacturing of traffic signals in the United States. Lots of green to be made there, with little chance of going in the red.

Morgan, a black inventor who lived in Cleveland, also came up with the first practical gas mask. It was originally used by firefighters in the early 1900s, and then by miners, before being put to grimmer uses on the European battlefields of World War I.

The earliest known traffic signal dates back to 1868 London, well before automobiles made an appearance. The signal, actually a revolving lantern that flashed red lights (for stop) and green lights (for caution), was illuminated by gas and operated by hand. It exploded on January 2, 1869, injuring the policeman-operator. In 1920, Detroit police officer William Potts devised the first four-way, three-color traffic light. It, too, was based on railroad signal technology. Within a year of the signal's appearance, the city of Detroit had installed fifteen of them around town.

But it was Morgan's version that caught the eye of GE. — *TL*

Also November 20:

1820: Whale Rams and Sinks *Essex*, Inspiring
  Moby-Dick
1984: SETI Institute Founded to Search for
  Extraterrestrial Intelligence

Also 1923:

February 2: Leaded Gasoline Goes on Sale
March 12: Talkies Talk ... on Their Own (see page 73)

November 21
1905: It Was a Very Good Year, If You Were Einstein

Albert Einstein publishes "Does the Inertia of a Body Depend Upon Its Energy Content?" It is the last in a series known collectively as Einstein's annus mirabilis (or "miracle year") papers.

The papers set forth the essentials of his theory of relativity and helped form the basis of modern physics. Einstein's final 1905 paper solidified his theory of special relativity, with its formula for mass-energy equivalence: $E = mc^2$.

Despite early speech difficulties, young Albert was clearly more than your basic precocious kid. By the age of twelve, he had mastered calculus and Euclidean geometry and possessed a solid grasp of deductive reasoning. Einstein wrote his first scientific paper in his teens, then got cocky and decided to skip the rest of high school. He attempted to enroll in Switzerland's Federal Polytechnic Institute, failed the entrance exam, and returned to high school.

Einstein completed the 1905 papers at age twenty-six. He hadn't landed a teaching job and was clerking at the Swiss Patent Office. Einstein won the 1921 Nobel Prize in Physics.

He was also a political man. A nonpious Jew who embraced a Spinozan view of God, he became a passionate Zionist who worked for a Jewish homeland. He loathed Soviet dictator Joseph Stalin, as he loathed all authority, yet he enlisted in many communist-inspired organizations, partly because of his equal disdain for capitalism. He helped develop the atomic bomb, then—after Hiroshima and Nagasaki—became an outspoken opponent of nuclear weapons. Einstein denounced McCarthyism and racism with the same fervor that he condemned European transgressions.

At the close of the twentieth century, *Time* magazine named Einstein Person of the Century. His birthday is March 14: 3/14, which is Pi Day.—*TL*

Also November 21:

1877: Thomas Edison Announces Invention of Phonograph
1968: Birth Defects Reveal Horrendous Chemical Pollution at New York's Love Canal

Also 1905:

October 22: Birth of Karl Guthe Jansky, Father of Radio Astronomy
November 7: Remote Control Wows Public (see page 313)

November 22
1963: Zapruder Films JFK Assassination

President John F. Kennedy is assassinated in Dallas.

Spectator Abraham Zapruder films twenty-six seconds of the assassination on his 8mm camera, contributing one of the twentieth century's earliest and most significant pieces of user-generated content. The events of the weekend that follows will be beamed by satellite worldwide in the first giant example of the global village.

Conspiracy theories abound, and half a century later, we're still not entirely sure what happened that day. But the assassination changed the political landscape of the United States. The aftermath changed the media landscape of the world. Zapruder sold publication rights to *Life* magazine, which ran his jarring, graphic still frames in its next issue. The sequence was not shown as a film clip on network television until 1975.

Where were the TV cameras that day? They were in studios: bulky and barely mobile, some the size of refrigerators. Mobile vans usually relied on landlines. The U.S. president was not yet under constant video watch. Compact TV cameras and microwave and satellite uplinks were still in the future. (Hundreds of witnesses carrying video-ready smartphones? Even further off.)

What was on television that November weekend was the return of Kennedy's coffin to Washington, his repose at the White House and lying in state in the Capitol, the funeral at St. Matthew's Cathedral, burial at Arlington, and the corteges linking these events. That and the live onscreen killing of suspected assassin Lee Harvey Oswald.

It was a harbinger of the media world to come. Transoceanic satellite links were new and expensive (see page 206), but for a story of such suddenness, importance, and personal drama, TV went full out. Millions around the world watched. Communications theorist Marshall McLuhan deemed the Kennedy funeral a founding instance of the global village, a media experience shared in real time across borders and continents. —*TL, RA*

Also November 22:

1910: Arthur Knight Patents Steel-Shaft Golf Clubs
1932: Robert Jauch Patents Gasoline Pump with
   Built-In Price Calculator

Also 1963:

July 19: Cracking the 100-Kilometer Altitude
   Barrier...in a Plane
December 7: Video Instant Replay Arrives
   (see page 343)

November 23
1889: SF Gin Joint Hears World's First Jukebox

The first jukebox is installed at the Palais Royale Saloon in San Francisco. It becomes an overnight sensation, and its popularity spreads around the world.

The Pacific Phonograph Company constructed the device around an Edison Class M phonograph, which was driven by electric batteries and played cylinder recordings. (Emile Berliner—see page 65—was still perfecting the disk record that would eventually sweep away the Edison cylinders.)

The jukebox phonograph itself was fitted inside an oak cabinet to which were attached four stethoscope-like tubes. (The batteries provided motive power only, not amplification.) The listening tubes operated individually, each activated by the insertion of a nickel, meaning that four different paying customers could hear the same song simultaneously. Towels were on hand, so a patron could hygienically wipe off the end of the tube before and after each song.

Louis Glass and William S. Arnold, the entrepreneurs who installed the new invention at the Palais Royale, originally marketed it as the nickel-in-the-slot player. (A nickel then had the buying power of about $1.20 today.) Glass and Arnold received a patent for their "coin-actuated attachment for phonographs" on May 27, 1890. The success of the jukebox eventually spelled the end of the player piano, then the most common way of pounding out popular music to a line of thirsty barflies.

But the automatic coin-operated phonograph did not come to be known as the jukebox until the 1930s; it picked up the name in the southern United States. The etymology of the word, though, remains a bit vague. It may derive from *juke house,* a slang reference to a bawdy house, where music was certainly not unknown. —*TL*

Also November 23:

1948: Frank Back Patents Zoom Lens for TV Cameras
1963: *Doctor Who* Materializes on BBC

Also 1889:

April 2: Hall's Aluminum Process Foils Steep Prices
September 23: Success Is in the Cards for Nintendo
(see page 268)
November 14: Journalist Nellie Bly Travels Around the World in Seventy-Two Days

November 24
1903: Starting Your Car Gets a Bit Easier

Clyde J. Coleman receives a patent for an electric automobile starter.

Coleman originally applied for the patent in 1899, but his early designs proved impractical. The need for this kind of starter for an internal combustion engine was obvious. Automobiles were getting larger, and hand-cranking—the method used to get the pistons moving in order to make ignition possible—was not only cumbersome but physically demanding and potentially injurious.

The hand cranks in use at the time were built with an overrun mechanism meant to disengage the crank from the spinning drive shaft, but it was designed to work in forward drive only. If the car backfired, the engine could slip into reverse, forcing the crank backward sharply. The result could be a broken thumb, or wrist, or hand, or arm, or—if you got tossed—a leg or even your skull.

The electric starter motor, when perfected, meant the end of the hand-cranked automobile. Coleman sold his patent to the Delco Company, which was taken over by General Motors. Charles Kettering, a Delco engineer who joined GM, did some tinkering with Coleman's design and received his own patent for an improved version. The 1912 model Cadillac proudly became the first car to replace the hand crank with an electric starter motor. Most automobile manufacturers switched over to the electric starter during the teens. The Ford Model T continued using the hand crank, offering an optional electric starter in 1919. With the exception of those old Model Ts, almost every American car on the road boasted an electric starter by 1920.

Now, even women could drive.—*TL*

Also November 24:

1859: Darwin Publishes *On the Origin of Species* (see page 184)
1974: Lucy Skeleton Discovered, First Recognizably Human Primate

Also 1903:

April 6: Birth of Harold Edgerton, High-Speed Photo Pioneer
December 17: Wright Brothers Fly at Kitty Hawk (see page 353)

November 25
1816: Theater Lighting—It's a Gas

Gaslight illuminates Philadelphia's Chestnut Street Theatre. Patrons are living in an age of wonders: lights that burn "without wick or oil."

Most lighting then relied on candles or whale-oil lamps. Merchant Charles Kugler wanted to bring to Philadelphia the modern marvel that illuminated posh London streets. But he felt that manufacturing gas from coal produced such a bad smell that it "could not, with propriety, be established but at a distance from the city." And coal was often expensive or scarce. Kugler replaced coal with pitch, which came from America's abundant trees. Pitch was also largely free of the rotten-egg smell of hydrogen sulfide. Turpentine (also from trees) dissolved the pitch, which was heated in a sealed chamber separate from the firebox beneath it. A chemical bath removed tars and odor-causing chemicals. The gas was collected under an adjustable hood that kept it under pressure to feed the lighting fixtures.

Kugler installed the furnace and storage tanks next to the auditorium of the theater building. Inspired by London's famous Covent Garden Theatre, the Chestnut Street Theatre was the young nation's first purpose-built theater. Its architects included Benjamin Latrobe, early designer of the U.S. Capitol.

Some Philadelphians denounced the gasworks as a danger to public health and safety, insisting it would emit an unpleasant, unhealthy stench. Furthermore, they said, gaslight would use up oxygen and affect the lungs of theatergoers. And an explosion would kill or maim people.

Amid this display of brotherly love, theater managers announced they were "happy to be the first to introduce this system of lighting theaters and flatter themselves that its superior safety, brilliancy and neatness will be satisfactorily expressed by the audience." And so it was. Until the Chestnut Street Theatre burned to the ground in 1820. It may have been arson. —*RA*

Also November 25:
1920: Gaston Chevrolet Dies in Race Crash
1948: Ed Parsons Installs First Cable TV (see page 215)

Also 1816:
January 9: Davy's Safety Lamp First Used in a Mine

November 26
1894: Cybernetics Pioneer Norbert Wiener Born

Norbert Wiener is born. A child prodigy, he goes on to become one of the twentieth century's most famous mathematicians and the founder of the discipline of cybernetics, the study of self-regulating systems.

Wiener graduated from high school at age eleven and from Tufts University at fourteen. He earned a PhD in math from Harvard at eighteen. Wiener joined the mathematics faculty across town at MIT in 1919 and remained there forty-one years.

Wiener made enormous contributions, including a mathematical explanation of Brownian motion. That led to modern probability theory and has implications in understanding many situations where countless tiny inputs produce a single output, from Dow Jones averages to line-noise distortions in electronic signals. His World War II work on automatically aiming antiaircraft guns led Wiener to a theory of cybernetics, also known as systems theory. Central to cybernetics is the feedback principle, whereby a system constantly adjusts itself based on feedback from the environment and from its prior adjustments. Wiener found the principle active not only in automation but also in living creatures.

Cybernetics derives from the Greek word *kybernetes*, meaning "helmsman." That also gives us (through Latin) words like *government, governor,* and *gubernatorial. Cybernetics* itself spawned a series of neologisms, including *cyborg, cyberspace, cyberpunk, cyberculture, cybersex,* and just plain *cyber.*

Cybernetic theory has been applied to biological systems (organisms), ecology, neuroscience, society, economics, and more, but it has arguably had its greatest impact in computers and robotics.

Wiener, though, was a critic of automation, warning it would cause widespread unemployment. He also feared the increasing power of computers would lead to a devaluing of human intellect. *Time's* 1964 obituary quoted him thus: "Render unto man the things that are man's, and unto the computer only the things that are the computer's." —*DT*

Also November 26:

1867: J.B. Sutherland Patents Refrigerated Railcar
1922: Archaeologist Howard Carter Enters King Tut's Tomb
1966: French President de Gaulle Opens World's First Tidal Power Station

Also 1894:

January 30: Charles King Patents Pneumatic Hammer
November 18: *New York World* Prints First Sunday Comics Section

November 27
1834: Electric Motor Gets Plugged In

Thomas Davenport invents the electric motor.

Vermont blacksmith Davenport heard about a machine that used an electromagnet to separate iron ores. He sold his brother's horse and some of his own possessions to buy an electromagnet. Which he promptly took apart, to find out how it was made. Davenport was soon making his own batteries and electromagnets, and in half a year he had come up with a motor powered by direct current from a galvanic wet cell (see page 81). Davenport's wife, Emily, maintained notes for him and suggested modifications and materials for the experiments. She also contributed strips of silk from her wedding dress to insulate the wires.

The brush-and-commutator scheme Davenport invented is still used in electric motors. Current flows through electromagnets mounted on a wheel, causing them to move toward fixed permanent magnets, rotating the wheel half a turn. That motion breaks the circuit powering the magnets and connects a new circuit with opposite polarity, reversing the polarity of the electromagnets and pushing each one away from the nearest magnet and toward the next, pushing the wheel another half turn. The process repeats, and the motor goes round and round.

Davenport patented his "Improvement in Propelling Machinery by Magnetism and Electro-Magnetism." He foresaw using it to power shop machinery and even locomotives, and he set up a workshop near Wall Street and published his own promotional newspaper on an electric-motor–powered printing press. But the business flopped, because batteries were still too weak, bulky, and unpredictable to provide reliable power. Dynamos (which reversed Davenport's motor to generate electricity from motion) were still decades off. Only then would electric motors become practical enough for everyday use.

Davenport died in poverty and poor health in 1851, a few days short of his forty-ninth birthday. —*DT, PG*

Also November 27:
1701: Birth of Anders Celsius, Inventor of Centigrade Temperature Scale
1895: Alfred Nobel Signs Last Will, Creating Nobel Prizes

Also 1834:
June 21: Cyrus McCormick Patents Reaping Machine

November 28
1895: Duryea Beats Benz in First U.S. Road Race

The first intercity U.S. automobile race runs from Chicago to Evanston and back.

The race was postponed from November 2 because most of the eighty-three entries weren't finished, broke down on the way to the race, or got busted for driving on city streets. Race organizers had to coax the city to pass an ordinance even allowing the newfangled vehicles to drive on the streets.

On November 28, Thanksgiving Day, the course was plagued by mud and snow-drifts. The three Benzes (see page 226) in the field were all three-wheelers. The only four-wheeled car to run was a "motorized wagon" from brothers Charles and Frank Duryea.

Two electric-powered cars got to the starting line. One of them couldn't start the race, and the other had to stop twice to replace the primitive batteries that were drained by the November chill. It covered only eleven of the course's fifty-four miles. The only other entries were two motorcycles, but they lacked the power to climb one of the hills and dropped out.

That left the Benzes and the Duryea, driven by Frank. The race also featured the first auto-racing accident. Shortly after the start, depending on whom you believe, either two cars vied for the same piece of road, or one Benz ran into a horse cart or was forced off the road by the horses. Whatever the cause, one Benz landed in a ditch. Then another Benz dropped out.

It was a nip-and-tuck battle between the last Benz and the Duryea. The Duryea led at the start, but the Benz passed it going into Evanston. The Duryea regained the lead on the return trip and crossed the finish line first.

The race proved how fast and how far automobiles could go. It was completed in just under eight hours, at a blistering average speed of 7 miles an hour. — *TB*

Also November 28:

1660: Britain's Royal Society Founded
1948: First Polaroid Instant Cameras Go on Sale
(see page 52)

Also 1895:

January 29: Steinmetz Patents Distribution of AC
Electricity (see page 29)

November 29
1972: Pong, a Game Any Drunk Can Play

Pong, the first popular video game, is released in its original arcade-game form.

If it seems crude by today's standards, well, it was crude then too. And it was meant to be. Pong was the brainchild of Nolan Bushnell, a founder of Atari, who was inspired to develop it after playing an electronic table-tennis game at a trade show. Bushnell had recently designed an arcade game he deemed too complicated because you had to read the instructions before you could play. So he strove for utter simplicity.

"I had to come up with a game people already knew how to play, something so simple that any drunk in any bar could play," Bushnell said later. The game, actually designed by Atari engineer Allan Alcorn, was Pong. It was indeed a game that drunks could play, and they did.

The first coin-operated Pong arcade game was installed at Andy Capp's, a tavern in Sunnyvale, California, where Atari was located. It was an instantaneous hit, confirming Bushnell's suspicions and vindicating, yet again, H.L. Mencken's famous dictum "Nobody ever went broke underestimating the intelligence of the American public."

Four months after Pong's appearance at Andy Capp's, there were upwards of ten thousand Pong arcade games scattered across the land. This caught the eye of Magnavox Odyssey, developer of the game that had inspired Bushnell to dream up Pong. A lawsuit followed, resulting in an out-of-court settlement in Magnavox's favor. By then, however, Pong had moved to a home-console model, which was very different from the original.

Bushnell cut a deal for Sears to act as Pong's exclusive retailer, and the 1975 Christmas shopping season was a lucrative one. This can fairly be said to have ushered in the era of home video-gaming. —*TL*

Also November 29:

1849: Birth of John A. Fleming, Inventor of Vacuum Tube (see page 322)
1975: First Use of "Micro-Soft" for Gates, Allen Partnership (see page 96)

Also 1972:

January 5: Nixon Okays "Low-Cost" Space Shuttle (see page 5)
February 17: VW Beetle Surpasses Ford Model T as Most Popular Car Ever

November 30
A St. Andrew's Day Salute to Scottish Inventors

It's St. Andrew's Day, the national day of Scotland. So we offer a toast to the great inventors whose Scottish ingenuity helped craft the modern world.

Some, like Alexander Graham Bell (see page 71) and James Watt (see page 56), are well known; others, like Arthur James Arnot, who moved to Australia and patented the electric drill, less so. "When we gaze out on a contemporary world shaped by technology, capitalism and modern democracy, and struggle to find our own place in it, we are in effect viewing the world through the same lens as the Scots did," Arthur Herman wrote in his 2001 book *How the Scots Invented the Modern World*.

Here, a small selection of the great Scottish inventors:

- Alexander Bain: electric clock
- Patrick Bell: reaping machine
- James Blyth: windmill electric generator
- Dugald Clerk: two-stroke engine
- Robert Davidson: electric locomotive
- James Dewar: Dewar flask (see page 265)
- Sir William Fergusson: surgical tools, including bone forceps, lion forceps, and vaginal speculum
- William Ged: stereotype printing
- Barbara Gilmour: Dunlop cheese, made from unskimmed milk of Ayrshire cows
- James Goodfellow: personal-identification-number tech
- Fleeming Jenkin: telpherage two-cable aerial tramway
- Charles Macintosh: waterproof fabrics (see page 365)
- James Clerk Maxwell: electromagnetic field theory
- John McAdam: improved road paving (see page 266)
- John Napier: logarithms
- James Nasmyth: steam hammer
- William Ramsay: discovered argon, krypton, neon, and xenon (see page 152)
- Robert Watson-Watt: radar

Also November 30:

1756: Birth of Physicist E.F.F. Chladni, Father of Acoustics

1858: John L. Mason Patents Mason Jar for Preserving Food

2004: Ken Jennings Finally Loses on *Jeopardy!*

Quite a list, eh, laddie? That's not the half of it.

Who invented whiskey? It may have been the Irish, but the Scots (see page 154) claim to have perfected it. *Sláinte!* —*LW*

December 1
1952: Ex-GI Becomes Blond Beauty

It's front-page news when George Jorgensen Jr. is reborn as Christine Jorgensen, the first widely known person to undergo a successful sex-change operation.

Jorgensen grew up, in her words, an "introverted little boy who ran from fistfights and rough-and-tumble games." Getting drafted only reinforced Jorgensen's belief that she was a woman trapped inside a man's body.

Upon discharge, Jorgensen learned about sex-reassignment surgery being performed in Sweden. Jorgensen began taking female hormones on her own, then headed to Europe. Stopping first in Copenhagen to visit relatives, Jorgensen met surgeon Christian Hamburger. He took the case and put his patient on hormone therapy as they prepared for surgery. Several operations were required, starting with castration, performed only after permission was obtained from Denmark's minister of justice. Hamburger did not give her an artificial vagina, so she remained "anatomically incorrect" for several years before undergoing a vaginoplasty in the United States.

Hormone therapy also changed Jorgensen's body. Fat was redistributed, and she began to take on womanly contours. Subsequent surgeries completed the process until she was ready to step into the spotlight. The sex change, possibly leaked by Jorgensen herself, hit the headlines on December 1, creating an international sensation. "Ex-GI Becomes Blonde Beauty" screamed Jorgensen's hometown *New York Daily News*.

She became an actor and nightclub singer, performing, predictably, "I Enjoy Being a Girl." But Jorgensen's world was not an enlightened one, particularly when it came to transgenderism. A first engagement fell through, and a second one failed as well when New York state refused to issue a marriage license. Her intended husband lost his job.

She later traveled on the lecture circuit, talking about her experiences and advocating for the nascent transgender cause. Jorgensen died of cancer in 1989, a few weeks short of age sixty-three. — *TL*

Also December 1:

1942: U.S. Rations Gasoline to Save Rubber
1999: *Nature* Publishes First Gene Sequence of a Human Chromosome

Also 1952:

May 7: Geoffrey Dummer Lectures on Integrated Circuit (see page 257)
July 22: Genuine Crop-Circle Maker Patented (see page 205)

December 2
1942: Nuclear Pile Gets Going
1957: Nuclear Plant Revs Up

Today is a double anniversary for nuclear energy: the first man-made sustained nuclear chain reaction and the first full-scale nuclear power plant.

1942: Enrico Fermi, Leo Szilard, and their colleagues achieve a successful controlled chain reaction in a squash court underneath the University of Chicago football grandstand. It lays the groundwork for the first atomic bombs (see page 220). The secret Manhattan Project brought top U.S. scientists to Chicago to create a nuclear chain reaction, starting with a *controlled,* nonexplosive one. Remarkably, the experiment was conducted within a big city.

On December 2, a three-man suicide squad was ready to douse the reactor in case it threatened to get out of control. Besides the main on/off switch, there was a weighted safety rod that would automatically trip if neutron intensity got too high, a hand-operated backup safety rod, and SCRAM—the safety-control-rod ax-man, who would cut a rope to drop the safety rod if all else failed. The suicide squad wasn't needed. The pile achieved a sustained nuclear reaction, and Fermi shut it down after twenty-eight minutes. It would be two and a half years before the world knew it had changed.

1957: The first U.S. nuclear plant goes to full power at Shippingport, Pennsylvania.

An experimental breeder reactor devised by Chicago Pile–1 veteran Walter Zinn had created the first nuclear-generated electricity in 1951, and the Soviets opened a small nuclear power plant in 1954 (see page 180). President Dwight Eisenhower broke ground that year for the first full-scale commercial plant, to be operated by Pittsburgh's Duquesne Light Company.

The Westinghouse-designed plant used nuclear fission to heat water, which converted the water in a secondary system into steam, which drove the turbine that created the electricity. Shippingport shipped its first power into the Pittsburgh grid on December 18, 1957. It was decommissioned in 1982.—*RA*

Also December 2:
1982: Barney Clark Gets First Artificial Heart

Also 1942:
January 8: Birth of Astrophysicist Stephen Hawking (see page 8)

Also 1957:
January 3: Debut of the Electric Watch, a Space-Age Marvel (see page 3)

December 3
1967: Patient Dies, but First Heart Transplant a Success

The first human-to-human heart transplant is performed. The operation is a success, but the patient dies after complications set in.

South African surgeon Christiaan Barnard prepared for this day by performing a number of experimental heart transplants involving dogs. Then his thirty-member surgical team implanted the heart of a young woman into fifty-three-year-old Louis Washkansky, a Cape Town grocer suffering from diabetes and incurable heart disease. Washkansky received the heart of twenty-five-year-old Denise Darvall, left brain-dead following an automobile accident the day before. She was removed from life support, and her father gave permission for the transplant.

The surgery at Cape Town's Groote Schuur Hospital was a success. Washkansky's body did not reject the heart, due in large part to immunosuppressive drugs he received. But those drugs also weakened his immune system, and he contracted pneumonia, which killed him eighteen days after the transplant. Barnard became an international celebrity (and reveled in it) as a result of the transplant. Over the next several years, he performed additional heart transplants, with the survival times for his patients gradually improving. One patient, Dorothy Fisher, survived twenty-four years after receiving a new heart in 1969.

Other surgeons, however, weren't as bullish on transplant surgery because of the high risk of organ rejection by the recipient. It wasn't until cyclosporine came into widespread use, in the early 1980s, that an effective means of reducing that risk was found. After that, organ-transplant surgery took off.

Barnard, meanwhile, became more interested in anti-aging research, and his reputation took a hit when he lent his name to Glycel, an anti-aging skin cream that in the end did nothing at all to slow the aging process. Barnard died in 2001. — *TL*

Also December 3:

1984: Poison Gas from Pesticide Factory in Bhopal, India, Kills Thousands

2001: Segway Starts Rolling (see page 97)

Also 1967:

January 12: It's Cold in Here — First Cryonic "Burial"

January 27: Three Astronauts Die in Apollo Capsule Fire

April 24: Cosmonaut Vladimir Komarov Killed When Soyuz 1 Parachute Fails

December 4
1996: GM Delivers EV1 Electric Car

General Motors begins delivery of the EV1, an electric vehicle that is a technical triumph for the time.

The technological innovations went well beyond the battery pack, inverter, and AC induction motor that propelled the EV1 without gasoline. The lead-acid battery pack could store only seventeen kilowatt-hours for the first generation, roughly equivalent to a half gallon of gasoline. So GM had to reduce the car's weight and aerodynamic drag to achieve a workable range.

To make the car aerodynamic, they adopted a teardrop shape with covered rear wheel wells. That yielded an unprecedented coefficient of drag of only 0.19, about half of other cars'. Unfortunately, it made the car half as good-looking. To reduce weight, the EV1 engineers used an aluminum chassis and plastic body panels. They also made the car a two-seater: a limiting factor in the marketplace but a necessary tradeoff. Efforts to reduce weight extended throughout the car, like using magnesium in the seat frames.

But GM was also working to dismantle the very regulations that necessitated the program: California's Zero-Emissions Vehicle mandate. GM succeeded in weakening the regulation in 2001, and CEO Rick Wagoner canceled the EV1 program shortly thereafter. In a move that angered EV1 lessees and enthusiasts, GM took back all the cars and sent them to the crusher, ostensibly to avoid having to provide parts and service for the following fifteen years.

GM returned to the electric market with the 2011 Chevy Volt hybrid plug-in. One can only imagine where the company, and EV tech in general, would be if GM had just moved forward with its technology advantage when it was well ahead of its competitors. —*DS*

Also December 4:

1858: Birth of Earmuff Inventor Chester Greenwood
1998: *Endeavour* Lifts Off with First U.S. Unit of
    International Space Station

Also 1996:

December 14: The $100 Million Christmas Bonus
    (see page 350)

1901: Disney, Heisenberg — Separated at Birth?

Animation pioneer Walt Disney and nuclear physicist Werner Heisenberg are born. If you've ever thought the uncertainty principle sounded goofy, you may be onto something.

Disney was born in Chicago. His animation innovations included the first sound-synch cartoon, *Steamboat Willie,* Mickey Mouse's 1928 debut. *Flowers and Trees,* the first full-color animated cartoon, won Disney the first of his thirty-two Academy Awards. *The Old Mill* (1937) was the first short to use the multiplane camera technique, with foreground, midground, and background on separate animation cels at different distances from the camera.

Fantasia (1940) combined live action with animation. Disney introduced time-lapse film photography to a wide public with films like *The Living Desert.* Southern California's Disneyland started the shift in 1955 from generic amusement parks to theme parks and included a futuristic sci-fi Tomorrowland. Disney conceived Florida's EPCOT, the Experimental Prototype Community of Tomorrow, as a showcase for technology that improves people's lives.

Heisenberg was born in Würzburg, Germany. Working in Copenhagen under Niels Bohr, he described a method for calculating the energy levels of "atomic oscillators" in a 1925 paper, "On Quantum Mechanical Interpretation of Kinematic and Mechanical Relations." It brought him immediate fame.

A second paper explained his famous uncertainty principle, which states that it's impossible to specify both the exact position and the exact momentum of a subatomic particle at the same time. Heisenberg received the Nobel Prize in Physics in 1932, the same year Disney won his first Oscar.

During World War II, while Disney was making military-training and civilian-propaganda films for the U.S., Heisenberg was director of Germany's uranium project, working on an atomic bomb. He was arrested in April 1945 and remained imprisoned in England until 1946. Heisenberg died in 1976, nine years after Disney. — *RA*

Also December 5:

1854: Aaron Allen Patents Theater Chairs with Fold-Up Seats

1951: Park-O-Mat Garage Uses Sixteen-Story Elevators to Store Seventy-Two Cars

Also 1901:

May 21: Connecticut Sets First Speed Limit at 12 MPH
December 12: Marconi Transmits Transatlantic Radio Signal (see page 348)

December 6
1850: The Eyes Have It, Thanks to the Ophthalmoscope

German physician Hermann von Helmholtz, who devoted much of his career to studying the eye and the physics of vision and perception, demonstrates his ophthalmoscope to the Berlin Physical Society. The invention revolutionizes ophthalmology.

Although von Helmholtz was not the first person to develop an ophthalmoscope nor the first to examine the interior of the eye, his device was the first to really be put to practical use.

The ophthalmoscope allows the examining doctor to look inside the patient's eye at the lens, retina, and optic nerve. It is the indispensable tool for diagnosing diseases of the eye, including glaucoma, and is used to screen for diabetic retinopathy, a condition in diabetics that can result in blindness. Caught early enough — and the ophthalmoscope is the method for pinning it down — the condition can be treated with laser surgery. The ophthalmoscopes most of us grew up seeing at the eye doctor's office are still in use as a basic diagnostic tool. For more complicated procedures, scanning laser ophthalmoscopy is available.

While the ophthalmoscope made von Helmholtz famous, he distinguished himself in a number of scientific disciplines. He measured the speed at which the nerves carry signals and wrote fundamental textbooks on physiological optics. Moving from sight to sound, he wrote on acoustics and the physiological perception of tone as the basis of musical theory.

He also contributed to the theory espousing the heat death of the universe as a necessary consequence of ever-increasing entropy. The *Encyclopaedia Britannica* wrote: "His life from first to last was one of devotion to science, and he must be accounted, on intellectual grounds, as one of the foremost men of the 19th century." — *TL*

Also December 6:

1957: Vanguard, First U.S. Satellite Launched, Blows Up on Launchpad
1998: Space Shuttle *Endeavour* Assembles First Two Pieces of International Space Station

Also 1850:

March 26: Birth of Edward Bellamy, Socialist Science-Fiction Author
July 14: What a Cool Idea, Dr. Gorrie (see page 197)

December 7
1963: Video Instant Replay Comes to TV

The college football game between Army and Navy marks the first use of video instant replay during a sports telecast. Many fans find it confusing.

Hockey actually had an earlier instant replay, but it wasn't on videotape: an unauthorized experiment in 1950 used a rush-processed kinescope film to replay a hockey goal on CBC's *Hockey Night in Canada*.

In the 1963 telecast, things didn't exactly go as planned. Various technical issues meant that the only time the CBS crew was able to successfully get a play rebroadcast over the air while the game was still under way was right at the end, when Army scored a touchdown to cut the deficit to six points. Fans at home then saw the Army quarterback run the same play and score again, right away. Lots of people phoned CBS to find out what happened.

Alas, Army had not scored again, broadcaster Lindsey Nelson assured the viewers, and the game ended with Navy knocking off Army, 21–15, sending the 102,000 people in attendance into jubilation or gloom. Navy quarterback Roger Staubach went on to win the Heisman Trophy before enjoying a Hall of Fame career with the NFL's Dallas Cowboys.

Instant replay has since been refined with slo-mo and freeze-frame and gone on to be embraced for official review of plays in many professional sports. The NFL gives each coach multiple challenges, so officials can stand in an isolated, covered area and look at multiple camera angles before correcting or confirming an on-field call. The NHL relies on a central nerve center in Toronto when goals are called into question. Major League Baseball limits reviews to a few specific calls. And soccer has been completely resistant. — *EM*

Also December 7:
1941: Attack at Pearl Harbor a Bold, Desperate
    Gamble for the Japanese
1999: Record Industry Sues Napster
    File-Sharing Service

Also 1963:
August 28: Road to Redmond Walks on Water
    (see page 242)

December 8
1993: GPS—It's Location, Location, Location

The U.S. secretary of defense opens the global positioning system to civilian use. It's about to change how people see where they are.

The U.S. Navy experimented with satellite navigation for its submarines in the mid-1960s. The Transit System used six satellites in circumpolar orbits, calculating the Doppler shift of radio signals to ascertain position.

The outlines of the current GPS were conceived at the Pentagon in 1973. The Navstar Global Positioning System was intended for military uses, like targeting missiles, and peacekeeping uses, like monitoring nuclear-bomb tests under the Nuclear Test Ban Treaty. But after Korean Airlines Flight 007 wandered into Soviet territory in 1983 and was shot down, with a loss of 269 lives, even the military thought there might be advantages to sharing GPS with civilians.

But it was 1993 before the Pentagon opened the GPS Standard Positioning Service (SPS) to the Transportation Department. The civilian SPS that finally became fully operational in 1995 provided an accuracy of one hundred meters. The military's Precise Positioning Service (PPS) was accurate to twenty-two meters. The disparity was artificially created by introducing random errors in the public signal.

Although civilian service has been as accurate as the military one since 2000, random errors (called selective availability) can still be introduced in combat zones so that only friendly forces get full accuracy. New techniques can now provide accuracy down to four inches. GPS is integrated into so many devices today, it's as ubiquitous as the cellphone.

We use GPS to navigate our car trips. It helps manage fleets of taxicabs, trucks, buses, and rental cars. First responders, parcel services, and airlines rely on GPS. Fishermen find their secret spots. Researchers track wildlife.

We even use it to trek down wilderness trails. Thoreau (see page 223) would be amazed.—*RA*

Also December 8:
1931: Coaxial Cable Patented

Also 1993:
April 22: Mosaic Browser Lights Up Web (see page 114)
September 24: Beautiful Myst Ushers in Era of
    CD-ROM Gaming

December 9
1968: The Mother of All Demos

Computer scientist Douglas Engelbart kicks off the personal computer revolution with a product demonstration that is so amazing it inspires a generation of technologists; they will call it "the mother of all demos."

The presentation included the debut of the computer mouse, which Engelbart used to control an onscreen pointer exactly the same way we do today. For a world used to thinking of computers as impersonal boxes that read punch cards, whir awhile, then spit out reams of teletype paper, this kind of real-time graphical control was amazing enough.

But Engelbart went beyond merely demonstrating a new input device—way beyond. His demo in San Francisco also premiered "what you see is what you get" editing, text and graphics displayed on a single screen, shared-screen videoconferencing, outlining, windows, version control, context-sensitive help, and hyperlinks. *Bam!*

What's more, it was likely the first appearance of computer-generated slides, complete with bullet lists and Engelbart reading aloud every word onscreen. Engineers at Stanford Research Institute in Menlo Park aided Engelbart in San Francisco through a microwave link and two high-speed 1,200-baud modem lines — about 0.3 percent the speed of a modern DSL line. The demo was the fruit of nearly ten years' work into ways computers might help ordinary people with tasks like writing memos, looking up information, filing things, communicating with others, persuading groups of people through presentations, and working collaboratively to solve difficult problems.

Engelbart called it augmented intelligence. While the tech industry enthusiastically adopted the mouse (see page 119) and many other innovations from his lab, few people carried forward the idea about making computers tools for collaborative problem-solving. President Bill Clinton honored Engelbart in 2000 with the National Medal of Technology. —*DT*

Also December 9:

1884: Richardson Patents Ball-Bearing Roller Skates
1921: Leaded Gasoline Shown to Reduce Knocking

Also 1968:

November 21: Birth Defects Reveal Horrendous
   Chemical Pollution at New York's Love Canal
December 24: Christmas Eve Greetings from Lunar
   Orbit (see page 360)

December 10
1845: The Pneumatic Tire, an Idea Ahead of Its Time

Robert W. Thompson, a Scottish engineer, receives a British patent for his new carriage tire. It has an inner tube inflated with air and encased inside a heavy rubber tire stretched around the rims. Behold: the world's first pneumatic tire.

Thompson's design was an improvement over the solid rubber tires that had been around for a while. It reduced vibration, which made for a smoother ride and improved traction. But his tire was not a commercial success, because it preceded the existence of both the pedal bicycle and the automobile. Thompson's tire was made for horse-drawn carriages, but it never really caught on and eventually faded from the scene.

It would be another forty years before the pneumatic tire reappeared, this time "reinvented" by John Boyd Dunlop, who claimed later to have had no knowledge of Thompson's work. The timing was right for Dunlop: the bicycle was popular, and his tire did a lot to enhance that popularity, again by providing a much smoother ride.

Innovations by Giovanni Battista Pirelli and the brothers André and Édouard Michelin made the new pneumatic tires more durable and improved their grip on the road. The Michelin brothers were the first to put pneumatic tires onto that new-fangled invention the automobile. They entered the famous 1895 Paris-to-Bordeaux road race, and though the tires may have made for a better ride, they didn't win the race.

Dunlop, Michelin, Pirelli: Any of those names sound familiar? Pneumatic tires soon became standard on automobiles. Some jurisdictions even banned solid tires because they wore out the roads. Not to mention what they did to riders' spines. —*TL*

Also December 10:

1626: Death of Edmund Gunter, Father of Cosine and
    Cotangent
1944: Death of Paul Otlet, Web Visionary (see page 237)

Also 1845:

March 17: Rubber Band Invented
March 26: A Sticky Application for an Old Problem
    (see page 87)
April 2: French Physicists Take First Photograph of Sun

December 11
1910: Neon Lights the City of Light

Georges Claude displays his neon lamps to the public at the Paris Expo. Electric advertising is about to take a colorful turn, not to mention quite a few twists.

Claude created his invention by combining an earlier one with a new discovery. German physicist Heinrich Geissler created the first geissler tube in 1855 by applying electricity to a tube filled with gas at low pressure. Then krypton ("hidden gas"), neon ("new gas"), and xenon ("strange gas") were all discovered in a matter of weeks in 1898 (see page 152). Around 1902, Claude, an engineer and chemist, started experimenting with neon as the filler gas for a tube. The red color of the light was distinctive. Claude was onto something.

After more tinkering, in 1910 he created two thirty-eight-foot long neon-tube lamps to show to the public. The "liquid fire" was pretty. But business turned out to be pretty slow. It was 1912 before a Paris barber put up the first neon advertising sign. The next year saw a breakthrough, as boulevardiers gazed upon three-and-a-half-foot neon letters spelling CINZANO.

The first city in the United States to get the neon treatment was Los Angeles. (Las Vegas was still a sleepy little desert town.) Earle C. Anthony imported two Packard signs for his LA auto dealership in 1923. Tokyo got its first neon signs in 1926. Neon advertising signs proliferated around the globe. Claude was nearly ninety when he died, in 1960, by which time Las Vegas had started its neon-aggrandized growth.

Neon tubes, strictly speaking, produce only red light. The other colors are produced by combinations of argon, helium, carbon dioxide, mercury, and specialized phosphor coatings on the insides of the tube.

That makes December 11 the brightest red-letter day in the calendar. —*RA*

Also December 11:

1844: First Use of Laughing Gas as Dental Anesthetic
1964: Martin Luther King's Nobel Address Warns
About Science Without Morality
1997: World Signs On to Kyoto Protocol on
Greenhouse Gases

Also 1910:

June 11: Birth of Jacques Cousteau (see page 164)
July 1: Automated Bread Bakery Opens in Chicago

December 12
1896: Marconi Demonstrates Radio
1901: Marconi Transmits Across Atlantic

Guglielmo Marconi amazes London in 1896 by demonstrating wireless communication across a room. Exactly five years later, he works wireless across an ocean.

Marconi started by replicating Heinrich Hertz's experiments on Hertzian waves, detecting sparks in one circuit with another circuit a few yards away. By 1895, he extended the range to over a mile. Marconi tried to interest the Italian government in transmitting messages without wires, but the *burocrati* weren't buying. Marconi's mother was Irish whiskey heiress Anne Jameson, so he used her connections in Britain to meet chief post office engineer W.H. Preece.

Preece arranged a public demonstration of Marconi's advanced apparatus. Marconi tapped a telegraph key in one part of the room, and Preece walked around with a receiver box. Every time Marconi hit the key, a bell rang. Look, Ma, no wires! The crowd was impressed.

Tickle me, Guglielmo. Marconi was twenty-two years old. He received the world's first patent for a system of wireless telegraphy and opened the world's first radio factory.

He sent radio signals twelve miles in 1897 and across the English Channel (twenty-one miles) in 1899. He patented "tuned or syntonic telegraphy" in 1900: different frequencies prevent simultaneous transmissions from interfering with one another. The improved signal quality also increased transmission range. Still, many people believed the curvature of the earth would limit radio to local use. Marconi proved them wrong.

Assistants telegraphed the letter *S* (three clicks in Morse code) from southwestern England to Marconi in Newfoundland on December 12, 1901.

By transmitting more than 2,100 miles across the Atlantic, Marconi demonstrated the practicality of worldwide wireless communication. He shared the 1909 Nobel Prize in Physics with Germany's Karl Ferdinand Braun, who'd strengthened Marconi's transmitters to make them practical. — *RA*

Also December 12:

1893: Cornele Adams Patents Mapmaking by Aerial Photography

2006: Baiji Yangtze Freshwater Dolphin Declared Extinct

Also 1896:

March 1: Becquerel Accidentally Discovers Radioactivity (see page 62)

Also 1901:

March 31: Wuppertal Monorail Opens (see page 92).

December 13
1809: Doc Decides to Remove Ovarian Tumor

Dr. Ephraim McDowell, a pioneer in abdominal surgery, examines his patient and makes the decision to attempt the first surgical removal of an ovarian tumor, earning him the sobriquet Father of Ovariotomy.

The forty-five-year-old patient, Jane Todd Crawford, had been misdiagnosed as being pregnant with twins. McDowell, who ran a surgical practice in Danville, Kentucky, had briefly studied medicine with the world-renowned medical faculty of Scotland's University of Edinburgh. He offered a different diagnosis: a large ovarian tumor. He decided to risk the previously untried surgery and set Christmas Day for the operation.

A reader today can only imagine Crawford's agony. McDowell, working without anesthetics (see page 275) or antibiotics (see page 294), which were then unavailable, removed a twenty-two-pound benign tumor. Crawford's suffering was rewarded, however: she made a complete recovery and lived until the ripe old age of seventy-eight.

McDowell's account of the operation, published in 1817, created a sensation in the medical world. He went on to perform eight more ovariotomies. (In 1951, under more modern conditions, Chicago surgeons removed a three-hundred-pound ovarian cyst from a fifty-eight-year-old in a four-day surgery.) Another notch in McDowell's distinguished medical bedpost occurred when he operated on future U.S. president James Polk, then a member of the Tennessee legislature. McDowell removed a gallstone and repaired a hernia.

Ironically, McDowell the abdominal expert died of a burst appendix in 1830. He was fifty-eight. Kentucky has honored his medical and surgical achievements with one of its two allotted statues in the National Statuary Hall Collection in the U.S. Capitol complex. The other is of a rather more famous historic Kentuckian: Henry Clay. —*TL*

Also December 13:

1816: Birth of Electrical Engineer Werner von Siemens, Founder of Siemens AG

Also 1809:

January 4: Birth of Louis Braille, Inventor of Alphabet for Blind

February 12: Birth of Charles Darwin (see page 363)

May 5: Hat-Weaving Tech Gets First U.S. Patent to a Woman

December 14
1996: The $100 Million Christmas Bonus

John Tu and David Sun, founders of Kingston Technology, share the wealth. They take $100 million from the sale of their privately held enterprise and give it to employees—a spontaneous gesture to those who had helped make the memory-module company a market leader.

Kingston's five hundred and fifty workers each received a median $130,000—almost $190,000 in today's money. Are you expecting a bonus anything like that this year?

Almost all the tech rags-to-riches stories have come from companies like Microsoft, AOL, Netscape, Yahoo, and Google. They're big public companies that made deals with workers on the way in—along the lines of "We'll underpay you, and here's some paper that may or may not be worth something someday."

Kingston, though, was a private company controlled by two guys. Nobody would have thought twice if they'd decided to keep all the money they got for selling their own company. Kingston had a history of honoring its employees. When it surpassed $1 billion in revenues in 1995, the company took out ads in local and national papers that read "Thanks a Billion!" and named every single employee.

Sun and Tu, both immigrants from Taiwan, sold 80 percent of Kingston to Softbank in August 1996 for $1.5 billion. And that would have been that, but for their decision to give $100 million to their workers...just before Christmas. Ironically, three years later, Sun and Tu bought the company back for just $450 million. It now has more than 4,000 employees and has thrived financially: *Forbes* listed Kingston in 2011 as the second-largest privately held tech-hardware company in the United States. *Fortune*, not surprisingly, lists it among the Best Companies to Work For in America.

Merry Christmas to all, and to all a good bonus. —*JCA*

Also December 14:

1900: Max Planck Gives First Paper on Quantum Mechanics

1962: Mariner 2 Reaches Venus, an Interplanetary First

1972: Last Humans Leave the Moon

Also 1996:

December 20: Carl Sagan Dies; Science Loses a Public Voice (see page 356)

December 15
2001: Leaning Tower of Pisa Reopens with New Angle

Italy's Leaning Tower of Pisa reopens its doors to tourists, leaning a little less.

The tower started tilting soon after construction began, in 1173. Work stopped and started often because of war and civic upheavals. Those interruptions probably kept the tower from collapsing before it was finished. The interruptions allowed the soft underlying soil to compact some, and masons corrected a little for the tilt. So the tower not only leans, it also curves.

After completion, around 1370, things were fine until nearby excavation work in the 1830s destabilized the base, and it began leaning a little more every year. After a tower collapsed in Pavia in 1989, the Italian government closed the Leaning Tower to tourists and looked for a rescue plan.

One scheme involved drilling ten thousand holes in the tower to reduce its weight. (That would almost certainly have caused the overstressed stone to crumble.) Another proposal was to build an exact replica of the tower leaning against it from the opposite direction. (Oh, sure.) Or maybe dismantle the tower stone by stone and rebuild it securely.

In the end, engineers stabilized the tower by pushing down on the high side of the foundation, loading its lip with one hundred tons of lead weights. Steel cables kept the tower from any further leaning, and forty-one corkscrew drills angled under the high side to remove soil underneath the eight-hundred-year-old foundation. It took three years, but it worked. The tower settled toward the high side, reducing its six-degree tilt by half a degree and its thirteen-foot overhang by seventeen inches. The temporary safety cables were removed.

The tower's tilt now approximates that of three centuries ago. Engineers say it won't need another overhaul for a few hundred more years. — *RA*

Also December 15:
1827: Boston Schools Require Vaccination

Also 2001:
January 28: Hey, Don't Tampa with My Privacy (see page 28)

December 16
1770: Birth of Beethoven Leads to Longer CDs

Ludwig van Beethoven is born. His Ninth Symphony will play a role in determining the length of the music CD. Exactly how big a role is a matter of debate.

The symphony, with its famous "Ode to Joy" finale, runs over an hour. The story goes that in 1979 and 1980, when Sony and Philips were negotiating a single industry standard for the audio compact disc, one or more of four people insisted a single CD be able to hold the entire Ninth. The four were the wife of Sony chairman Akio Morita, campaigning for her favorite music; Sony VP Norio Ohga, who studied at the Berlin Conservatory; Mrs. Ohga (her favorite piece too); and conductor Herbert von Karajan. The longest recording they found was a 1951 performance conducted by Wilhelm Furtwängler: a languorous seventy-four minutes.

But Philips engineer Kees A. Schouhamer Immink, who participated in the technical negotiations, says that's not the whole story. Yes, there was pressure to fit the Ninth on a single CD, but Sony also knew that Philips had a factory prepared to make 115mm CDs, and Sony pushed for the eventual 120mm standard to erase Philips's head start. Maximum playing length was set at seventy-four minutes and thirty-three seconds. But the effective limit was seventy-two minutes, the length of the U-Matic videotapes then used for audio masters. The Furtwängler Ninth wasn't released on a single CD until new technology made that possible in 1997.

So, there's a hole in our story, just like the hole in the middle of the CD. That hole, by the way, matches the size of an old Dutch coin. So even if the Japanese prevailed on disc size, the Dutch called the shots on the hole.

Happy birthday, Ludwig. Your flame still shines. —*RA*

Also December 16:

1832: Birth of Towering Engineer Gustave Eiffel
1884: W.H. Fruen Patents Coin-Operated
   Liquid-Vending Machine
2004: France's Millau Viaduct Opens

Also 1770:

April 15: Chemist Joseph Priestly Recommends Using
   Rubber to Erase Pencil Marks

December 17
1903: Bicycle Brothers Make Airplane Work Wright

Orville Wright successfully flies in a heavier-than-air machine that takes off from level ground under its own power and is controlled during flight. It's the first airplane.

Several people, including the brothers themselves, had already managed to get aloft in some sort of device. What the Wrights finally did in 1903 was make the airplane workable.

Orville and Wilbur Wright made bicycles in Dayton, Ohio. Their experience building light, strong machines proved valuable when they turned to flight. They also learned from birds. By lowering a wingtip, a bird causes that wing to rise and begins banking in the opposite direc-

The first flight.

tion. The Wrights used this wing-warping technique on their early gliders, along with a rudder to facilitate the turn and an elevator to control pitch up and down. These three control ideas led to success.

The *Flyer I* had a wingspan of forty feet, four inches. To fly the plane from the dunes of Kill Devil Hills, North Carolina, the brothers laid a wood track down on the sand, and the airplane sat on a little wheeled trolley. With five witnesses present from the nearby lifesaving station, Orville lay down at the controls of the *Flyer I* at 10:35 a.m. on December 17. With the engine running wide open, the tethering rope was released, and Orville rolled down the rail, lifted off the ground, and began the first-ever controlled flight of a powered, heavier-than-air machine.

The flight lasted twelve seconds and covered 120 feet. The brothers made two flights each that day. The longest was Wilbur's fifty-nine-second flight of 850 feet, which ended with a small crash. Wilbur was fine, but the plane needed repairs. They packed up their machine and went home to refine their designs. By 1908, the brothers gained worldwide fame for their achievements. —*JP*

Also December 17:
1790: Aztec Calendar Stone Discovered
1935: First Flight of the DC-3

Also 1903:
January 4: Edison Fries an Elephant to Prove His Point
(see page 4)

December 18
1878: Let There Be Light — Electric Light

Joseph Swan demonstrates the electric bulb to the Newcastle Chemical Society in northern England...a year before Edison's bulb.

In 1809, English chemist Humphry Davy connected two wires to a battery and inserted a charcoal strip between the other ends of the wires. The strip glowed, making it the first electric lamp. Inventor Warren De la Rue, about ten years later, enclosed a platinum coil in an evacuated tube and passed electric current through it to make it glow.

Swan attempted in 1860 to make an electric light using a carbon filament inside an evacuated glass chamber. But the carbon deposited a dark layer of soot inside the chamber, obscuring the light, and it blew out quickly. Carbonized paper lasted longer: thirteen and a half hours, enough time to signal a real future for this invention.

Swan kept working. Better pumps created better vacuums in the bulb, so the filament could glow without catching fire. After the initial demonstration to the chemical society, Swan did another presentation, in February 1879, with more than seven hundred people in the audience. His lamp then burned for forty hours: good but still not commercially workable.

Edison (see page 296) realized the bulb would need a filament with high electrical resistance so that longer lines and less power could be used to heat it. His bamboo-filament bulbs could last fifteen hundred hours. But Swan had filed for a patent in 1861, revising it in the next decade when he improved the design. The patent was strong enough for Edison Electric to go for a merger with the Swan Electric Light Company, creating the British firm Edison & Swan United.

Swan also developed a method of drying wet photographic plates, and he patented bromide paper, still used for photographic prints. — *PG*

Also December 18:

1987: Perl Simplifies the Complexity of Computer Programming
1997: Nine-Mile Tokyo Bay Bridge-Tunnel Opens

Also 1878:

February 12: A Face-Saving Invention from Harvard (see page 43)

December 19
1974: Build Your Own Computer at Home!

The Altair 8800 microcomputer goes on sale. It's the small start of a big trend toward small things.

A small New Mexico company — with the big name of Micro Instrumentation and Telemetry Systems and the small name of MITS — manufactured the Altair as a do-it-yourself kit. Founder Ed Roberts got the name Altair from the star destination in a *Star Trek* episode.

The Altair's heart was an Intel 8080 microprocessor with the remarkable capacity of 8 bits, or 1 byte. The kit offered a 256-byte memory, enough to contain one sentence of text. The Altair's open, 100-line bus structure evolved into the S-100 standard. Keyboard? That was a few years in the future. Input was accomplished through the Sense Switches, eight toggles on the front panel. Monitor? Nope. Output was accomplished through LEDs on the front panel: high tech for 1974.

The Altair 8800 kit sold for just under $400 (more than $1,800 in today's money). Without the case, the kit was under $300. Or you could order everything fully assembled for $595. To soup it up, MITS offered a few peripherals: a video card, a serial card for connecting to a terminal, a 64KB RAM-expansion card, and an 8-inch floppy drive. The floppies stored 300 KB each.

The Altair excited Paul Allen and Bill Gates, who wrote the first microcomputer BASIC for the 8800 and, within months, went on to found Microsoft (see page 96). MITS sold more than 2,000 Altairs by the end of 1975. The Commodore PET and the Apple II, complete with keyboards and monitors, both debuted in 1977. Altairs went out of production in 1978.

Microsoft and Intel are still around. You noticed? — *RA*

Also December 19:

1863: Walton Patents Improvement to His Invention of Linoleum

1958: Eisenhower's Greeting Is First Recorded Message from Space

Also 1974:

June 26: By Gum! There's a New Way to Buy Gum (see page 179)

September 1: Jet Flies from New York to London in Under Two Hours

December 20
1996: Science Loses Its Most Visible Public Champion

Carl Sagan dies.

Calling Carl Sagan a scientist is a little like calling the Beatles a rock band. Sagan was certainly a scientist (astronomer, biologist, and astrophysicist, to be precise). But he was also science's most visible public advocate, a secular humanist, a fervent believer in extraterrestrial life, teacher, author, television host, and political activist.

Accurately fixing the surface temperature of Venus and positing the presence of seas on Jovian and Saturnian moons are among his contributions to astronomy, but his lasting contribution to humanity was popularizing the natural sciences for hundreds of millions of people. Through his PBS series *Cosmos: A Personal Voyage*, Sagan reached an enormous worldwide audience. He also authored books that further helped popularize the natural sciences.

Reaction to Sagan from the scientific fraternity was mixed—not everyone appreciated his eager embrace of celebrity—but his professional credentials were sound, and many colleagues were pleased to see the sciences rise in the public consciousness. Sagan the scientist is perhaps best remembered as a proponent of the search for extraterrestrial life, but he enjoyed a distinguished, wide-ranging career. Assisting the U.S. space program, he was involved in everything from devising mission experiments to briefing Apollo astronauts before their moon landing (see page 203).

Sagan was also deeply involved in more earthbound matters and was often concerned about technology's impact on civilization. He worried about our growing ability to annihilate ourselves with nuclear weapons, and he warned against our casual abuse of the planet. He vigorously opposed President Ronald Reagan's Star Wars proposal for a space-based weapons program, and he sounded early warnings about global climate disruption.

Sagan was sixty-two when he died from the complications of myelodysplasia, a blood disorder linked to anemia and leukemia.—*TL*

Also December 20:

1880: Charles Brush Lights Broadway with Electric Arc Lamps — the Great White Way

1901: Birth of Robert Van de Graaff, Inventor of High-Voltage Van de Graaff Generator

1909: The Volta, Ireland's First Cinema, Opens Under Direction of James Joyce

Also 1996:

February 10: Chess Champ Loses to Computer (see page 41)

December 21
1898: The Curies Discover Radium

Radium is discovered by the husband-and-wife team of Pierre and Marie Curie.

Physicist Pierre Curie was a professor at the School of Physics in Paris when student Marie Sklodowska caught his eye. They wed in 1895 and began a happy marriage and fruitful professional collaboration. Spurred by the work of Wilhelm Roentgen (see page 314) and A.H. Becquerel (see page 62), the Curies succeeded in isolating element 84, polonium (named for Poland, Marie's homeland), and then element 88, radium.

The Curies and Becquerel shared the 1903 Nobel Prize in Physics for their radioactivity research. Marie Curie, who had just won her doctorate, became the first female Nobel laureate, nosing out Bertha von Suttner's Peace Prize by two years. After Pierre was killed in a wagon accident in Paris in 1906, Marie continued their work. She took over Pierre's physics professorship, then became director of the University of Paris's Radium Institute.

The French Academy of Sciences rejected Curie's membership application in 1911. One academy member stated flatly: "Women cannot be part of the Institute of France." Vindication came later that year, when she received a second Nobel Prize, this time in Chemistry. Curie spent the rest of her life in science, founding a radiation laboratory in Warsaw and promoting the healing properties of radium. She died of leukemia, possibly radiation-induced, in 1934. The Curies' daughter and son-in-law, Irene and Frédéric Joliot-Curie, won the 1935 Nobel Prize in Chemistry.

Curium, element 96, is named in honor of Pierre and Marie Curie. Francium, element 87, is named for France, site of the Curie Institute where it was discovered. The curie is an international unit of measurement for radioactivity. — *TL*

Also December 21:

1824: Death of Dr. James Parkinson, Author of "Essay on the Shaking Palsy"
1842: Birth of Peter Kropotkin, Anarchist and Darwin's Detractor

Also 1898:

February 18: Birth of Enzo Ferrari
May 30: Krypton Discovered Before Superman (see page 152)

December 22
1882: Looking at Christmas in a New Light

An inventive New Yorker finds a brilliant application for electric lights and becomes the first person to use them as Christmas tree decorations.

Edward H. Johnson, who toiled for Thomas Edison's Illumination Company, used eighty small red, white, and blue electric bulbs, strung together along a single power cord, to light the Christmas tree in his New York home. Edison himself may have strung electric lights as Christmas decorations around his lab in 1880. But sticking them on the tree was Johnson's idea. It was a mere three years after Edison had demonstrated that lightbulbs were practical at all (see page 296).

The idea of replacing the Christmas tree's traditional wax candles (customary since the mid-seventeenth century) with electric lights didn't, umm, catch fire right away. Stringed lights were mass-produced as early as 1890, enjoying a vogue with the wealthy and in shopwindows, but they didn't become popular in humbler homes until a couple of decades into the twentieth century.

In 1895, President Grover Cleveland supposedly ordered the family's White House tree festooned with multicolored electric lights. General Electric began selling Christmas-light kits in 1903.

Another New Yorker is generally credited with popularizing indoor electric Christmas lights. According to the story, Albert Sadacca, whose family sold ornamental novelties, became a believer in 1917 after reading the account of a bad fire caused by a candlelit tree bursting into flames. Whether or not that's the reason, Sadacca began selling colored Christmas lights through the family business. More homes were wired for electricity, so the timing was right, and sales took off.

With his brothers, Sadacca later started a company devoted solely to the manufacture of electric Christmas lights. He roped some competitors into a trade association, which then dominated the Christmas-light industry until the 1960s. —*TL*

Also December 22:
1666: French Academy of Sciences Founded

Also 1882:
March 24: Koch Pinpoints TB Bacillus (see page 85)
April 29: Trackless Trolley Starts Rolling

December 23
1947: Transistor Opens Door to Electronic Future

John Bardeen and Walter Brattain demonstrate the transistor they built with William Shockley at Bell Laboratories in Murray Hill, New Jersey.

It's been called the most important invention of the twentieth century. The transistor is a semiconductor device that can amplify or switch electrical signals. It was developed to replace vacuum tubes (see page 322), which were bulky, unreliable, and consumed too much power.

Shockley had worked on the concept for nearly a decade but couldn't build a functioning prototype. He asked Bardeen and Brattain to step in. They tinkered with Shockley's design and replaced silicon with germanium, which increased amplification by about three hundred times. A few modifications later, Brattain had a gold metal point extended into the germanium. That resulted in better ability to modulate amplification at all frequencies. The final design of a point-contact transistor had two gold contacts lightly touching a germanium crystal that was on a metal plate connected to a voltage source. This little plastic triangle became the first working solid-state amplifier.

Bardeen and Brattain demonstrated the device to Bell Lab officials on December 23, 1947. Shockley was reported to have called it "a magnificent Christmas present," but he wasn't present and was said to be bitter about missing the big day. Shockley refined the idea and came up with the superior bipolar, or junction, transistor in early 1948. Bell Labs publicly announced the transistor at a New York press conference on June 30, 1948.

The transistor replaced vacuum tubes and mechanical relays, revolutionizing electronics. It became the basic building block upon which all modern computer technology rests. Shockley, Bardeen, and Brattain shared the 1956 Nobel Prize in Physics, but the trio never worked together again after the first few months of their initial creation of the transistor. — *PG*

Also December 23:

1953: First Successful Kidney Transplant from Living Donor, an Identical Twin

1970: Construction Workers Place Highest Steel on World Trade Center

Also 1947:

February 21: *Take a Polaroid* Enters the English Language (see page 52)

December 24
1968: Christmas Eve Greetings from Lunar Orbit

The crew of Apollo 8 delivers a live, televised Christmas Eve broadcast after becoming the first humans to orbit another space body.

Apollo 8 transmitted the first televised view of an earthrise.

Frank Borman, Jim Lovell, and William Anders made their now-celebrated broadcast after entering lunar orbit on Christmas Eve, which might help explain the heavy religious content of the message. After announcing the arrival of lunar sunrise, each astronaut read from the book of Genesis. The astronauts were clearly moved: "The vast loneliness is awe-inspiring, and it makes you realize just what you have back there on Earth," Lovell said during another broadcast. (There were six broadcasts in all from the crew.)

But admiring the vastness of space was not Apollo 8's primary mission. This was a pivotal step on the way to the ultimate goal of landing a man on the moon, achieved less than a year later (see page 203). During a flight lasting six days and including ten orbits of the moon, the Apollo 8 astronauts photographed the lunar surface in detail, both the near and far side, and tested equipment for later missions. The Apollo 8 command module is now on display at Chicago's Museum of Science and Industry.

The very first Christmas Eve broadcast was 1906 and is generally considered the first public voice-over-radio broadcast of any kind. Canadian inventor Reginald Fessenden was promoting his alternator-transmitter to potential buyers of his patent rights. Like Apollo 8's, Fessenden's broadcast was of a pious nature. There was a reading from Luke, chapter 2, and Fessenden himself played "O Holy Night" on the violin. Fessenden's program was available to anyone with a receiver within range of his transmitter in Brant Rock Station, south of Boston. Unlike Apollo 8's, his audience was pretty small: primarily shipboard radio operators off the coast. — *TL*

Also December 24:

1936: Radioactive Isotope First Used to Treat Cancer

Also 1968:

August 31: First Simultaneous Multiorgan
Transplant — One Donor, Four Recipients
December 9: The Mother of All Demos (see page 345)

December 25
2004: Next Stop, Titan, Saturn's Largest Moon

The Huygens probe begins its descent from the Cassini orbiter to Titan, Saturn's largest moon.

The Cassini-Huygens mission, an international effort involving NASA, the European Space Agency (see page 167), and the Italian Space Agency, represents the first thorough survey of Saturn and Titan. The Huygens probe was named for seventeenth-century Dutch astronomer Christiaan Huygens, Titan's discoverer.

The probe was designed by the ESA to take atmospheric measurements during its descent through Titan's dense cloud cover. Huygens was also equipped with cameras to photograph the moon's topography as it broke through the clouds. The probe reached Titan's surface on January 14, 2005, plopping down unceremoniously in a muddy patch. Despite the sloppy ground, the landing was flawless, smoother in fact than mission control had anticipated.

The soft landing produced a bonus: Huygens transmitted physical data from the moon's surface for about an hour, until its batteries ran out. Those images show evidence of rivers and streams flowing full of methane.

The Cassini spacecraft was named for seventeenth- and early-eighteenth-century Italian-French astronomer Giovanni Domenico Cassini. It was assembled at NASA's Jet Propulsion Laboratory, launched in October 1997, and entered orbit around Saturn in June 2004.

Cassini's twelve instruments have sent back volumes and volumes of data about Saturn, its mysterious rings (which may be far older than originally believed), and moons like the giant Titan and the very bright Enceladus. The orbiter is powered by radioisotope thermoelectric generators, which turn the heat of natural radioactive decay into electricity. Now in its second mission extension, Cassini is slated to continue observing through 2017. — *TL*

Also December 25:

1868: Death of Lock Inventor Linus Yale
1999: Space Shuttle *Discovery* Astronauts Release Hubble Space Telescope After Repairs

Also 2004:

February 4: You've Got a Friend in the Facebook (see page 35)
February 13: Largest Pulsating White Dwarf "Diamond Star" Announced
April 1: Google Unveils Innovative Webmail Service, Gmail

December 26
1982: *Time*'s Top Man? The Personal Computer

The personal computer is selected as *Time* magazine's Man (or, in this case, Machine) of the Year.

It marked the first time that the editors selected a nonhuman recipient for the award (the entire planet, Endangered Earth, would be second, in 1988), which *Time* has bestowed annually since 1927. The magazine's essay is now a quaint reminder of the era's dawning awareness of the computer as a force in modern life. In 1982, 80 percent of Americans expected that "in the fairly near future, home computers will be as commonplace as television sets or dishwashers"!

But the primitive PCs of 1982 were doing remarkable things, things that the big mainframes had already done to transform the workplace. Once the silicon chip became the industry standard, computers dramatically shrank in size, and their moving to the home front was only a matter of time. In 1980, according to *Time*, 724,000 personal computers were sold in the United States. The following year, with more companies joining the frenzy, that number doubled to 1.4 million. In 1982, the number doubled again.

In winning the nod from *Time*, the PC beat out some formidable competition, including President Ronald Reagan (who would be named twice), Britain's Margaret Thatcher, and Israel's Menachem Begin. But as the magazine opined: "There are some occasions, though, when the most significant force in a year's news is not a single individual but a process, and a widespread recognition by a whole society that this process is changing the course of all other processes."

As we sit here typing this, that's a hard argument to refute. —*TL*

Also December 26:

1791: Birth of Computer Pioneer Charles Babbage
   (see page 158)
1967: Wham-O Patents the Frisbee

Also 1982:

September 9: 3-2-1 ... Liftoff! First Private Spaceship
   Launched (see page 254)
October 1: Sony Sells First CD Players

December 27
1831: *Beagle* Sets Sail with a Very Special Passenger

The Royal Navy brig sloop HMS *Beagle* sets sail from Plymouth, England, on its second voyage as a survey vessel.

On board, at the invitation of *Beagle* captain Robert FitzRoy, is young biologist Charles Darwin. His account of the journey, *The Voyage of the* Beagle, published in 1839, will establish him as one of the foremost naturalists of his time.

But it was a serendipitous sequence that gave Darwin the opportunity to change our view of the world. FitzRoy recruited a naturalist by asking Captain Francis Beaufort, who asked a Cambridge professor, who had the perfect candidate, who was unavailable. So the professor asked another professor, who recommended Darwin: four degrees of separation. Darwin first had to overcome the objections of his father, who saw the proposed two-year voyage as a chance for Charles to continue idling.

What became a five-year voyage was intended primarily to explore the coastline of South America, much of which remained uncharted. Darwin spent lots of time ashore as the *Beagle* sailed near South America and the Galapagos Islands. The ship also visited numerous islands in the South Pacific and Indian Oceans. The young naturalist was fascinated by the seemingly endless variety of plant and animal life he encountered. His ruminations on the source of that variation resulted in his theory of evolution by natural selection, published in an 1858 paper (see page 184) and a year later in his paradigm-shaping *On the Origin of Species.*

The *Beagle* returned to England on October 2, 1836. It undertook another survey voyage and served as a coastguard vessel before being laid up. It was broken up in 1870. The following year, Darwin published *The Descent of Man,* based on his observations of the natives of Tierra del Fuego and the South Seas four decades earlier. — *TL*

Also December 27:

1571: Birth of Astronomer Johannes Kepler
1654: Birth of Mathematician Jacob Jacques Bernoulli
1822: Birth of Chemist and Microbiologist Louis
   Pasteur

Also 1831:

August 20: Birth of Geologist Edward Suess, Coiner of
   the Word *Biosphere*

December 28
1973: Endangered Species Get a Helping Hand

President Richard Nixon signs the Endangered Species Act into law.

The act came into being as Americans grew increasingly aware of the damage being done to the environment—and the threat posed to specific animal and plant species—by rampant economic growth. Declaring existing conservation policies inadequate, Nixon tasked Congress with devising new legislation aimed specifically at protecting species and their ecosystems threatened by economic encroachment. The result was the Endangered Species Act.

The wide-ranging act forbids any government agency from funding or participating in any action that may "jeopardize the continued existence" of any endangered or threatened species. There are also provisions for private citizens to take legal action against the government to make sure the act is enforced.

Individual plant and animal species are added to a protection list and categorized as either endangered or threatened based on a number of factors, including population figures and the condition of the habitat. If conditions improve significantly, a species can be downgraded from outright endangered to merely threatened, or even removed from the lists. Currently, there are nearly two thousand plant and animal species on the lists.

Two agencies, the U.S. Fish and Wildlife Service and the National Oceanic and Atmospheric Administration, are responsible for administering the act. There have been a number of well-publicized legal challenges under provisions of the Endangered Species Act, resulting in the delay of some projects and the complete scrapping of others.

The ESA suffered considerably under the George W. Bush administration, while the Barack Obama administration has received higher marks from environmental advocates. —*TL*

Also December 28:

1879: "Badly Designed, Badly Built and Badly Maintained" Tay Bridge Collapses in Scotland, Killing Seventy-Five

1923: Death of a Towering Figure in Engineering, Gustave Eiffel

Also 1973:

April 3: Motorola Calls AT&T ... by Cellphone
(see page 95)

December 29
1766: He Put the *Mac* in *Mackintosh*

Charles Macintosh is born in Glasgow, Scotland. He will be remembered as the inventor of rubberized waterproof clothing, especially the raincoat that bears his name.

Son of a dye maker, Macintosh developed an early interest in chemistry and science. By age twenty he was already running a plant producing ammonium chloride and Prussian blue dye, introducing new techniques for dyeing cloth. In partnership with Charles Tennant, Macintosh developed a dry bleaching powder that proved popular, making a fortune for both men. The powder remained the primary agent for bleaching cloth and paper until the 1920s.

Macintosh began experimenting with waterproofing fabric, using waste by-products from the dye process. That included coal tar, which could be distilled into naphtha. By joining two sheets of fabric with dissolved India rubber soaked in naphtha, he made the first truly waterproof fabric that was supple enough for clothing. He patented it in 1823.

When the waterproof fabric was made into a raincoat, the garment quickly became known as the mackintosh. (The *k* comes from the spelling preferred by the chief of clan Mackintosh.) The coat came into widespread use by the British army and the general public. The fabric got stiff when cold, and sticky when warm. And wool's natural oils caused the rubber cement to deteriorate. But the process was improved over time and was effective enough to be used for one of Sir John Franklin's Arctic expeditions.

But Macintosh was no one-trick pony. Besides developing rubberized fabric and bleaching powder, he also helped devise a hot-blast process for producing high-quality cast iron. — *TL*

Also December 29:

1800: Birth of Charles Goodyear, Inventor of Vulcanized Rubber

1952: First Transistor Hearing Aids Go on Sale

Also 1766:

September 6: Birth of Physicist-Chemist John Dalton, Father of Elemental Atomic Theory

December 30
1924: Hubble Reveals We Are Not Alone

Astronomer Edwin Hubble announces that the spiral nebula Andromeda is actually a galaxy, and our own Milky Way is just one of many galaxies in the universe.

The Andromeda galaxy.

Before Copernicus and Galileo, humans thought our world was the center of creation. After learning that the sun and planets did not revolve around the earth (see page 256), we discovered that the sun was not the center of the universe or even a major star in its galaxy. Then, astronomer Edwin Hubble knocked us off our egotistical little pedestal once again.

He trained the powerful new hundred-inch telescope at Mount Wilson in Southern California on spiral nebulae. These fuzzy patches of light were generally thought to be clouds of gas or dust within the Milky Way, which was presumed to include almost everything in the universe.

Hubble found not only stars in Andromeda but Cepheid variable stars. He used Henrietta Leavitt's formula for variable stars to calculate that Andromeda was 860,000 light-years away: eight times the distance to the farthest stars in the Milky Way. This conclusively proved that the nebulae are separate star systems and that our galaxy is not the universe. This cosmic news didn't make the front page of the *New York Times*. The paper did notice the following February 25 that Hubble and a public health researcher split a $1,000 science prize ($13,400 in today's money).

Hubble discovered another couple of dozen galaxies, and by analyzing the Doppler effect on the spectra of receding stars, he established that red shift is proportional to distance. Hubble was honored by being the first to use the two-hundred-inch Mount Palomar telescope (see page 278) and by NASA naming its space telescope after him. —*RA*

Also December 30:
1854: First Petroleum Company, Pennsylvania Rock Oil Company, Incorporated

Also 1924:
February 8: First Execution by Gas Chamber

December 31
1938: Set 'Em Up, Joe... for a Breath Test

Cops in Indianapolis put an early version of the breathalyzer to its first practical test — on New Year's Eve. It proves a success.

Indiana University biochemist Rolla N. Harger invented the drunkometer in 1931. He patented his device in 1936 and helped draft the act that made it a legal method for helping to establish blood-alcohol level.

Harger's drunkometer, a model of simplicity, was the first tool to successfully measure alcohol levels using breath analysis. The subject being tested blew into a balloon. The captured air was then mixed with a chemical solution, which changed color if alcohol was present. The darker the solution, the more alcohol in the breath. From there, the level of alcohol in the person's bloodstream was estimated using a mathematical formula, which Harger also developed. As he pushed for his patent, Harger also pushed to outlaw drunk driving, which was a growing problem after the repeal of Prohibition.

Attempts to measure alcohol levels by breath content date back to the late 1700s. Prior to the drunkometer, the only effective method was the direct testing of blood or urine samples. Both methods were cumbersome and costly — not to mention completely irrelevant in terms of preventing trouble. The beauty of Harger's method was that police could pull drunk drivers off the road before accidents occurred.

Inebriation is apparently a subject of some interest in Indiana. In 1954, the breathalyzer, the tool that eventually replaced Harger's drunkometer, was invented there by Dr. Robert Borkenstein, a laboratory technician with the Indiana State Police.

No device can measure actual impairment, however, because a variety of factors determine how alcohol affects individual drinkers. Hence the expressions *hollow leg* and *cheap drunk*. — *TL*

Also December 31:

1805: 10 Nivôse An XIV, Last Day of French
   Revolutionary Calendar (see page 267)
1879: Edison Demonstrates Incandescent Lights to
   Public (see page 296)
1999: Horror or Hype? Y2K Arrives and World Trembles

Also 1938:

April 6: Teflon Gets Off to a Slippery Start (see page 98)

ACKNOWLEDGMENTS

This book would not have been possible without all my colleagues at *Wired*, especially Tony Long, who wrote all the first year's *This Day in Tech* entries and most of the second year's. Editor in chief Evan Hansen also deserves special mention, as well as all the other writers who contributed to this book (see page vii). Thanks also to the Wired.com art department for help in locating photos; Ron Ferrato and his staff in the *Wired* kitchen; my book agent David Fugate; and editor John Parsley, copyeditor Tracy Roe, production editor Deborah Jacobs, and all the other pros at Little, Brown. Special thanks to Tom Morrison, who rescued and returned the stolen page proofs.

Outside the office, I received encouragement and support from my reading companions in Le Petit Écolier book club; buddies in the OMG (Old Media Guys) lunch group; classmates on the Y-1967 e-mail listserv; my sisters, Susan Mann and Darryl Forman; and my partner, Ben Shum. Thanks to all.

Finally, most of these entries are shorter than their original counterparts, some far shorter. The authors did their work, but I must take responsibility for errors or distortions caused by my rigorous compression of the originals down to single pages. — *RA*

WANT MORE INFO?

Mad Science is by no means encyclopedic, nor are the individual stories definitive. These short tales are meant to illuminate and entertain. We selected one story per day for this book, but most of the "also this day" and "also this year" listings (at the bottom of each day's main entry) are searchable online in *This Day in Tech*. You'll also find expanded info on this book's mini-histories, plus links to their source material. Just run a web search on Wired.com plus the headline or some key words.

INDEX

Italic page numbers refer to illustrations.

Aqua-Lung, 164
Arago, François, 233
arc lamps, 356
ARCO Juneau (oil tanker), 153
Arkus-Duntov, Zora, 183
Armstrong, Neil, 203, 259
Arnold, Kenneth, 177
Arnold, William S., 329
Arnot, Arthur James, 336
ARPANET, 99, 104, 123, 144
artificial snow, 319
Ashen Light of Venus, 9
Ashley, Maurice, 41
Aspaigu (driver in automobile rally), 21
aspirin, 67, *67*
Association for Computing Machinery, 260
asteroids, 1, 13
at sign (@), 126
AT&T, 95, 182, 206, 302, 321
Atari, 293, 295, 335
Atlantis space shuttle, *50,* 167
ATMs (automated teller machines), 247
Atomic Age, 199
atomic bombs, 32, 37, 59, 220, 293, 327, 338
Atomic Energy Act, 180, 244
atomic nuclei, 42, 106
atomic symbols, 248
atomic tests, 243
atomic theory, 250, 365
atomic-hydrogen welding torches, 31
Atoms for Peace program, 236
auto clubs, 307
autobahns, 220
Automats, 162
automobiles, 20, 25, 29, 42, 65, 67, 71, 102,
    122, 124, 209, 213, 258, 306, 310, 311,
    340; assembly lines for, 14; auto shows,
    290, 309; and diesel, 6, 317; electric start-
    ers for, 330; first road trip, 226;
    flathead V-8 engines, 92; racing of, 21,
    43, 120, 259, 334; safety belts for, 193;
    tires/wheels for, 143. *See also* Ford Motor
    Company; General Motors (GM); motor-
    cycles; Rosemeyer, Bernd; Tucker, Pres-
    ton; *names of individual makes/models*
aviation, 13, 46, 49, 78, 88, 146, 162, 168,
    200, 202, 205, 208, 262, 276, 280, 298;
    altitude barrier, 202, 328; breaking of
    sound barrier, 140; Canary Island run-
    way collision, 88, 153; DC-3 first flight,
    353; in-flight movies, 202; instrument
    flying, 45, 270; pilot loops the loop, 227,
    324; round-the-world air service, 170;
    rugby team plane crash, 288;

stewardesses, 137; transpacific airplane
flight, 287; wind tunnels, 149. *See also*
aircraft; Earhart, Amelia; jets;
Lindbergh, Charles; *Spruce Goose;*
Wright brothers (Orville and Wilbur)

B-29s, 70
B-50 aircraft, 63, 215
Babbage, Charles, 129, 158
Babcock and Wilcox nuclear reactor, 236
Back, Frank, 329
Baird, John Logie, 112, 171
Baker, Philip, 83
ballistic missiles, 161
balloons, 7, 47, 54, 64, 82, 157, 185, 198, 225,
    230, 231, 233
ballpoint pens, 163
Band-Aids, 87
Banks, Joseph, 81
Banting, Frederick, 11
barbed wire, 178
Bardeen, John, 359
Bareev, Evgeny, 41
Barnard, Christiaan, 339
barometers, 164
Barry, James (first woman physician), 208
Barry, William, 175
baseball, 43, 146, 240
BASIC computer language, 96, 123, 150,
    217, 355
Basov, Nicolay, 138
batteries, 73, 115, 214, 241
battles (war), 36, 70, 157, 170, 188, 201,
    240, 272
Baumgartner, Felix, 230
Beach, Alfred Ely, 58
Beagle (HMS), 184, 363
Beam Gun, 268
Beauchamp, George, 196, 224
Beaufort, Francis, 363
Bechet, Sidney, 181
becquerel, 62
Becquerel, Antoine Henri, 62, 357
beer, 24, 311
Beethoven, Ludwig van, 352
Bel Geddes, Norman, 122
Bell Laboratories, 95, 138, 206, 359
Bell X-2 rocket plane, 206
Bell, Alexander Graham, 27, 65, 71, 218,
    229, 336
Bellamy, Edward, 87, 342
Bellanca Skyrocket aircraft, 280
Bendix race, 140
Bennett, Elizabeth, 14

Bennett, Jesse, 14
Benz, Berta, 226
Benz, Eugen, 226
Benz, Karl, 226
Benz, Richard, 226
Berlin, Irving, 162
Berliner, Emile, 65, 329
Berliner, Henry, 169, 329
Berners-Lee, Tim, 99
Bernoulli, Jacob Jacques, 363
beryllium, 198
Bessemer steelmaking process, 292
Bessemer, Henry, 292
Best, Charles, 11
Betamax video recorders, 160
Beutler, Jules, 21
bicycles, 48, 280, 314, 353
Bierce, Ambrose, 264
Bierey, Gottlob Benedikt, 125
Big Ben, 289
"Big Ear" Observatory, 229
Bigelow, Henry, 275
Bikini Atoll, 143, 271
bikini bathing suits, 188
Bingham, Hiram, 27, 207
Binnig, Gerd, 112
Biological Weapons Treaty, 5, 102
biosphere, 234
Birmingham (USS), 18
Bíró, Georg, 163
Bíró, László, 163
birth control, 131, 291
Bismarck (German battleship), 149
black boxes, 78
Black Death, 240
blackouts, 153, 196, 241
Blackwell, Elizabeth, 273
Blenkinsop, John, 177
Bleriot, Louis, 208
blind, alphabet for, 4, 349
blitzkrieg, 246
blood circulation, 156, 320
blood transfusions, 320
blue jeans, 142, 216
Blundell, James, 320
Bly, Nellie, 320, 329
BMW automobile, 120
BOAC (British Overseas Airways
   Corporation), 279
Boeing aircraft, 40, 64, 86, 149, 198,
   202, 203
Boeing, Bill, 64
Boggs, David, 144
Bohlin, Nils, 193

Bohr, Niels, 120
Bollée, León, 129
Bolton, John, 219
Boltzmann, Ludwig, 250
Bond, George, 273
Bond, James, 105
Boole, George, 125, 308
Boolean logic, 308
Borchers, W., 227
Borden, Gail, Jr., 315
Borkenstein, Robert, 367
Bosch, Carl, 241
Boston, Massachusetts, 15, 64, 113, 145, 191,
   210, 258, 275, 283, 287, 290, 351
Bowser, Sylvanus, 250
Boyle, Robert, 25
Bradley, James, 142
Brahe, Tycho, 222, 317
Braille, Louis, 4, 349
brain death, 164
Branson, Richard, 8
Brattain, Walter, 359
bread factories, automated, 184, 347
Brearley, Harry, 227
breathalyzers, 367
bridges, 2, 65, 101, 112, 149, 163, 196, 242,
   313, 325, 354, 364
Britain, 1, 106, 131, 140, 163, 219, 250, 260,
   276, 288, 313, 320, 322, 329, 365;
   adhesive postage stamps used in, 128;
   appoints Royal Commission on
   pollution, 201; and aviation, 13, 78; and
   Brunel engineering, 101; calendar of,
   61; and coronation of Elizabeth II on
   television, 155; instant coffee in, 225;
   and krypton, 152; and periodic table of
   the elements, 38; on pollution, 222; and
   polyethylene, 88; postal service of, 20;
   and radar invention, 58; red telephone
   boxes of, 17; and satellites, 206, 241; and
   scurvy, 142; shrapnel shells and army of,
   74; television in, 207; and transatlantic
   cable, 210; in World War I, 36; and zebra
   crosswalks, 306. *See also* England;
   Royal Society
Brown, Arthur Whitten, 168
Brown, Louise Joy, 208
Brownian motion, 332
Brunel, Isambard Kingdom, 101
Brunel, Marc, 101
Brush, Charles, 356
Buchanan, Sharon, 179
Buck Rogers (comics), 191
Budding, Edwin, 263

bulletin-board systems (BBSs), 47
Bunsen, Robert, 230
Burney, Leroy, 270
Burt, William, 176
bus service, 79
Bushnell, Nolan, 335
Byron, Lord, 59, 158
bystander effect, 74

C++ computer language, 289
cable cars, 216
Caesar, Julius, 61, 172
calculating machines, 129, *129*, 282, 318
calculus, 304
calendars, 61, 267, 353
California, 24, 32, 55, 91, 92, 106, 110, 140,
    143, 151, 185, 209, 216, 219, 231, 267,
    301
Callaghan, Jim, 306
calliopes, 284, 292
Caltech, 31, 278
Calypso (research vessel), 164
cameras, 16, 124, 233, *233*
Cameroon, 235
Campbell, Sir Malcolm, 248
Campbell-Swinton, Alan Archibald, 171, 185
can openers, 277
Canary Island runway collision, 88, 153
cancer, 31, 190, 270, 286, 360
canister shots, 74
canning, 315, 323
Cannizzaro, Stanislao, 38
cannon, 59, 118, 240
Canter, Laurence, 103
Cape Canaveral, 37, 207
Čapek, Karel, 25
Capone, Al, 45
carbon 14, 59, 115, 140
carbon microphones, 65
carbon paper, 101, 282
cardiac imaging/catheterization, 305
Carlisle, Anthony, 81
Carlson, Chester, 297
Carothers, Wallace, 60
Carpenter, Scott, 243
Carré, Ferdinand P. E., 197
Carrier, Willis Haviland, 200
Carroll, Charles, 263
Carroll, Lewis, 14
Carson, Rachel, 34
Carter Handicap, 16
Carter, Garnet, 40
Carter, Howard, 95, 332
Carver, George Washington, 5

cash registers, 100, 310
Casino Royale (Fleming), 105
Cassini spacecraft, 9, 300, 361
Castro, Fidel, 105
catcher's masks, 43
Catholic Church, 256
Cavalli, Earnest, 127
cavalry charges, 306
CB Simulator, 269
CBS News, 310
CD players, 276, 362
cellphones, 5, 17, 95
cellular pathology, 250
Celsius, Anders, 333
centigrade temperature scale, 333
Ceres (asteroid), 1
CERN, 99
Cernan, Eugene, 203
cesarean sections, 14
Challenger (space shuttle), 50, 103
Chambers, Stan, 32
champagne, 218
Charles (oil tanker), 132, 213
Charles II (king), 175
Charlston, Abraham, 59
Chemical Bank, 247
chemical plant explosion, 211
Chemie Gruenenthal, 276
chemotherapy, 245, 307
Chen, Steve, 46
Cheng, Regan, 295
Chernobyl nuclear plant, 50, 118, 180
Chertok, Boris, 279
Chesebrough, Robert, 66
chess, 41
Chevrolet, 183
Chevrolet, Gaston, 245, 331
Chicago, Illinois, 45, 47, 113, 146, 155, 184,
    202, 222, 225, 293, 307, 334, 338, 341, 349
Chicago "L," 155
Chicago Cubs, 146
China, 179, 180, 187, 194, 241, 290
chiropractic, 263, 314
Chladni, E. F. F., 266, 336
chlorine gas, 110, 114
chocolate, 190
cholera, 85, 253
Christensen, Ward, 47
chronometers, 187, 218
Churchill, Winston, 155, 260
CIA (Central Intelligence Agency), 105
Cinématographe (movie camera/projector), 44
Civil War. *See* American Civil War
Clark, Barney, 264, 338

Cuban missile crisis, 173, 297
Cugnot, Nicolas-Joseph, 177
Curie, Marie, 23, 62, 357
Curie, Pierre, 62, 137, 357
Curry, Adam, 35, 227
Curtiss biplanes, 18
Cushing, Harvey, 100, 132
Custer, George Armstrong, 71, 178
Cuyahoga River, 175
cybernetics, 332
cyberspace, 252
cyclotron subatomic particle accelerator, 51

da Vinci, Leonardo, 107
Dacron, 32, 130
Daguerre, Louis-Jacques-Mandé, 9, 233
daguerreotypes, 233
Daily Source Code, 35, 227
Daimler, Adolf, 244
Daimler, Gottlieb, 65, 124, 244
Dalton, H. C., 251
Dalton, John, 248, 365
Daniell, J. F., 73, 214
Danish Observatory, 222
Dartmouth Oversimplified Programming
    Experiment/DOPE, 123
Dartmouth Simplified Code/Darsimco, 123
Dartmouth Time Sharing System (DTSS),
    123
Darwin, Charles, 39, 184, 265, 357, 363
Davenport, Emily, 333
Davenport, Thomas, 333
Davidson, Jim, 215
Davis, Harry, 274
Davy, Humphry, 81, 331, 354
Dawson, Clyde, 179
day of darkness, 141
Day, Horace Harrell, 87
de Forest, Lee, 73
de Gaulle, Charles, 332
de Jouffroy, Claude-François-Dorothée, 198
De la Rue, Warren, 354
De Mestral, George, 258
Declaration of Independence, 187
Deep Blue computer, 41
Deep Junior (software program), 41
Deep Space Network (NASA), 166
deep-sea exploration, 23, 27
Defense Advanced Research Projects Agency
    (Pentagon), 99
deHavillands (aircraft), 279
Denys, Jean-Baptiste, 320
Derringer, Paul, 146
Descartes, René, 161

detective genre, 112
Dewar, James, 265, 336
Dewey Decimal System, 7, 237
Dewey, Melvil, 7, 237
diabetes, 11, 97
Diamond Sutra, 133
Dick, Albert Blake, 222
Dickson, Earle, 87
difference engine, 129
Digital Millennium Copyright Act, 303
dinner knives, 135, 161
diphtheria vaccine, 253
direct current (DC), 4, 29, 66
Director tape computer programs, 69
dirigibles, 67, 75, 248, 313
Discovery space shuttle, 51, 361
disk records, 65
Disney, Walt, 341
Disneyland, 165, 169, 341
DNA (deoxyribonucleic acid), 117, 192, 193,
    208, 255, 265, 309
Docuteller machine, 247
Dodgson, Charles. *See* Carroll, Lewis
Dolby noise reduction, 18
Dolby, Ray, 18
Dolly the sheep (cloned mammal), 54, 188
Dolphin (USS), 210
domain names, 76, 93, 176, 316
Donora Zinc Works, 301
Donora, Pennsylvania, 301
Doolittle, James Harold "Jimmy," 45, 270
Dorsey, Jack, 82
draisine, 48
Drake, Sir Francis, 96
drawing/quartering, 278
drills, 336
Drinker, Philip, 287
drive-in movie theaters, 159
drug experiments, 105
Drysdale, Charles R., 270
Dulles, Allen, 105
Dummer, Geoffrey, 257
DuMont, Allen B., 29, 112, 171
Dunlop, John Boyd, 346
DuPont company, 32, 60, 98, 130, 163,
    211, 299
Durand, Peter, 323
Duryea, Charles, 334
Duryea, Frank, 334
Dutch elm disease, 79, 302
Dvorak, August, 16, 134
Dvorak, John C., 26
Dylan, Bob, 111, 208
dynamite, 296

Eagle lunar module, 203
Earhart, Amelia, 157, 185
Earl, Harley, 183
earmuffs, 184, 289, 340
Earth, 7, 31, 142, 172, 256, 366
Earth Day, 109, 114
earthquakes, 26, 110, 143
Eastman, George, 124, 249
Eastman-Kodak Company, 27, 124, 297
Eckert, J. Presper, 310
Edgerton, Harold, 98, 330
Edgeworth, Richard, 266
Edison, Thomas, 4, 29, 52, 66, 228, 249, 322;
   and "hello" telephone greeting, 229;
   birth of, 42; and carbon microphone, 65;
   and electric lighting, 19, 296, 354, 358;
   and mimeograph, 222; and movies, 32,
   168; and phonograph, 327; and Tesla
   (Nikola), 192; and typewriters, 176
Edsel, 3, 249
Eiffel, Gustave, 352, 364
Einstein, Albert, 120, 151, 192, 317, 327
Einthoven, William, 143
Eisenhower, Dwight D. "Ike," 180, 212, 236,
   244, 271, 310, 338, 355
ejection seats, 8, 13, 42, 193
electric eels, 81
electric meters, 228
electric motors, 333
electricity, 4, 17, 19, 81, 90, 142, 151, 156,
   180, 204, 228, 241, 249, 268. *See also*
   alternating current (AC); direct current
   (DC)
electrocardiograms, 143
electrocutions, 220
electrolysis, 81
Electro-Motive Diesel, 213
electronic news strip, 312, *312*
electrons, 122, 165
Elephant Man, 250
elevators, 84, *84*
Ellul, Jacques, 6
Ellwood refinery (California), 55
Elsener, Karl, 165
Ely, Eugene, 18
e-mail, 269, 361
Emerson, Jack, 287
emoticons, 264
Endangered Species Act, 364
Endeavour space shuttle, *5,* 243, 340, 342
Enewetak Atoll, 307
Engel, Joel, 95
Engelbart, Douglas, 119, 345
Engine Company 21, 113

England, 6, 8, 77, 124, 163, 240, 272, 280,
   292, 348; bicycle time trials in London,
   314; Big Ben in, 289; cholera epidemic
   in, 253; electric lightbulbs in, 296; first
   woman physician in, 273; gaslight in,
   28; and Greenwich mean time, 324;
   Luddites in, 59; and smoking, 270;
   subway service in, 10; traffic signals in
   London, 326; world's fair in, 7. *See also*
   Britain; Royal Society
English Channel, 7, 208
Eniac computer, 310
Enigma Cipher Machine, 131
Enterprise (space shuttle), 248
Enterprise (*Star Trek* starship), 253
Enterprise (USS), 131, 269
Environmental Protection Agency, 34, 175
environmentalism, 34, 223
epidemiology, 253
Eratosthenes (Greek astronomer), 172
Erie Canal, 283
Essex (whale ship), 326
Estonia, 182
Etch A Sketch, 195
ether, 91, 275
Ethernet, 95, 144
European Launcher Development
   Organisation (ELD), 167
European Space Agency (ESA), 167, 361
European Space Research Organisation
   (ESRO), 167
evolution, 27, 184, 204, 255
executions, 153, 255, 278, 366
Explorer 1, 31, *31*
Exxon Valdez spill, 85, 153, 303
Eyak language, 21

Fabricius, Hieronymus, 156
Facebook, 35, 46, 82
facial-recognition surveillance cameras, 28
Fahlman, Scott, 264
Fahrenheit, Daniel Gabriel, 261
Fancher, W. H., 170
Faraday, Michael, 267
Farnsworth, Philo T., 112, 171
Fawcett, Eric, 88
FBI (Federal Bureau of Investigation), 46,
   286
FCC (Federal Communications
   Commission), 72, 184, 206, 215, 321
Fender Electric Instrument Manufacturing
   Company, 224
Fender Precision Bass (P-Bass), 224
Fender, Clarence "Leo," 224

development in, 13; and Garros (Roland), 110; Koch (Robert) of, 85; Krupp (Alfred) of, 59; and Maxim gun, 36; monorail in, 92; and movie sound, 73; and nuclear power plants, 180; and rockets, 77; and Steinmetz (Charles Proteus), 29; television in, 83; and use of poison chlorine gas, 114; and use of tanks, 260; V-1 missile of, 136, 166; West Germany on vehicle safety belts, 193. *See also* U-boats

Gernsback, Hugo, 169, 230
Gibson guitar company, 196, 224
Gibson, Reginald, 88
Gibson, William, 78, 252
Giger, H. R., 36, 115
glass bottles, 100
Glass, Louis, 329
Glenn, John, 51
Glennan, T. Keith, 212
Glidden, Joseph, 178
global positioning system (GPS), 45, 344
Gloster Meteor jet fighter, 201
Gmail, 46, 93
Goddard, Robert, 77
Godzilla, 121
gold, 24, 91
Goldberger, Joseph, 110
Goldman, Sylvan, 157
Goldmark, Peter Carl, 174
Goldschmidt, Hans, 227
golf, 40, 328
Goodall, Jane, 27
Goodwin, Hannibal, 124
Goodyear, Charles, 365
Google, 35, 46, 93, 252, 350, 361
Gorrie, John, 197
Gossamer Condor (aircraft), 237
Gould, Gordon, 138
Gould, Stephen Jay, 255
Grand Coulee Dam, 151
Grandclément, Catherine, 157
Granjean, Arthur, 195
gravity, 107
Gray, Elisha, 71
Great Britain (ship), 101
Great Eastern (ship), 101, *101,* 210
Great Western (ship), 101
greenhouse gases, 42, 47, 347
Greenwich mean time, 288, 324
Greenwich meridian, 288
Greenwich Observatory, 175
Greenwood, Chester, 184, 289, 340
Gregorian calendar, 1, 267

Gross, Samuel, 87
Grosvenor, Gilbert, 27
Guillet, Leon, 227
guillotines, 153, 255
guitars, 196, *196,* 224
Gulf of Mexico, 156
Gumper, Jake, 250
Gunter, Edmund, 346
Gutenberg, Johannes, 126, 133

H.L. Hunley (submarine), 48
Haagen-Smit, Arie, 209
Hagia Sophia, 129
Haig, Douglas, 260
HAL 9000 computer, 12
Hale Telescope, 278
Hale, George, 278
Hall, Charles Martin, 94, 329
Halley's Comet, 91, 141, 167, 172
Hallidie, Andrew, 216
Hamburger, Christian, 337
hanafuda, 268
Hardart, Frank, 162
Hare, Robert, 315
Harger, Rolla N., 367
Harrison, John, 187, 218
Hart, Guy, 196, 224
Harvard Computation Laboratory, 221
Harvard Mark I/II computers, 134, 150, 221
Harvard University, 35, 43, 96, 141, 221, 290, 302
Harvey, William, 148, 156, 320
hat-weaving, 127, 349
Hawking, Stephen, 8
Hawkins, Coleman, 181
Haynes, Elwood, 227
Hazard, Cyril, 219
hearing aids, 365
heart pacemakers, 283
heart surgery, 128, 251
heart transplants, 301, 339
heart-lung machines, 128
hearts, artificial, 264, 338
heat-work relationship, 48
Hedley, William, 177
Heinkel He-280 jet fighter, 13
Heisenberg, Werner, 120, 341
helicopters, 11, 169
heliocentrism, 256
heliopause, 44
helium, 152, 232, 237
"hello" telephone greeting, 229
Helwig, Frank, 69
Henle, Jakob, 85

Hennébique, François, 199
Henry the Navigator, Prince, 319
Henson, Matthew, 98, 307
Herculaneum, 238
heredity, 39
Herman, Arthur, 336
Herndon, Hugh, Jr., 280
Herschel, John, 233
Herschel, William, 74
Hershman-Leeson, Lynn, 158
Hertz, Heinrich, 54, 84, 348
Hewlett, Bill, 232
Hewlett-Packard Company, 232
Heyerdahl, Thor, 109, 139
high-density polyethylene (HDPE), 88
high-mobility multipurpose wheeled vehicles
    (the HMV), 83
high-rises, steel-frame, 123, 288
high-voltage Van de Graaff generators, 356
highway interchanges, 267
Hill, Rowland, 128
Hillary, Sir Edmund, 151
Hindenburg, 128
Hitchhiker's Guide (Adams), 147
Hobbit, The (Tolkien), 196, 266
Hoffmann, Felix, 67
Hofmann, Albert, 108
Hollerith punch-cards, 61, 154
Hollerith, Herman, 61
Hollingshead, Richard, Jr., 159
Home Insurance Building (Chicago),
    123, 288
Homebrew Computer Club, 66, 96
Hooke, Robert, 56, 148
Hoover Dam, 151, *151*
Hopkins, Samuel, 53, 214
Hopkirk, Paddy, 21
Hopper, Edward, 162
Hopper, Grace, 150, 221
Hormel, 188
Horn, Joe, 162
horse racing, 16
hospitals, 42
hotlines, 244
Houdini, Harry, 63
Hounsfield, Godfrey, 239
Howe, Elias, 181, 255
Hubbard, Eddie, 64
Hubbard, Gardiner Green, 27
Hubble Telescope, 44, 103, 116, 142, 167, 187,
    361, 366
Hubble, Edwin, 278, 366
Hughes, Howard, 149, 308
humanities/science, 129

Hunt, Walter, 102
Hunt, William, 178
Hurley, Chad, 46
Hutton, James, 107
Huygens, Christiaan, 156, 361
hydroelectric power, 41, 192, 202
hydrogen bombs, 7, 17, 143, 202, 234, 271,
    305, 307, 312
hygrometers, 261

IBM, 12, 22, 26, 41, 48, 76, 96, 99, 119, 123,
    144, 154, 179, 217, 221, 226, 282, 297,
    316
ICANN (Internet Corporation for Assigned
    Names and Numbers), 93, 263
ice cream sundaes, 95
ice makers, 197, *197,* 258
ice plows, 258
Id Software, 127
Ideal X (tanker), 118
immunology, 323
Indiana legislature, 36, 68
Industrial Light and Magic, 136
Industrial Revolution, 56, 59, 223
Indy 500, 152
infectious diseases, 85
information theory, 122, 291
infusion pumps, 97
inkblots, 94
insect zappers, electric, 314
instant replay, 50, 72
insulin, 11, 251
insulin pumps, 97
integral sign, 304
integrated circuits, 111, 257
Intel, 111, 257, 355
intellectual-property law, 80. *See also*
    copyright; patents
intercontinental ballistic missiles
    (ICBMs), 37
internal combustion engines, 65, 228, 244
International System of Units, 130
Internet, 6, 78, 87, 89, 93, 99, 104, 148, 177,
    263, 286, 322; broadband connections,
    72; concerts, 235; cyberspace, 252;
    domain names, 76, 93, 176, 316; dot-
    coms, 71, 72; hackers, 38; households
    with, 231; and intellectual-property law,
    80; kiosks, 17; as news medium, 256;
    service providers, 269; at sign (@) used
    for, 126; spam, 123; voting, 182. *See
    also* Facebook; Google; social media;
    webcasts; World Wide Web; YouTube
Internet Explorer, 114, 273

interstate highways, 122, 271
Inventors Hall of Fame, 206
inversion layers, 301
iPhones, 182, 298
iPods, 298
iron lung artificial respirator, 287
irrigation systems, 205
I-17 submarine, 55
Ishi (of Yahi American Indian tribe), 86, 260
Isle of Man road race, 259, 322
isotopes, 49, 190, 227, 360
iTunes Music Store, 120, 241
Iwatani, Toru, 285
Ixtox 1 oil well, 156

Jackson, Charles, 275
Jacquard, J. M., 190
James IV (king of Scots), 154
Janney, Eli, 121
Jansky, Karl Guthe, 297, 327
Janssen, Zacharias, 277
Japan, 37, 55, 70, 81, 106, 127, 180, 220, 225, 242, 268, 280, 285, 318, 343, 347, 354
Jauch, Robert, 328
Jefferson, Thomas, 53
Jeffreys, Alex, 255
Jenkins, Charles Francis, 185
Jenner, Edward, 136, 323
Jennings, Ken, 336
jets, 13, 40, 103, 124, 201, 203, 246, 279, 355
Job Control Language (JCL), 69
Jobs, Steve, 93, 136, 182, 248, 261, 298
John Paul II, Pope, 256
Johnson, Edward H., 358
Johnson, Robert Wood, 87
Joint Long Range Proving Ground, 207
Joliot-Curie, Frédéric, 357
Joliot-Curie, Irene, 357
Jorgensen, George, Jr. (Christine Jorgensen), 337
Joule, James, 48
Joyce, James, 356
jukeboxes, 329
Julian calendar, 1, 61
Jupiter, 56, 240
Jupiter (balloon), 231

Kael, Pauline, 12
Kahn, Julius, 236
Kaiser, Henry J., 308
Kaiserwagen, 92
Kalashnikov, Mikhail, 189
Kamen, Dean, 97
Karim, Jawed, 46

Kasparov, Garry Kimovich, 12, 41
Kato-Coffee Company, 225
Kato, Satori, 225
Keck I telescope (Hawaii), 9, 278
Keeler, Leonarde, 33
Kellogg, John, 68
Kellogg, Will, 68
Kelly, Kevin, 237
Kelly, Michael, 178
Kelly, William, 292
Kemeny, John G., 123
Kennedy, John F., 53, 105, 138, 147, 155, 202, 206, 234, 257, 271, 328
Kenyon, David, 113
Kepler, Johannes, 277, 292, 363
Kettering, Charles, 330
keyboards, 16, 134, 176, 186
Khrushchev, Nikita, 207
kiddie seats, 157
kidney transplants, 359
Kilby, Jack, 257
Killian, James, 212
Kinemacolor, 191, *191*
King, Charles, 30, 332
King, Martin Luther, Jr., 15, 347
Kingston Technology, 350
Kittinger, Joe, 230
Klussmann, Friedel, 216
Knapp, Michelle, 284
Knight, Arthur, 328
Knight, Margaret, 45
Knott, Max, 112
Knowles, Red, 145
Koch, Robert, 85
Koch's postulates, 85
Kodak film, 124
Komarov, Vladimir, 116, 339
Kon-Tiki, 120
Kornei, Otto, 297
Kozo, Nishino, 55
Krakatau volcano, 240, 324
Krikalyov, Sergei, 83
Krolopp, Rudy, 95
Kropotkin, Peter, 265, 357
Krupp, Alfred, 59, 118
krypton, 152, 347
Kubrick, Stanley, 12
Kugler, Charles, 331
Kurtz, Thomas E., 123
Kyoto Protocol, 42, 47, 347

Laennec, René, 69
Lake Mead, 151
Lake Nyos, 235

Mariner spacecraft, 197, 205, 243, 350
Maritime Administration, 236
Mars, 50, 167, 197, 235, 240, 243, 248
Mars Observer, 235
Martin, George, 306
Martin, Harry, 163
Marvel, Carl, 88
masers, 138
mason jars, 336
Mason, James, 39
Mason, John L., 336
mass communications, 204
Master Lock, 63, *63*
matchbooks, 272
matrix mechanics, 120
Matthews, Thomas, 219
Mauchly, John, 310
Mauermann, Max, 227
mauve dye, 49, 73
Max Planck Institute, 11, 286
Maxim, Hiram, 36
Mayer, Martin, 174
Maynard, John, 87
McAdam, John, 266
McCandless, Bruce, II, 38
McCollum, Elmer, 64
McCormick, Cyrus, 46, 333
McCullough, David, 2
McDonald, Charlie, 252
McDowell, Ephraim, 349
McKim, Mead & White, 290
McKinley, William, 225
McLean, Malcom, 118
McLuhan, Marshall, 204, 328
McMurry, Erin, 141
McNichol, Tom, 192
Me-262s (aircraft), 13
Mead, Elwood, 151
mechanical grain reaper, 46
mechanism, as scientific instrument, 139
medical devices, 97
Meitner, Lise, 42, 122
Méliès, Georges, 246
Melissa (Internet worm), 87
Mendel, Gregor, 39, 184
Mendeleyev, Dmitri, 38, 233
Mercator, Gerardus, 70
Mercury, 235, 243
Mercury space program, 51, 101, 243
Mergenthaler, Ottmar, 186
Messerschmitt 262 fighter, 201
Metcalfe, Bob, 144
Metcalfe, John, 266
Metcalfe's law, 144

meteorites, 270, 284
Metius, Jacob, 277
metric system, 130, 267
Michaux, Ernest, 48
Michelin brothers, 346
Micro Instrumentation and Telemetry
    Systems (MITS), 96, 355
microphones, 65
microscopes, 112, 148, *148*, 253
Microsoft, 26, 44, 96, 144, 148, 220, 286,
    350, 355
microwave ovens, 300
MileHiCon, 303
milk bottles, 100, *100*
Milky Way, 366
Mill, Henry, 176
Millau Viaduct, 352
Mille Miglia Race, 120
Miller, Phineas, 75
Miller, Stanley, 137
Millikan, R. A., 315
milling machines, 108
mimeograph, 71, 222
mind-control tests, 105
Mini Cooper, 21
Mir space station, 50, *50,* 83, 178
Miss Veedol (aircraft), 280
MIT (Massachusetts Institute of Technol-
    ogy), 69, 76, 93, 191, 286, 332
Mitchell, Billy, 285
Mitnick, Kevin, 46, 286
MK-ULTRA project, 105
Möbius, A. F., 214, 323
Model 0 Flatbed Duplicator, 222
molasses explosion/flood (Boston,
    Massachusetts), 15
Molina, Mario, 286
Monier, F. Joseph, 199
Monitor (USS), 70, 170
Monnartz, P., 227
monoplanes, 228
monorails, 92, *92*
Monsanto House of the Future, 165
Monte Carlo Automobile Rally, 21
Monterey Bay Aquarium, 295
Montgolfier brothers (Joseph and Jacques),
    157, 198
Moog synthesizers, 309
Moog, Robert, 309
moon: of Earth, 9, 37, 77, 147, 202, 203,
    235, 239, 243, 257, 259, 282, 350, 356,
    360; of Neptune, 235, 285; of Saturn,
    167, 300, 356, 361; in science fiction,
    246

Moore, Gordon, 111
Moore's Law, 111
Moran, George "Bugs," 45
Morgan, Garrett, 326
Morse code, 173, 179, 264
Morse, Samuel F.B., 20, 173
Morton, William, 275
Mosaic web browser, 114, *114*, 302
Moseley, Henry, 38
Mothra, 121
Motograph News Bulletin ("zipper"), 312
motorcycles, 244, 259
Motorola, 95
Mount Everest, 151
Mount St. Helens, 140
mouse (for computers), 119, 345
movies, 12, 20, 29, 32, 69, 73, 76, 92, 99, 100,
     109, 136, 152, 158, 159, 168, 191, 202,
     232, 246, 281, 311, 322, 356
Mozilla, 114
"Murders in the Rue Morgue" (Poe), 112
Murray, Matthew, 177
music, 10, 65, 111, 125, 174, 181, 182, 196, 208,
     215, 224, 298, 301, 309, 329, 343, 352
Mustang automobiles, 109, 123
Muybridge, Eadweard, 168
Mylar, 163
MySpace, 35
Myst (game), 269, 344

Nansen, Fridtjof, 285
Napoléon, 74, 81, 267, 323
Napster, 343
NASA (National Aeronautics and Space
     Administration), 5, 31, 37, 42, 44, 50, 51,
     83, 101, 109, 116, 166, 167, 194, 203,
     206, 212, 235, 243, 248, 254, 279, 361
National Advisory Committee for
     Aeronautics (NACA), 212
National Aeronautics and Space Act, 212
National Broadcasting Company (NBC), 321
National Center for Supercomputing
     Applications (NCSA), 12, 114
National Geographic Society, 27
National Highway Traffic Safety
     Administration, 42
National Inventors Hall of Fame, 193, 239
National Museum of American History, 162
National Oceanic and Atmospheric
     Administration, 40
National Safety Council, 288
National Science Foundation, 99
National Theater (Prague, Czechoslovakia),
     25

Nature Research Society, 39
Nautilus (USS), 21, 180
NCSA Mosaic 1.0 web browser, 99, 114
Neanderthals, 193
Nelson, George, 103
neon, 152, 312, 347
Neptune, 21, 235, 239, 268, 275, 285
Nescafé, 225
Nessler, Karl Ludwig, 283
Netscape browser, 114, 273, 302
Netscape company, 223, 350
neutrinos, 274
neutron, 59
New York, 10, 58, 67, 84, 100, 102, 122, 153,
     162, 169, 171, 177, 185, 196, 231, 258,
     283, 290, 291, 293, 302, 309, 312
Newcomen engines, 198
Newcomen, Thomas, 56
Newlands, John, 38
Newmark, Craig, 89
newspapers, 20
newsreels, 20
Newton, Isaac, 39, 77, 107, 188, 261, 304
NFL (National Football League), 50, 72
Niagara Falls, 41, 91, 192, 202
Nicholson, William, 81
nicotine, 29, 138
Nièpce, J. N., 233
Nightingale, Florence, 134
911 calls, 47
1984 (Orwell), 22, 161, 215
Nintendo Entertainment Systems, 293
Nintendo Koppai, 268
nitroglycerin detonator, 38
Nixon, Richard M., 5, 138, 203, 207, 271, 364
Nobel Prizes, 11, 62, 85, 112, 117, 120, 138,
     152, 239, 257, 265, 286, 294, 296, 314,
     327, 333, 341, 347, 348, 357, 359
Nobel, Alfred, 38, 296, 314, 333
North American Radio Broadcasting
     Agreement (NARBA), 90
North Pole, 27, 98, 307
North Side Gang (of George "Bugs" Moran),
     45
North, Simeon, 108
notation of differentials, 304
Noyce, Robert, 257
nuclear chain reaction, 159, 257, 338
nuclear power plants, 50, 65, 89, 118, 180,
     338
Nuclear Test Ban Treaty, 219, 344
nuclear weapons, 37, 192, 236, 254, 280, 300,
     327, 344, 356. *See also* atomic bombs;
     hydrogen bombs

Nufer, Jacob, 14
nylon, 60

Occupational Safety and Health Act, 34
ocean liners, 8
oceans, 164
Odeo (Silicon Valley company), 82
Odyssey spacecraft, 50
Ohio Art Company, 195
oil, 10, 148, 153, 156, 213, 231, 241, 292, 366
Olympic Games, 16, 139
ophthalmoscope, 342
Oppenheimer, J. Robert, 42
optics, 39
organ transplants, 245, 291, 301, 339, 359
Origin of Species (Darwin), 184, 363
Orwell, George, 22, 161, 215
Osterhoudt, J., 277
Otis, Elisha Graves, 84
Otis, William, 233
Otlet, Paul, 237
Otto, Nicolaus, 65, 228
Ötzi (Alpine iceman), 264
ovarian tumors, 349
Owen, Robert, 136
oxygen, 215
oxyhydrogen blow-pipe, 315
ozone, 23, 47, 200, 209, 286

Pacific Phonograph Company, 329
Packard, Dave, 232
Packard, James Ward, 213
packet-switching, 72, 99
Pac-Man, 285
padlocks, 63
Palace Theater (New York), 322
paleontology, 53
Pall Mall (London), 28
Palmer, Daniel David, 263, 314
Palomar Observatory, 278
Pan Am, 170
Panama Canal, 126, 229, 322
Pangborn, Clyde, 280
Panthéon (Paris), 7
paper, 72
paper-bag-making machine, 45
parachutes, 54, 60, 116, 230
Parker, Charlie, 181
Parkes Radio Telescope, 219
parking meters, 135, *135*
Parkinson, James, 103, 357
Parkinson's disease, 103, 357
Park-O-Mat garage, 341
Parmelee, Henry S., 225

Parsons, Ed, 215
Pascal, Blaise, 79, 129
Pasteur, Louis, 85, 189, 244, 363
patents: for designs (U.S.), 315; first (U.S.),
    214; first to a woman (U.S.), 349; laws
    granting/protecting (U.S., Venice), 53,
    80; numbering of (U.S.), 57
Patrick Air Force Base, 207
Patrick Henry (SS), 272, *272*
Paul, Ron, 46
Pauling, Linus, 117
Paxton, Joseph, 217
peanut butter, 5
Peary, Robert, 27, 98, 307
pedal boats, 182
pellagra, 35, 110
Pemberton, John, 130
pendulums, 7, 156
Penguin publishing, 213
penicillin, 68
Pennington, Mary Engle, 283
Pennsylvania (USS), 18
pens, 163
Pentagon, 83, 99, 150, 212, 244, 344
Pepys, Samuel, 320
Pérignon, Dom Pierre, 218
periodic table of the elements, 38, 152
Perkin, W. H., 49, 73
Perl computer language, 354
Perlman, Louis Henry, 143
"permanent wave" hair styling, 283
Perrin, M. W., 88
Persil (household detergent), 211
pesticides, 34
Phi Beta Kappa, 187
Phillips head screws/screwdrivers, 190, *190*
Phillips, Henry, 190
phonautograph, 101
Phonofilm movie process, 73
phonograph cylinders, 65
phonographs, 327, 329
photocopiers, 297
photography, 9, 20, 27, 28, 52, 94, 98, 117,
    124, 168, 233, 249, 330, 346, 348
physicians, women, 273
pi, 36, 68, 75
Piazzi, Giuseppe, 1
Picasa photo-sharing site, 28
Piccard, Auguste, 23
Piccard, Jacques, 23, *23*
Pielke, Roger, Jr., 5
Pink Floyd, 134
Pioneer 10, 166, *166*
Pirate Bay website, 153

Pirelli, Giovanni Battista, 346
Pitman, Isaac, 4, 108
Pixar animation studio, 136
Planck, Max, 120, 167, 290, 350
Planetary Observer project, 235.
planets, 104, 113, 238. *See also* moon; *names of individual planets*
Plankalkül computer language, 134
Planned Parenthood, 291
plastic beverage bottles, 299
plastics, 88
playing cards, 268
plows, 170, *170*
Plunkett, Roy, 98
Pluto, 21, 49, 166, 238
plutonium, 55, 220, 312
pneumatic hammers, 30, 332
pneumatic tires, 346
podcasts, 35, 227
Poe, Edgar Allan, 112
poison gas, 339
Polacolor film, 52
Polaris missile, 203
Polaroid instant camera, 52
polio, 118, 281, 287
pollution, 175, 201, 222, 301, 327, 345. *See also* smog
Polyakov, Valery, 50, 83
polyester film, 163
polyethylene, 88, 211
polygraph machines, 33
polytetrafluoroethylene (Teflon), 98
Pompeii, 238
Pong, 295, 335
pony express, 106, 299
Post, Wiley, 88, 205
postage stamps, 128
postal services, 20, 51, 64, 106, 128, 206, 231, 267, 299
potassium carbonate, 53
Potts, William, 326
presidential elections/inaugurations, 20, 310
Priestly, Joseph, 215, 352
prime meridian, 288
primordial-soup hypothesis, 137
Principia Mathematica (Newton), 188
printing, 27, 70
Project Excelsior, 230
Project Explorer, 212
Prokhorov, Aleksandr, 138
Psychodiagnostics (Rorschach), 94
punch-card tabulators, 61
Pusey, Joshua, 272

quadrophonic surround-sound, 134
quagga (zebralike species), 226
quantum mechanics, 290, 350
quasars/quasi-stellar objects (QSOs), 219
Quinlan, Karen Ann, 164

R.F. Mosley (knife-making firm), 227
R.U.R. (Čapek), 25
Ra II (boat), 109, 139
rabies vaccine, 189, 244
radar, 58
radiation belts, 31
radio, 20, 98, 115, 155, 192, 221, 245, 274, 282, 300, 313, 320, 321, 322, 348, 360
radio astronomy, 297, 327
radio compasses, 189, 274
radio frequencies, 90
Radio Shack, 269
Radio Shack TRS-80 computer, 96, 217
radio signals, 67
radioactive isotopes, 190, 360
radioactivity, 62
radio-dating, 115
radio-frequency identification chips, 309
radium, 137, 357
railroads, 20, 48, 66, 101, 121, 128, 132, 169, 177, 204, 216, 223, 243, 253, 263, 272, 283, 324, 332
raincoats, 365
RAM (random access memory), 97, 133
ramen instant noodles, 239
Ramsay, William, 152, *152*
Ramstein Air Show, 242
Ranger 7 lunar probe, 214
Ray-O-Vac, 115
Raytheon, 300
RCA (Radio Corporation of America), 65, 73, 86, 112, 174, 215, 297, 309, 321
Reagan, Ronald, 84, 356, 362
reaping machines, 333, 336
Rebok, Jack, 98
red telephone boxes, 17
Reddi-wip, 193
Redstone ballistic missile, 31
Reeves, George, 169
refrigerated railcars, 332
refrigerators, 317
Regional Agreement for the Medium Frequency Broadcasting Service in Region 2, 90
Reichenbach, Henry, 124
Reid, George, 113
Reilly, Frank C., 312
reimplantations/replantations, 145

remote controls, 313
repeating rifles, 102
revolvers, 57
Reye's syndrome, 67
RFC (request for comments), 99
Riccioli, Giovanni, 9
Richards, Patti, 93
Richardson, Levant M., 345
Richelieu, Cardinal, 135, 161
Rickenbacker, Adolph, 196, 224
Ride, Sally, 26, 171
Ringley, Jennifer, 106
Ritty, James, 100
RMS *Queen Mary 2* (ocean liner), 8
road-building, 266
Robert E. Peary (SS), 272
Roberts, Ed, 355
Robertson, Morgan, 275
robotics/robots, 25, 97, 103, 192, 303
rockets, 31, 37, 77, *77,* 147, 207, *207,* 234,
    254, 299
Rockville Centre, New York, 97, 247
Rodgers, Cal, 262
Roebling, Emily Warren, 2
Roebling, John Augustus, 2
Roebling, Washington, 2
Roentgen Society of London, 171
Roentgen, Wilhelm, 62, 314, 357
Rogers, Buck, 313
Rohrer, Heinrich, 112
roller coasters, 169
roller skates, 345
Roper, Sylvester, 244
Rorschach, Hermann, 94
Rose, Henry, 178
Roselle, New Jersey, 19
Rosemeyer, Bernd, 28
Rosetta Stone, 256
Ross, Doug, 69
rotary engines, 227
Rougier, Henri, 21
Rowland, F. Sherwood, 286
Rowley, Charles, 102
Royal Air Force, 163
Royal Canadian Air Force F-86 Sabrejet, 140
Royal Navy, 142, 149, 208, 260, 363
Royal Observatory, 288
Royal Society, 38, 39, 81, 142, 148, 304, 334
rubber, 352, 365
rubber bands, 78, 346
rubber-asphalt compound, 252
rubberized waterproof clothing, 365
Rubik, Ernö, 30
Rubik's Cube, 30

Ruska, Ernst, 112
Russell, Bertrand, 192
Russian Federation, 167, 180
Rutgers University, 294

S.F. Bowser Company, 250
Sabine, Wallace Clement, 290
Sadacca, Albert, 358
Sadtler, Philip, 301
safety belts, 193
safety lamps, 9, 331
safety pins, 102
safety razors, 321
Sagan, Carl, 356
sailing, 180
Sakharov, Andrei, 22
Salisbury (ship), 142
Salyut space stations, 111, 208
Sandage, Allan, 219
sandblasting, 293
San Francisco, California, 18, 86, 110, 194,
    216, 272, 293, 306, 329, 345
Sanger, Larry, 15
Sanger, Margaret, 291
Santorio Santorio, 261
satellites, 31, 45, 47, 93, 103, 147, 167, 179,
    195, 206, 212, 234, 241, 279, 309, 328,
    342, 344
Saturn, 167, 240, 300, 361
Saturn 5, 37
Savannah (ship), 236, *236*
Savery, Thomas, 56
Savitskaya, Svetlana, 208
Sax, Adolphe, 181, *181*
saxophones, 181
scanners, 138, 239
Schaefer, Vincent, 319
Schaffert, Roland, 297
Schatz, Albert, 294
Schawlow, Arthur, 138
Schenck, Helmut, 13
Schick, Jacob, 79
Schlitt, Karl-Adolf, 106
Schmidt, Maarten, 219
Schouhamer, Kees A., 352
Schrödinger, Erwin, 120
science fiction, 266, 307
science/humanities, 129
Scientific American, 87, 213, 242, 315
scientific method, 161
Scopes, John, 204
scotch tape, 253, 317
Scotch whisky, 154
Scott, Giles Gilbert, 17

thermionic emission, principle of, 322
thermometers, 261
thermos, 265
Thermos GmBH, 265
Thing, the, 309
Thompson submachine guns, 45
Thompson, LaMarcus Adna, 169
Thompson, Leonard, 11
Thompson, Pat, 303
Thompson, Robert W., 48, 346
Thomson, Elihu, 228
Thomson, J. J., 122, 165
Thoreau, Henry David, 223, 253
Thornton, Stan, 194
Three Mile Island, 89, 180
3M Company, 253, 317
THX (sound reproduction), 136
tidal power stations, 332
Tihanyi, Kalman, 171
Tilghman, Benjamin Chew, 293
tire wheels, 143
Titan (moon), 361
Titan 1 (ICBM), 37
Titan 2, 37
Titanic (RMS), 107, 247, 275
tobacco. *See* smoking/tobacco
toilet paper, 208
toilets, 27, 106
Tokheim, John, 250
Tolkien, J. R. R., 266
Tombaugh, Clyde, 238
Toole, Betty, 158
toothbrushes, 179
Torres-Quevedo, Leonardo, 313
Torricelli, Evangelista, 164
Townes, Charles, 138
traffic signals, 326
trains, 20, 242, 263
Trans-Alaska Pipeline System, 153
transatlantic cable, 210
transistor radios, 221, 300
transistors, 111, 282, 359, 365
transplanted embryos, 34
Travers, Morris, 152
Treaty of the Meter, 130
Trevithick, Richard, 177
tricycles, 210
Trieste (deep-diving research vessel), 23, *23*
Trinity atomic test, 199
Triton, 235, 285
Triton (USS), 132, 230
Trojan horse software, 116
Trojan War/Horse (of Greeks), 116
trolleys, 121, 358

Trowbridge, A. E., 205
TRS-80 computers, 217
Tu, John, 350
tuberculin, 85
tuberculosis, 85, 294
Tucker, Preston, 22, 207
Tudor Ice Company, 258
Tudor, Frederic, 258
Tufte, Edward, 253
Tunguska explosion, 183
tunnels, 101, 111, 199, 250, 287, 325, 354
Tupper, Earl S., 211
Tupperware, 211
turbines, electric, 294
Turing, Alan, 176
Turner, Edward, 191
Tuscany (ship), 258
Tutankhamen, 332
TV Guide, 167
TWA (Trans World Airlines), 202
Twain, Mark, 229
Tweed, William M. "Boss," 58
Twilight Zone (television show), 123, 172
Twitter, 20, 35, 82
2001: A Space Odyssey (film), 12
2010 (Clarke), 12
Tyng, James, 43
typesetting machines, 186
typewriters, 176, 222
Typhoid Mary (Mary Mallon), 132, 268

U.S. Air Force, 37, 207, 230
U.S. Army, 31, 40, 108, 262, 318
U.S. Army Air Corps, 140
U.S. Army Corps of Engineers, 77
U.S. Census, 216
U.S. Congress, 40, 53, 80, 104, 132, 151, 200,
    212, 214, 275, 324
U.S. Department of Energy, 153, 218
U.S. Naval Reserve, 150
U.S. Navy, 14, 18, 23, 70, 88, 189, 221, 230,
    272, 274, 344
U.S. Patent and Trademark Office, 54, 57,
    67, 187
U.S. Pioneer mission, 9
U.S. Post Office, 64
U.S. Public Health Service, 301
U.S. Supreme Court, 65, 291, 322
U-boats, 106, 122, 131, 201, 220, 272, 308
UFOs, 177, 191, 269
Unabomber (Theodore Kaczynski), 95
United Kingdom. *See* Britain
Univac (Universal Automatic Computer),
    310, *310*

388   INDEX